T0212296

Earth and Environmental Sciences Library

Series Editors

Abdelazim M. Negm, Faculty of Engineering, Zagazig University, Zagazig, Egypt

Tatiana Chaplina, Antalya, Turkey

Earth and Environmental Sciences Library (EESL) is a multidisciplinary book series focusing on innovative approaches and solid reviews to strengthen the role of the Earth and Environmental Sciences communities, while also providing sound guidance for stakeholders, decision-makers, policymakers, international organizations, and NGOs.

Topics of interest include oceanography, the marine environment, atmospheric sciences, hydrology and soil sciences, geophysics and geology, agriculture, environmental pollution, remote sensing, climate change, water resources, and natural resources management. In pursuit of these topics, the Earth Sciences and Environmental Sciences communities are invited to share their knowledge and expertise in the form of edited books, monographs, and conference proceedings.

Mario Tribaudino · Daniel Vollprecht ·
Alessandro Pavese

Editors

Minerals and Waste

 Springer

Editors

Mario Tribaudino
Department of Earth Sciences
University of Torino
Torino, Italy

Daniel Vollprecht
Institute of Materials Resource
Management
Augsburg University
Augsburg, Germany

Alessandro Pavese
Department of Earth Sciences
University of Torino
Torino, Italy

ISSN 2730-6674   ISSN 2730-6682 (electronic)
Earth and Environmental Sciences Library
ISBN 978-3-031-16137-7   ISBN 978-3-031-16135-3 (eBook)
https://doi.org/10.1007/978-3-031-16135-3

This Springer imprint is published by the registered company Springer Nature Switzerland AG
The registered company address is: Gewerbestrasse 11, 6330 Cham, Switzerland

Preface

Environmental issues are at stake in every moment of our life. A huge community of scientists is involved, investigating a wide range of topics, from health issues to waste disposal and recycling.

Environmental sciences are typically interdisciplinary, with the contribution of medical, chemistry, ecology, engineering and earth sciences investigators. Within this community, mineralogy provides significant contributions: as an example, we may recall investigations in asbestiform minerals, in biomineralogy and in the recycling of mine tailings, all of which are a subject of extensive research.

However, in the general field of waste disposal, recovery and recycling, the contribution of the mineral sciences is underrepresented. Indeed, most wastes, not only mining or demolition wastes but also industrial residues from ceramic activities, are made of minerals. Also municipal wastes, after removing the organic part in incinerators or after long-time disposal, are just an assemblage of minerals.

Moreover, although human-made, waste deposits resemble natural counterparts in many aspects, and we must apply what we know from natural systems, to understand waste assemblages, to predict their fate, at short and long times, and to suggest the best for high-temperature recycling. This complies with the drive for the reuse and recycle of wastes as much as possible, which is the goal of waste management.

Mineral behaviour in wastes was reported in several papers focused on various issues of waste management, but a comprehensive work on the different aspects of the relations of mineralogy and wastes is still missing.

This book comes to fill the gap between mineral sciences and the field of waste investigation and management. It comes as a product of the MINEWA School, held from 20 to 24 June 2022, in Bardonecchia (Italy), by the European Mineralogical Society (EMU) and the Società Italiana di Mineralogia e Petrologia (SIMP). It is meant to give in a book the full background information on what are wastes and how they are managed, the basics of the high- and low-temperature geochemistry involved, and a sense of all that is going on in different fields of waste investigation and how they are linked with minerals. Mineral tailings, incinerator bottoms and fly ashes, metal slags and residues of the ceramic industry are here addressed, as well as sanitary issues and waste management from a global perspective.

This book comes as a tool for teaching in applied mineralogy, geology, chemistry and environmental engineering courses. It is meant not only as an introduction to waste management problems, for graduate and undergraduate students, but also as a reference book to researchers in the field.

Torino, Italy   Mario Tribaudino
Augsburg, Germany   Daniel Vollprecht
Torino, Italy   Alessandro Pavese

Contents

Minerals and Wastes, an Overlooked Connection 1
Mario Tribaudino

What is Waste, and How We Manage in Europe 21
Mariachiara Zanetti and Deborah Panepinto

Thermodynamics and Kinetics of HT-Processes 39
Alessandro Pavese

Bio-mineral Interactions and the Environment 67
Giovanni De Giudici, Daniela Medas, and Carlo Meneghini

Metals: Waste and Recovery 117
Gilberto Artioli

Mineralogy of Metallurgical Slags 135
Daniel Vollprecht

**Bottom Ash: Production, Characterisation, and Potential
for Recycling** 155
Jacques Rémy Minane and Raffaele Vinai

**Spatialising Urban Metabolism: The Supermarket as a Hub
for Food Circularity** 213
Emma Campbell, Greg Keeffe, and Seán Cullen

A Brief Glance on Global Waste Management 227
Astrid Allesch and Marion Huber-Humer

Waste, Environment, and Sanitary Issues: Are They Really at Odds? ... 259
Maura Tomatis, Jasmine Rita Petriglieri, and Francesco Turci

Minerals and Wastes, an Overlooked Connection

Mario Tribaudino

1 Introduction

A common view is that wastes are something very different from natural objects. Wastes are the end-of-life of products: products are manufactured objects; we keep products as long as they are useful, and then we discard products that become waste. The difference between a useful product and a waste is indeed something regarding property: the objects have owners, whereas wastes take their origin from dismissing property. It is all a manmade game field. The most obvious definition of minerals holds instead as a topic point that minerals are nature made objects (e.g. [19]). Minerals are nature made, wastes are man-made. We will go through a definition of minerals more suitable for an application on wastes in this chapter, but first we will find how they are related.

A first step in relating minerals to wastes may simply come from an internet search. If you search "minerals and wastes" you will find several links involving disposal of quarry residuals. Natural minerals, which are discarded during quarrying, accumulate, forming dumps. It is obviously a deposit coming from the human activity, but made by minerals. The minerals that form the dumps have the same chemical and physical properties of natural minerals, just they are ground to some extent. Still no doubt quarrying deposits are wastes, although these materials are not covered by the waste definition of the European Waste Framework Directive.

Another step forward: mining residues. Here again we have natural minerals, which form deposits mimicking natural sedimentary rocks. They are made from the original rocks, after separation of the ore, but also from newly formed minerals, which were not present in the mined rock. These minerals form by reaction of the tailings with the atmosphere and the biosphere, following oxidation, leaching, adsorption and

M. Tribaudino (✉)
Dipartimento di Scienze della Terra, University of Torino, Via Valperga Caluso, 35, 10125 Torino, Italy
e-mail: mario.tribaudino@unito.it

precipitation processes. In some cases, they originate from the industrial processes to concentrate the ore from the gangue. The transformation of sulphides in sulphates enhances the release in the environment of metals, and acid mine drainage, a process originating from a transformation from a natural ore mineral to an environmental risky tailing [11]. An example is in gold mining, whose impact in terms of As and heavy metal in the environment has been widely discussed [1].

We have also minerals made by the inorganic chemistry processes, involving reactions of something that in nature was formerly present, but now has been recycled in the human processes. A major example is concrete, where we obtain a mixture of non-crystalline and crystalline phases, together with the addition of common silicates and carbonates, starting from natural minerals. In the case of Portland cement we have clay minerals and calcite, as a start, but the clinker is by no means natural. Still, at the end we have minerals, most of them have a natural counterpart, and, most important, they can be studied exactly as the other minerals. Other examples are the ashes from the incineration of the municipal solid waste, residues from coal plants, from foundry, metal scrapes, and so on.

So, mining and quarry wastes, our first internet definition of minerals and wastes largely underestimates the inherencies of mineralogy to the waste management. There are several fields where minerals are involved, covering the field of inorganic wastes. A definition of minerals in the broader sense of natural and analogues of natural ones must be introduced: minerals are solid state inorganic compounds, mostly crystalline, but not necessarily, which can be natural or having a natural counterpart. This because natural minerals and counterpart do have the same properties: same reactivity, same structure, same phase stability. And waste related systems can be studied likewise.

If minerals are the first player, wastes are the second one: minerals are an obvious by-product in smelting ores, a less obvious one in construction waste, and an almost surprising one in municipal solid waste. There is a lot of artificial in the process of making concrete, but the final products of concrete do not escape thermodynamic and kinetic constrains, and although man-made, they are mineral analogues. In municipal waste one would not expect minerals, we do not throw in bins solid minerals, but paper, plastic, wood, glass. Already these products contain minerals as fillers, like calcite and clay minerals, albeit not as main constituents. However, as a result after incineration, or in landfill disposal given enough time to reduce and react organic compounds, we have an assemblage of minerals. In it, incineration ashes are a material with substantial criticalities in recycling, but they are criticalities that can be described in the field of mineral behaviour.

Time is a major difference between mineralogical assemblages from natural processes alone, and those observed in waste residues. We are very good in promoting strongly off equilibrium reactions to obtain materials, which eventually will degrade in the environment. We are not so good in mastering the degradations timing. Sometimes this will occur in too much of a short time, as in the leakage of batteries, sometimes in a time exceedingly long for us, like in the consumption of plastics. In general, manmade assemblages are far from equilibrium. For instance, glass is a widespread product, but glass is also a solid phase far from thermodynamic equilibrium. Far from equilibrium mineral assemblages are more reactive; potentially toxic elements could be included in the minerals and released in the environment as the mineral reacts to the environment, generally by leaching.

How common are minerals as waste constituents? As long as some mineral can be found in almost any solid waste, we have several wastes where minerals are the main constituent: mining and quarrying activities, construction and demolition residues, industrial waste, mostly in ceramics, ashes from municipal waste incineration, metallurgical slags and residue from carbon based plants. The amount is huge: in the European Union in 2020 [8] the total generated waste by all economic activities and households amounted to 2 277 million tonnes. Construction contributed to 35.7% of the total and was followed by mining and quarrying (26.3%), manufacturing (10.7%), waste and water services (10.2%) and households (8.2%),the remaining 9.4% was waste generated from other economic activities, mainly services (4.4%) and energy (3.5%). Construction and demolition, and mining and quarrying can be safely classified as mineral wastes, and account for two third of the total. Of the remaining waste we have several contributions from mineralogy, as for instance in municipal solid waste incineration bottom ashes. Each of the above fields will be a subject of the following chapters.

2 A Geochemical Perspective: Continental Crust as a Benchmark

Why so many mineral wastes? Since Palaeolithic, the primary materials of man craft come from the exploitation of the biosphere, lithosphere and atmosphere of the Earth. At first just stone was used; objects from stone carving were used to machine wood and animal skin or furs. Then, we took advantage of ores, which are minor, rather exceptional occurrence in the Earth crust, and went into the metal ages. Later on, oil and the outbreak of plastics introduced new materials. Now we are at the edge of a new era, with the incoming green economy. This did occur not by a complete replacement: the stone, in cases crushed and/or reacted to make bricks or cement, is still now and in the future a basis of our economy. Ores give metals, which are still crystalline materials, but also a wealth of mineral tailings as wastes. Plastic primary material are hydrocarbons obtained by the refinery of oils, which should not be considered minerals, but, as an industrial product, during disposal will release elements, like chlorine, which are found in several mineral products.

Stones come from the lithosphere, which is the slice of about 100 km involved in plate tectonics. On top of it is the Earth continental and oceanic crust. Continental and oceanic crust differ in thickness and mineral composition. Oceanic crust is hardly exploitable, there are many metals in the oceanic floor, but their mining is yet to come. Therefore, we get the primary minerals from the continental crust.

The continental crust has an average thickness of about 30 km; it geographically corresponds to the continents and the floor of shallow internal seas. The continental crust is highly heterogeneous. In it during geological times several processes concentrated elements and compounds, in such a way that now we can use them. The Earth crust changes its nature by internal processes, but also with an interaction from the biosphere and atmosphere, following geochemical paths. The human activity is a geologically very recent source of strong interaction between the lithosphere and the biosphere. It creates new deposits from the exploitation of earth crust minerals. This is what we call the Anthropocene. However, although the Anthropocene sedimentary deposits in most cases do not have a natural counterpart, the chemical reactions which transform the man-made deposits in an often unwanted way are the same that transform minerals on earth surface. They are the clue to understand and assess the environmental concerns of a given waste assemblage and its treatment.

Wastes are the typical Anthropocene deposit. Waste composition gives information on the products of the human activity, and its interaction with the earth crust. As an example, we may compare the composition of the average continental crust and that of two major mineral wastes: demolition wastes and bottom ashes from municipal solid waste incinerators (MSWI) (Fig. 1). For both we have to average over highly heterogeneous compositions in the crust and in the wastes. To check for possible inconsistencies due to different sampling, for MSWI bottom ashes we plotted three different estimates of the most common elements.

In MSWI bottom ashes we find that:

(1) the oxides of most elements, namely Al, Fe, Si, Ti, K and Na are present in bottom ashes to an extent very close to that of the continental crust;
(2) just two elements show strong enrichment in bottom ashes, i.e. Ca and P. Another difference is in the higher loss on ignition (LOI) content in bottom ashes respect to the average crust;

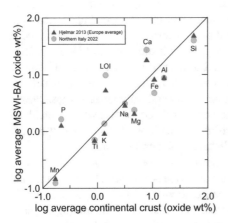
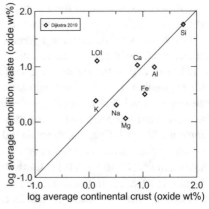

Fig. 1 **Left**: Average composition of the oxides of most common element (more than 1wt%) in MSWI bottom ashes vs the composition of the average continental crust (CRC Handbook 2016). LOI: loss on ignition. Data from: NI: average of the composition of the bottom ashes in 5 incinerators in Northern Italy (unpublished results); Hjelmar 2013: compilation for European Incinerators [10]. **Right**: Average composition of the oxides of the most common elements in demolition wastes [7] versus continental crust

(3) three elements account for most of the composition in bottom ashes: Si, Ca and Al, with Fe at a lower extent. Trying to simplify the picture, the phase relations between the oxides, SiO_2–CaO–Al_2O_3, described in the CAS system, will give a general description of the system once we take into account the effect of Fe impurities;

(4) The values shown in Fig. 1 come from an average of the reported compositions from a number of MSWI plants. Different waste input, different incinerator procedures, different weathering time and intensity may change significantly the composition from a given incinerator. For instance, the degree of separation of the metals done in each plant could account for a higher Fe content.

As concerns demolition waste [7] we find some similarity with bottom ashes. Again, Ca, Si and Al are the most common oxides, with Fe and Mg significantly less: the CAS system is even more representative. Again, LOI is much higher than in earth crust, but Ca does not show an enrichment. P, Mn and Ti, present in MSWI bottom ashes in significant quantities, are here negligible.

In the next paragraph, we will recall some basic notion of mineralogy, preliminary to an interpretation of the above results. For a more detailed discussion, the reader can consult the mineralogical textbooks in the references.

3 Elements and Mineral Phases

The average compositions give only a part of the picture. Chemical analyses are made after complete digestion of the phases. This occurs after an aggressive attack, which is far beyond the dissolution that may occur in natural environment. After complete dissolution, we have no information whether the content of a given element is in a form environmentally critical or potentially recyclable. The mineral form in which the element is encased is the clue to predict its environmental destiny. We will here discuss the mineral form in which major elements are present in natural minerals, how we can predict the behaviour of less common element in minerals, and last compare it to what we know from the average composition of less common elements in incinerators.

Few elements are the main constituents of the solid earth: one is most abundant, and takes the form of an anion, oxygen, and the others are cations, in the order of abundance Si, Al, Ca, Fe, Mg, K, Na. Together they make more than 99 wt% of the solid earth crust. These elements are also the most common in mineral wastes.

The form in which we find the minerals that compose most of the earth is that of a complex oxide, in which the different cations form a prevalently ionic bond with oxygen. The only common mineral made by a simple oxide is quartz, SiO_2, where the two most common elements bond together. Although we know more than 5000 different mineral species, only few are widespread. Si is by far the most common cation, and therefore most rock forming minerals are silicates. Bonding in silicates is a strong one, and a considerable energy is needed to break it.

The most common mineral phases are a combination of these elements: quartz, feldspars, which are the main reservoir of Si, Al, Ca, K and Na in the continental crust, and a few mineral families, pyroxenes, amphiboles and micas, which are silicates containing also Fe and Mg. Other minerals form from an interaction of the rocks exposed at the surface with the atmosphere: most important are clay minerals, occurring after hydration of the above minerals, and carbonates, among which calcite, which mostly occur after reaction of leached Ca with atmospheric CO_2.

Although most common elements are present as silicates or, less commonly, carbonates (Ca and Mg, calcite $CaCO_3$ and dolomite $CaMg(CO_3)_2$) and sulphates (Ca, in gypsum $CaSO_4.2H_2O$), other less abundant, but economically important minerals occur by combination of common and less common elements. For instance, Zn occurs mostly as a sulphide (ZnS, sphalerite).

The elements within a given mineral make ordered sequences of atoms and atom clusters (Fig. 2). Being oxygen the most common element, we have that a number of oxygen atoms surrounds each cation. In silicates the bonding is essentially ionic, which means that we may consider atoms as charged particles of an almost spherical shape. The number of oxygen atoms near a given cation depends on the ratio between the cation and the anion; being oxygen much bigger than the other ions, and negative, an ordered repetition of oxygen clusters with cations in the interstices makes the atomic structure of crystalline silicates. We can represent the clustering of oxygen surrounding a cation by polyhedra with the centre of the oxygen atoms at the vertices. In a synthetic system, in which only few elements are present, we have that cations fill the polyhedra in a stoichiometric ratio with oxygen. In natural samples, different cationic species may fill the interstices created by oxygen polyhedra (Fig. 2).

The most common polyhedron is the SiO_4 tetrahedron, in which four oxygen at the vertices of a tetrahedron make an interstitial space which is filled by the Si atom. The repetition of this polyhedron gives rise to different mineral silicate families. Other cations, like Fe, Mg, Al, fit better in larger polyhedra with 6 oxygen at the vertices, other like Ca, K, Na with 8 or more.

Less common elements can form their own phase, in which they are concentrated, but in most cases simply enter one or more polyhedra fitting the element size in the structure of the most common minerals. There are obviously limits, dictated by the cation size, which make a given atomic species unable to fit within a polyhedron in a given mineral.

The result is that minor elements, among which are Potentially Toxic Elements (PTE), like Cu, Zn, Pb, Cr… can form different minerals, with different solubility, and different Ph-Eh stability, but also be present as an impurity locked into a silicate structure. In silicates they may hardly leach, whereas in sulphates or carbonates the leaching, with all relevant sanitary issues is easier. This limits how the analyses, made after dissolution of all the cations, are useful to predict potential leaching, exploitation and recycle of the PTE elements. Leaching of MSWI bottom ashes shows for instance that only a very minor fraction, between 10^{-6} and 10^{-8} of the available atoms, is leached. This likely corresponds to the fraction that is mostly soluble, that is likely present in a mineral, which may occur at very small amount. The analysis of an element has little use, if it is not related to the determination of

Fig. 2 Different representations of the mineral enstatite, $Mg_2Si_2O_6$, showing the pyroxene structure. Top: actual size of the oxygen atoms (red) respect to other atoms. Oxygen fills almost completely, the structure, leaving only interstitial gaps where other atoms enter. Middle: representation of the same structure with reduced atom is size to enhance the bonding relations within polyhedra. Bottom: same view, but showing the polyhedral where Si (blue polyhedron) and Mg (orange) enter the structure. Within a given polyhedral other elements, if sized to fit, may enter

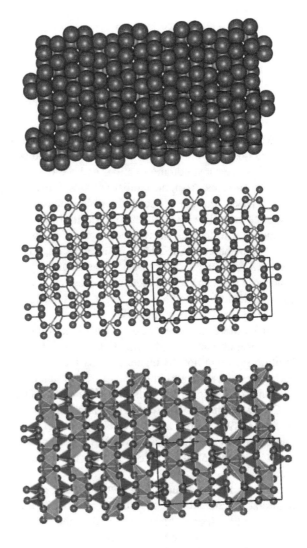

the phases, where an element is present. Within this book several examples will be made.

The polyhedral structure of silicates is present in crystalline or amorphous silicates. In a crystalline structure we have a regular repetition of atoms related in three dimensions by geometrical vectors, whereas in the amorphous structure of glasses, we have an atomic framework unrelated by geometrical vectors, but still showing a strong atomic bonding within the framework. What is said for silicate crystals can be repeated for silicate glass, which is a major product in several wastes. As for mineral phases, also different glass may coexist, each with its own leaching behaviour. For instance, in MSWI ashes we have alkali richer or poorer glassy phases, formed during incineration, together with glass unreacted from the original waste (Fig. 5).

This occurs by the heterogeneous waste material, but also by the thermal gradients within the burning chamber.

4 The Fate of Major and Minor Elements

We may now re-examine the element concentration in waste and in earth crust. We find that most common elements are present in continental crust and in wastes. Silicon is the most common cation, and we expect that also mineral wastes are made by silicates. Also, the similar compositional scaling of the major elements in bottom ashes, and the not so different in demolition waste, confirms that mineral wastes homogenize the products of industrial activity to obtain something close in composition to the natural continental crust. In spite of the large heterogeneity in both continental crust and MSWI bottom ashes, when averaged for most elements the local differences compensate. In geochemistry, when local spikes of element concentration occur, as for instance in ores, we say that fractionation has occurred. Here, in demolition and MSWI wastes, most major elements underwent little or no fractionation respect to earth crust.

Some difference however exists. The higher Ca and LOI is found as Ca is present in the form of calcium carbonate, $CaCO_3$, which loses the CO_2 component during ignition, on a significantly higher quantity than in the earth crust. The presence of Ca carbonates and sulphates in construction and in a number of industrial applications accounts for the higher concentration of Ca-carbonate. However, a significant amount of Ca is also present in glass and crystalline silicates. Moreover, the higher amount of P in MSWI bottom ashes is likely due to the burning of organic residuals. In demolition wastes, the significantly lower content of Fe and Mg is possibly related to the abundance of construction materials in wastes, generally impoverished in these elements.

Less common elements, however, do fractionate. As an example of the kind of anthropic fractionation of the minor elements in waste, in Fig. 3 we have plotted a number of minor elements averaged in the MSWI bottom ashes respect to the continental crust. Being the continental crust mostly silicate, most minor elements will go into the silicate structure, make their own minerals interspersed at a negligible amount within the silicate matrix or, occasionally, will be enriched in ores. The ore enriched portion is a small fraction of the available budget. We can then compare the anthropic pressure on a given element like one alternative geochemical process, which may or may not concentrate an element.

A first group of elements, among which Co, Zr, Li, Hf, Ti, Rb, F, V, Mn, Sr, Ga, U, Th shows an average composition in bottom ashes very close to the earth continental crust. Among them, we find a number of elements with large industrial concern, like Co, Li, Mn, F, V, and we would expect them to be enriched in manmade craft. In some cases they are object of an active recycling, but in general they are present in silicate structures as trace elements. In the geochemical classification of Goldschmidt [13] they belong to the group of lithophile elements, those who are most present as

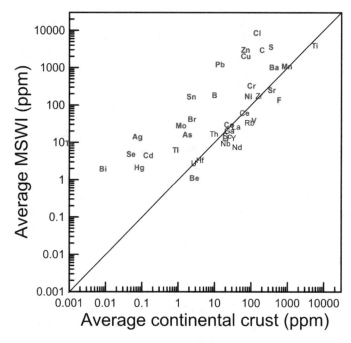

Fig. 3 Minor element composition in MSWI bottom ashes and in average continental crust. Data from: red Hjelmar et al. [10], black Northern Italy incinerators, unpublished data

substituents in the structure of silicates and oxides, e.g. Sr in plagioclase, Rb in K-feldspars and Ti and Mn in ferromagnesian minerals. These elements as a rule are locked in silicates, glass or crystals, which are most abundant also in bottom ashes. In an analysis the fraction of these elements coming from anthropic concentration, like metals, and the fraction present as impurities in silicates are merged: being the latter more abundant, we cannot show a significant enrichment of the element in the ashes. Although the overall budget is bound in silicate structures, and not prone to leaching, we still must consider that even a very small portion in a soluble phase could result in a critical leaching.

A second group is made by some elements, Cl, Zn, Cu, S, Pb, Br, Mo and Sn, which are significantly enriched in bottom ashes, between one and two orders of magnitude. Some of them are present with a concentration higher than 1000 ppm (Cl, Zn, Cu, Pb, S). These elements are concentrated by human activity: their higher abundance in manmade craft resides in that they are present in relatively large deposits, but in a mineral phase in which they are not bonded strongly. In nature Cl is found in the highly soluble mineral halite (NaCl), Zn and Pb in sulphides, ZnS (sphalerite) and PbS (galena), B in borates, which make the element very easy to extract. This enables massive exploitation and subsequent industrial concentration in the plastic (Cl), wires (Cu) or sulphates (S).

At a lower concentration level, we see a number of elements, which are present in earth crust at concentration well below 1 ppm. They are found in bottom ashes at concentration more than two orders of magnitude. Some of them are renowned for toxicity, like Hg and Cd. Again, we find these elements in easy to process and very concentrated phases, like HgS cinnabar for Hg, or impurities within the galena in Ag. Like in nature, also in wastes these elements will prefer non-silicate structures.

As an example we may consider the case of copper [9]. Copper earth crust average concentration is 50 ppm. As a result, the upper 3 km of the earth's continental crust contain about 60,000 billion metric tons of copper. However, in common rock copper concentration goes from 1 to 100 ppm, whereas a budget, yet minor, can be found in few deposits with ore grade concentration (8000 ppm minimum). In the ore copper is present as a sulphide, like $CuFeS_2$, chalcopyrite, or carbonate, like azurite and malachite. In average bottom ashes from MSWI, the average concentration is not far from the ore grade (3500 vs 8000 ppm), supporting a concentration, or a very limited recycling, of this element by human activity. In the waste however copper is generally found as a metal.

The above discussion shows how the mineralogical composition affects the waste composition, and its environmental concern. The same element can occur in stable or in highly reactive non-equilibrium materials, which are likely to release potentially toxic elements, in dependence of the kind of fractionation occurring. Another example showing how the mineral species hosting an element can change during anthropic processing is zinc. In nature, zinc is found mostly as a sulphide; when extracted zinc is worked to make metal coatings, but in an incinerator, it is not found as a metal, but most frequently as an impurity within a glass, or in spinel oxides, more rarely as a hydroxide (Fig. 4, [12]). Much of it concentrates in the fly ashes, which are Zn enriched [17]. This changing phase makes a difference in the environmental approach to the recycling of Zn bearing wastes, following the principle that environmental science looks at minerals as long as they are reactive.

Fig. 4 Ca–Zn hydroxide (elongated crystals, light gray) and $Zn(OH)_2$ (small white spots) in the matrix of bottom ashes. Such phases are overlooked in XRD but have a potential in leaching Zn (unpublished)

100µm

5   A Mineralogical Perspective: What is Mineral?

It appears now that minerals are major constituents in wastes. We define minerals as the crystalline phases that have a natural counterpart, used as such or manmade, but also amorphous silicate glassy phases, which will react as minerals, and the synthetic phases, which share crystal structure and reaction processes with natural minerals. This is a broader definition: man-made phases and glass are not minerals in a strict sense [19]. We shall make some example of the different solid phases that we may call minerals.

(1) We may have minerals as natural crystalline phases that we find in waste. Although we are outside Stone Age, we still use a lot of stone in our products. A stone, made by minerals at a scale from the metric blocks used for breakwaters to the micrometric size of superfine quartz in coatings, is the starting material for several products. In this case, the mineral, resistant to leaching and reactions during manufacturing, will stay as such in the whole product cycle. As an example quartz, which, owing to the high stability at high temperature, and the resistance to hydrolysis, is a refractory mineral. To be defined as a natural mineral, through the extraction to waste cycle of a product it is sufficient that it does not experience conditions in which it reacts. An example is inert material added in concrete.

(2) We have also minerals that are manmade, but are the same as natural ones. In the process of manufacturing we make reactions occur, some of them have an analogue in nature, some of them not, but the resulting phases are the same we find in nature. An example is the widespread occurrence of mullite in a number of ceramic materials. Mullite is a natural mineral, occurring in shales included and heated by volcanic magmas. It is a mineral found only in few occurrences in nature, whereas as a product of ceramic industry it is widespread. We could tell the difference between natural and synthetic mullite only after detailed structural and chemical analysis. Another example is silicon. Silicon is almost invariably bound with oxygen. As a metal, it is an exceptional natural material, found just in few locations where reducing conditions occur, and where Si is in excess to form silicide phases. On the other side, it is an important phase in alloys and in electronics. What matters more, synthetic and natural Si and mullite samples show the same chemical properties in waste and in nature.

(3) We will consider minerals also a number of synthetic products, most common in material science, which are chemically different but structurally equivalent to the natural counterparts. A number of chemical substitutions were done starting from the crystal structure of the mineral apatite, or the perovskites, to improve the technological properties (e.g. [5]). Although such processes are economically expensive, and we do not expect massive perovskite waste, these are potentially target to recovery the PTE. A number of alloys is the major addition to this group. Structurally they are not different to some native mineral like iron, gold or silicon, but have a very different composition (Fig. 5).

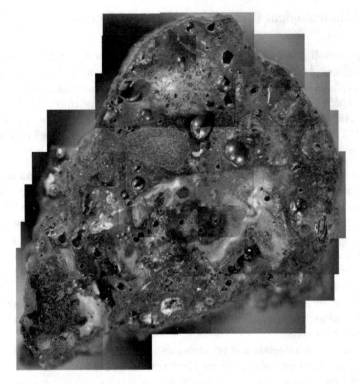

Fig. 5 MSWI bottom ash mineral grain. Note the highly heterogeneous aspect, with voids for the CO_2 emission during heating, the smeared contact between grey, red and yellow areas, the molten rounded metals, the grey metal grains. Grain size 2x2 mm^2

(4) We will add to our definition also amorphous phases, like glass or bitumen. They may react to form crystalline phases, and play a major role in reactions in wastes. In concrete, ceramics and MSWI waste, amorphous phases are often the dominant ones.

(5) Any other solid crystalline or amorphous phase could be included in our definition, no matter if yet the natural counterpart was discovered. Properties and reactions follow the same principles. However, not every solid object is a mineral, some solid waste, like plastics, rubber and generally, organic compounds, do not fall in the field of mineralogy, and so wood and wood made products. Also, bacteria and other livings have their internal properties by cellular organization and reaction which are outside the methods of mineral investigation. However, it will be important to understand the interaction between minerals and non-minerals, as they may be crucial for waste processing, increasing or decreasing the waste behaviour, in a way wanted or unwanted.

6   A Mineralogical Perspective: Which Minerals?

Let us now focus on the formation of minerals in wastes. The products of the human activities come from processes at high or low temperature that in several cases mimic those occurring in nature. Also, when they do not, we may still have minerals as a product.

Let us take as an example the bottom ashes from incinerators: in a natural environment, such a mixture of organic and inorganic constituents at high temperature and for a short time, in oxidizing conditions, is something exceptional. Still, the resulting material is a forge of minerals, crystals and amorphous glass, each of which has a natural counterpart (Fig. 6). The overall assemblage is a non-equilibrium one, with reduced metals coexisting with silicate oxides and glass. Incidentally, such a metal and glass coexistence, albeit with different phases, is found in natural chondritic meteorites [15].

A number of investigations were done on the mineralogy of MSWI residues, and a number of crystalline phases together with a significant or preponderant glass was found. The glass comes from two sources, residual from glassy wastes, like bottles, or newly formed during incineration. Some interspersed organic residual is also present. The crystalline phases found by X-ray diffraction, present at least as 1% in weight, almost invariably are minerals according to points 1 and 2 of the above definition.

A list of minerals found in MSWI bottom ashes is reported in Table 1. This list is far from complete, and the quantity of each phase may vary between incinerators. Electron microscope analysis shows a wealth of other phases, among which some metals and lower concentration silicates.

The composition is reported when clearly identified by major elements. Melilite is a solid solution between akermanite and gehlenite $Ca_2Mg(Si_2O_7)$–$Ca_2Al(AlSiO_7)$; plagioclase between albite and anorthite ($NaAlSi_3O_8$–$CaAl_2Si_2O_8$); pyroxene mostly diopside ($CaMgSi_2O_6$) with exchange of Ca for Mg and Na, and Mg for Al and Fe^{3+}; FeO_x, undefined iron oxide.

Few points are apparent: at first, the glass is the most present. We find that it is most concentrated in larger sized fraction of the ashes [12], where unreacted glassy pieces are easily detected by optical observation. However, also in the smaller portions we have up to 50% glass, which in part comes from melting of the formerly crystalline phases. Unreacted plastics or rubbers also make a portion of the amorphous share.

We can classify crystalline minerals as inert, like quartz, formed in incineration, like melilite, or grown in weathering, like calcite. The primary inert phase is quartz. Quartz was present before incinerations and survived the process. Quartz is a ubiquity mineral, found by crystallization in magmatic and metamorphic environment, and concentrated in sedimentary rocks as a fraction less prone to alteration. More rarely, if burning does not occur up to 900–1000 °C, residual muscovite from clay mixtures may still be present. Muscovite ($KAl_2(OH)_2(AlSi_3O_{10})$) is a common mica mineral found in a variety of rocks and is used in paints, plastics or cosmetics as a filler.

More minerals are formed during incineration: most common melilite, which is a solid solution of gehlenite and akermanite. In nature, melilite is not extremely

Fig. 6 Mineral textures in MSWI: top: spinel grains, crystallized with sharp edges. In one grain a darker core is related to exchange of Cr^{3+} and Al for Fe and Ti. Further growth of spinel occurs as small grains at the edges; middle: dendritic wollastonite growth, with white drops of Fe with P as an impurity to about 7 wt%, in a glassy matrix, with higher Si content.; bottom: kirschsteinite ($CaFeSiO_4$), showing skeletal growth features in a glass

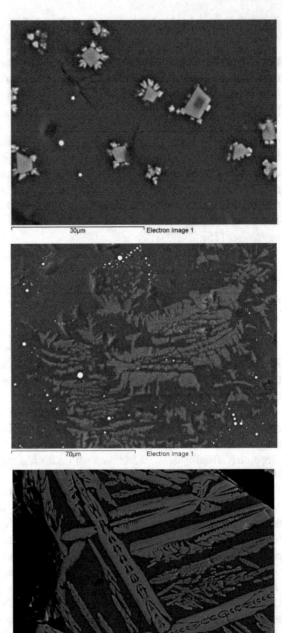

Table 1 Compilation of minerals in bottom ashes from different incinerators; ss: solid solution between different end member compositions. Data from: [2, 3]; NI 1 and NI 2: unpublished data from 5 incinerators from Northern Italy, The N1 column averages the mineralogy for the three amorphous richer products, NI 2 from the two amorphous poorer; Bay. fresh and Bay. aged, [4] analyses of fresh and aged ashes, respectively

Minerals		Alam	Assi	NI 1	NI 2	Bay. fresh	Bay. aged
Quartz	SiO_2	13	4	6	10	31	9
Melilite	ss	5		2	5	3	2
Plagioclase	ss	5	2	4	8	1	2
Pyroxene	ss	4		2		4	2
Muscovite	$KAl_3Si_3O_{10}(OH)_2$					2	3
Amphibole	ss					1	4
Calcite	$CaCO_3$	13	8	6	17	1	5
Vaterite	$CaCO_3$	2	2				
Magnesite	$MgCO_3$		2			–	–
K-carbonate	K_2CO_3					1	2
Rutile	TiO_2	1				1	1
Hematite	Fe_2O_3	4	1	1	1	2	2
Spinel (magnetite + hercynite)	$Fe_3O_4 + FeAl_2O_4$	9		0	1	5	3
Wuestite	FeO	1				1	0
FeO_x[a]	ss	14					
Cristobalite	SiO_2	1	2	1		0	0
Anhydrite	$CaSO_4$		1			3	2
Ettringite	$Ca_6Al_2(SO_4)_3(OH)_{12} \cdot 26H_2O$	1		4	10	1	7
Stratlingite	$Ca_2Al_2SiO_7 \cdot 8H_2O$		1				
Hydrocalumite	$Ca_4Al_2(OH)_{12}(Cl,CO_3,OH)_2 \cdot 4H_2O$			3	5	1	15
Apatite	$Ca_5(PO_4)_3OH$	6	6			–	–
Halite	$NaCl$	1				–	–
Larnite	Ca_2SiO_4				7		
Portlandite	$Ca(OH)_2$		1		5		
Chalcocite/covellite	Cu_2S/CuS					2	1
Amorphous		36	67	72	30	33	31

common, and it is found in high temperature Al-rich volcanic rocks. As a rule in MSWI bottom ashes, gehlenite, quartz and glassy phases are the dominant place for silicates. Alkali elements go in glassy phases. During incineration magnetite and hematite are also formed. Other important high temperature minerals which form during incineration are feldspars, reported with one of the several names given to their different compositions (albite, anorthite, bytownite, plagioclase, K-feldspar), and pyroxenes. Pyroxenes are often reported as diopside, a name which should be limited to the pyroxene with the specific composition $CaMgSi_2O_6$. Also wollastonite ($CaSiO_3$) forms by high temperature crystallization in melt.

During weathering other phases form. Carbonates, mostly as calcite, but also vaterite and dolomite, form by reaction of Ca oxides with atmospheric CO_2. Carbonates are a newly formed phases, as at 700 °C the carbonates present in large amount before incineration decompose in free lime (which may hydrate to portlandite after cooling and contact with water) and CO_2. The reverse reaction occurs during weathering. Other phases, like ettringite, form during washing of the ashes; stratlingite and hydrocalumite are hydration products of Al-rich cements [14, 17], whereas Ca-sulphates, as anhydrite, gypsum or bassanite, according to the H_2O content in the crystal structure, occur from sulfation. Except carbonates the above phases are quite soluble, posing environmental concerns for the leaching of PTE.

7   Mineral Textures

A last point will focus on mineral textures. Textures in minerals from wastes can be interpreted in the same way as they do in natural systems. We can define mineral textures as the shape of the minerals coexisting in relation to other phases. The key point in mineral texture is the crystal growth. We may obtain an information on mineral textures by microscopic observation at the optical and electron optical scale.

Mineral textures are revealing on the mineralogical history of a given assemblage. In natural rocks which underwent long lasting (millions or billions of years) cooling, equilibrium textures are obtained. We find large crystals, compositionally homogeneous, often with intergranular angles by 120°. Chemical reaction due to metamorphic transformation or low temperature hydrolysis promote growth of other newly formed phases, generally at the boundaries. Dendritic, skeletal textures form instead due to rapid crystallization; the evidence that a phase grew in non-equilibrium, during extremely fast cooling, comes from the observation of dendritic or skeletal textures of the phase in a glass. On the other side, evidence that the phase underwent a resorption is proven by the observation of gulfs or rounded edges at the boundaries of the crystalline phase in a glass (Fig. 6). Zoning will also indicate a non- equilibrium cooling. This can be observed in several examples in Fig. 6.

8   Conclusions

Similarities between natural minerals and mineral waste are apparent. Mineralogical systems, and all the research on them, can provide a basis to interpret and predict the behaviour of wastes. However when doing so, we should be aware of the differences between natural and anthropic systems.

One difference is in time: human activities are on a short or very short time scale compared to natural. As a rule, manmade products do not crave equilibrium conditions: there is trade-off between product resistance and its consumption as a waste. We wish that a product keeps its properties as long as we use it, but it should also magically dissolve properly after disposal. If it does not, we are overwhelmed by garbage accumulation, like for plastics in the oceans; if it does too soon, we may have the release of toxic elements. Fast processes are generally non-equilibrium ones. The Ostwald rule, which states that the first phases are generally kinetically favoured but less stable, provides a basis for the lower stability of anthropic products.

Another difference is in composition: metals are a major presence in human activities. We have them in several wastes, in spite of our effort for recovery. Metals are typically a reduced chemical composition, and they are major constituents in wastes.

This said we might examine a given waste deposit at several levels in a mineralogical approach: the first one is the geochemical composition, compared to the average continental one. We have compared an average with an average, both from data with strong variability. We may expect grouping of different waste data, for instance different MSWI bottom ashes from different plants, indicating different policies of municipal garbage collection, national usages, plant configuration, and even changes with time. For a composition very similar in major elements, we could have a different pattern for minor elements. Composition will be obtained by standard geochemical techniques, as ICP optical or mass spectroscopy for minor elements, and X-ray fluorescence (XRF) for major ones.

Moreover, we should seek for mineralogy, which provides the basis for the interpretation and prediction of the waste in recovery. This will give further information on the materials that made the product before it becomes a waste, on how it was machined, and on the reactions it underwent after weathering. X-ray diffraction, together with Rietveld refinement to determine glass and other phases quantitative content, will be the technique to obtain the average mineralogy. SEM–EDS will complement the data, to show the mineralogical composition and seek for phases at lower abundance.

Last, we should look for the textures, which will show how the mineral got to the non-equilibrium conditions of the waste as an assemblage of minerals, so to understand the reactions occurring between the different phases. Optical and electron optical microscope analysis will be the technique with which we will observe the mineral textures.

Clearly, this general picture will have different outcome with the different kind of wastes: for the different wastes we will have different average composition, different minerals, different reactions and different textures. This will be discussed in detail in the following chapters.

References

1. Abdul-Wahab S, Marikar F (2012) The environmental impact of gold mines: pollution by heavy metals. Central Eur J Eng 2:304–313
2. Alam Q, Schollbach K, van Hoek C, van der Laan S, de Wolf T, Brouwers HJH (2019) In-depth mineralogical quantification of MSWI bottom ash phases and their association with potentially toxic elements. Waste Manage 87:1–12. https://doi.org/10.1016/j.wasman.2019.01.031
3. Assi A, Bilo F, Federici S, Zacco A, Depero LE, Bontempi E (2020) Bottom ash derived from municipal solid waste and sewage sludge co-incineration: First results about characterization and reuse. Waste Manage 116:147–156
4. Bayuseno AP, Schmahl WW (2010) Understanding the chemical and mineralogical properties of the inorganic portion of MSWI bottom ash. Waste Manage 30:1509–1520
5. Bhalla AS, Ruyan GR, Roy R (2000) The perovskite structure—a review of its role in ceramic science and technology. Mat Res Innovat 4:3–26
6. CRC Handbook of Chemistry and Physics (2016–2017) Abundance of elements in the earth's crust and in the sea, John Rumble Ed. Taylor and Francis, 97th ed, pp 14–17
7. Dijkstra JJ, Comans RNJ, Schokker J, Van der Meulen MJ (2019) The geological significance of novel anthropogenic materials: deposits of industrial waste and by-products, Anthropocene, 28, 2019. ISSN 100229:2213–3054
8. EUROSTAT (2020) Energy, transport and environment statistics. Edition 2020. Luxembourg Publication Office, pp 1–192. ISBN 978-92-76-20736-8
9. Henckens MLCM, Worrell E (2020) Reviewing the availability of copper and nickel for future generations. The balance between production growth, sustainability and recycling rates. J Clean Prod 264:121460. ISSN 0959-6526
10. Hjelmar O, Van Der Sloot HA, Van Zomeren A (2013) Hazard property classification of high temperature waste materials. ECN Publishing. http://resolver.tudelft.nl/uuid:9216fae6-9a3d-4896-8d6b-6bfb6b7b4592
11. Lemos M, Valente T, Marinho Reis P, Fonseca R, Delbem I, Ventura J, Magalhães M (2021) Mineralogical and geochemical characterization of gold mining tailings and their potential to generate acid mine drainage (Minas Gerais, Brazil) Minerals 11(1):39. https://doi.org/10.3390/min11010039
12. Mantovani L, Tribaudino M, De Matteis C, Funari V (2021) Particle size and PTE speciation in MSWI bottom ash. Sustainability 13(4):1911
13. Mason B, Moore CB (1982) Principles of geochemistry, 4th edn. John Wiley & Sons Inc., New York
14. Moon J, Oh JE, Balonis M, Glasser FP, Clark SM, Monteiro PJM (2011) Pressure induced reactions amongst calcium aluminate hydrate phases. Cem Concr Res 41:571–578
15. Papike JJ (2018) Planetary materials. De Gruyter, Berlin, Boston. https://doi.org/10.1515/978 1501508806
17. Tandon K, Heuss-Aßbichler S (2021) Chapter 12 Fly ash from municipal solid waste Incineration: from industrial residue to resource for zinc. In: Pöllmann H (ed) Industrial waste: characterization, modification and applications of residues. De Gruyter, Berlin, Boston, pp 379–402. https://doi.org/10.1515/9783110674941-012

Bibliographic References

A basic acquaintance in mineralogy is assumed in this chapter and in the following. For the interested reader we suggest the following books

18. Wenk HR, Bulakh A (2016) Minerals their constitution and origin, 2 ed. Cambridge University Press Cambridge. ISBN: 9781107514041

19. Klein C, Philipotts A (2016) Earth materials. Introduction to mineralogy and petrology. Cambridge University Press. ISBN: 9781316608852
20. Dyar MD, Gunther ME (2012) Mineralogy and optical mineralogy. ISBN 978-1-946850-02-7

What is Waste, and How We Manage in Europe

Mariachiara Zanetti and Deborah Panepinto

1 Introduction

Strategies of waste management: reduction at the source, end of pipe treatment.

Waste management in the European Union is regulated by the Waste Framework Directive (WFD—Dir. 2008/98/EC) [1]. Into the Italian legislation, this Directive is transposed in the Consolidated Environmental ACT (TUA, Italian Legislative Decree 152/2006) [2]. The WFD is based on a lot of waste management concepts, such as:

- Reducing the use of resources;
- Consideration of the entire life cycle of materials/products;
- Achieving the best overall environmental result;
- Application of the "polluter pays" principle through the mechanism of extended producer responsibility.

The hierarchy of operations that the Directive defines for waste mechanism is as follow:

- Prevention;
- Reuse;
- Recycling;
- Recovery (i.e. energy recovery);
- Final disposal in landfill.

This hierarchy specifies the priority for the application of management operations, according to which prevention of waste generation is preferable to any other management method. Anyway prevention is an action of the production of the waste so the first really management intervention is the second, the reuse.

M. Zanetti (✉) · D. Panepinto
Politecnico Di Torino, Corso Duca degli Abruzzi 24, 10129 Torino, Italy
e-mail: mariachiara.zanetti@polito.it

D. Panepinto
e-mail: deborah.panepinto@polito.it

Indeed, when the waste has been produced, it is preferable to reuse the material (into the same process that produce it) and secondly to recycle it. If none of these options are sustainably viable, then it is preferable to use it for recovery (e.g. as energy recovery in thermal treatment plants) before send it to final disposal in landfill.

The WFD also specifies that it is possible to deviate from this general hierarchy in particular cases if the greatest environmental benefit can be demonstrated on the basis of a Life Cycle Assessment (LCA).

The last amendment of the WFD was introduced at the end of May 2018 with the introduction of the so-called "Circular Economy Package". This package reinforces some concepts that were already contained in the WFD, and clarifies some definitions also in order to improve the collection and processing of statistical data on waste management. Particular attention is paid to the target of keeping materials as long as possible within the cycle of production and consumption, so as to minimise the need of virgin materials and the amount of waste to send to the final disposal.

From the viewpoint of the average waste production in the European countries the amount is more or less stable in the year as reported from the Eurostat website [3].

In Fig. 1 is reported the amount of waste production in the EU countries in the year 2018 and in Fig. 2 the total waste generation by economic activities and household.

EU = European Union	DE = Germany	ES = Spain	NO = Norway
EE = Estonia	PL = Poland	LT = Lithuania	UK = United Kingdom
FI = Finland	FR = France	SK = Slovakia	RS = Serbia

(continued)

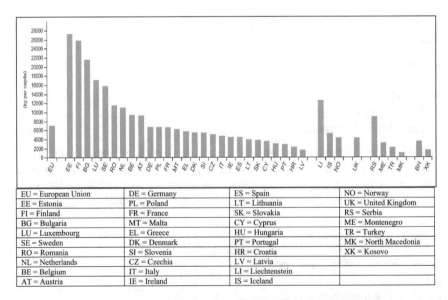

EU = European Union	DE = Germany	ES = Spain	NO = Norway
EE = Estonia	PL = Poland	LT = Lithuania	UK = United Kingdom
FI = Finland	FR = France	SK = Slovakia	RS = Serbia
BG = Bulgaria	MT = Malta	CY = Cyprus	ME = Montenegro
LU = Luxembourg	EL = Greece	HU = Hungary	TR = Turkey
SE = Sweden	DK = Denmark	PT = Portugal	MK = North Macedonia
RO = Romania	SI = Slovenia	HR = Croatia	XK = Kosovo
NL = Netherlands	CZ = Czechia	LV = Latvia	
BE = Belgium	IT = Italy	LI = Liechtenstein	
AT = Austria	IE = Ireland	IS = Iceland	

Fig. 1 Total Wastes generated in EU in the year 2018. Modified from [3]

(continued)

EU = European Union	DE = Germany	ES = Spain	NO = Norway
BG = Bulgaria	MT = Malta	CY = Cyprus	ME = Montenegro
LU = Luxembourg	EL = Greece	HU = Hungary	TR = Turkey
SE = Sweden	DK = Denmark	PT = Portugal	MK = North Macedonia
RO = Romania	SI = Slovenia	HR = Croatia	XK = Kosovo
NL = Netherlands	CZ = Czechia	LV = Latvia	
BE = Belgium	IT = Italy	LI = Liechtenstein	
AT = Austria	IE = Ireland	IS = Iceland	

In Fig. 3 the situation concerning the total waste management operated in the year 2018 in EU are reported. In this picture it is possible to see deep differences about the most advanced Countries, located in the left side of the graph, where the final disposal in landfill is almost completely over and the other countries (in the right side of the graph) where the landfill management method is still widely used.

2   Reuse, Recycling

(a)   Technological issues for reuse, recycling

On 30 May 2018, four new European directives were approved that constitute the so-called "Circular Economy Package" and amend six pre-existing directives, which identify a series of measures to make the transition to a stronger and more circular economy, in which resources are used according to an economic system that can be regenerated. The strategic objectives pursued by the Circular Economy Package focus on strengthening the waste hierarchy, according to the following priority:

– Upstream reduction of waste (prevention and ecodesign);
– Reuse of material;
– Recycling in the form of matter;
– Recovery in the form of electrical and/or thermal energy.

The first action, the most virtuous, is related to prevention or to the reduction of waste production and is therefore configured as an intervention on production. The first action on waste management is therefore reuse (that it means the reuse of a material in the same production cycle) followed by recovery (of matter or energy, therefore in this case the recovery of matter or energy from a material once it has reached the end of its life by means of applied treatments) and the final storage in landfills.

Fundamental is the byproduct definition and distinction from waste. More recently, the European Commission, through its Communication COM(2007)59 of 21/2/2007, has intended to provide some useful guidelines to distinguish, in the context of residues deriving from a production process, what is waste from what is

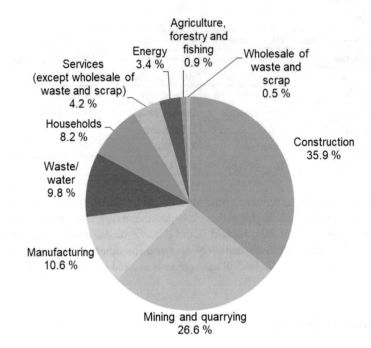

Fig. 2 Waste generation by economic activities and household, EU, 2018 (% share of total waste). Modified from [3]

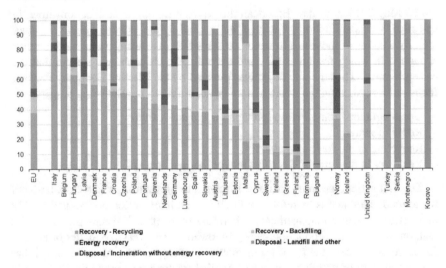

Fig. 3 Waste treatment by type of recovery and disposal, 2018 (% of total treatment) [3]

Fig. 4 Simplified diagram related to heat treatments [8]

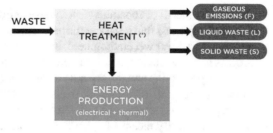

*Including waste pre-treatment, combustion, treatment, etc.

not. In the need to give an answer on how to make a "distinction between materials that are not the primary objective of a production process, but that can be considered by-products not similar to waste, and materials that must instead be treated as waste", the Commission has adopted the three conditions indicated by the Court of Justice of the European Communities, so that byproduct distinction can be made as precise as possible, consisting of:

- the reuse of the material is certain and not only possible;
- the material may be reused without being previously processed;
- the preparation of the material for its reuse takes place during the production process.

A waste that has been treated may be reused or recycled if comply with the following requirements:

- the substance or article is intended to be used for specific purposes;
- there is a market or demand for that substance or object;
- the substance or object meets the technical requirements for the specific purposes and complies with existing legislation and standards applicable to products.

On the grounds of the above mentioned definitions the substantial difference between a byproduct or waste definition is that the first may be reused without any treatment, normal industrial practice excepted.

The unit for the separation and processing of wastes are designed to accomplish the following:

modify the physical characteristics of the waste so that the components can be removed more easily;
to remove specific, useful components and contaminants from the waste stream;
to process and prepare the separated materials for subsequent uses.
Physical treatments

Physical treatments for the separation of constituents coming from a solid residue exploit physical properties such as: Dimension, mechanical resistance, shape, elastic properties, density, surface properties, magnetic and electrical properties, reflecting power and thermal properties.

The main physical treatments are:

- Sieving: By means of apparatus such as sieve or trommel. Size separation or classi-fication is a process that separates the particles into fractions according to their size and shape, in dry or wet environment. In solid waste processing, the most common process is dry screening, that is targeted to: remove the oversized materials from the waste processing stream, separate the feed into proper size fractions according to the requirements of following treatments, selectively sort waste components into separate products in case that selective shredding is accomplished.
- Size reduction: By means of apparatus such as shredder (rotary, blade, impulse) or mill (hammer, flail). Size reduction is a mechanical process that is used to obtain a uniform product of reduced size compared with that of the initial waste stream, either for direct use and/or storage or to make the materials useful for further separation processes. Materials with high mechanical resistance will be less reduced than materials characterized by low mechanical resistance.
- Fluid media separation: By means of apparatus such as column classificator, centrifuge classificator (cyclon) or density classifier with callow cone and rotating drum employing different fluids (mainly air and water). Fluid media separation is based on the dimension, shape and density of waste constituents.
- Elastic properties: particles undergo to impact and according to their weight and/or elasticity follow different trajectories. Relative apparatus are ballistic separators.
- Surface properties. Surface strength of particles indicates low or high affinity for waters (hydrofobe or hydrophilic particles). This properties may be exploited by means of the flotation process (i.e. preferential adhesion of hydrophobic particles to air) to achieve the separation of constituents. Therefore in the flotation cells the growth of air bubbles is enhanced and the hydrophobicity of materials may be enhanced by means of collector agents and or surfactants.
- Magnetic separation: By means of apparatus such as magnetic drums, suspended belt magnets and magnetic plates exploiting the different magnetic susceptibility of waste constituents.
- Electrical separation: By means of apparatus such as electrostatic separator and eddy current separator that exploit the electrical conductivity (proper of metals) of waste constituents.
- Reflecting power: These systems use sensors to detect specific optical, chemical, or physical properties of single particles. Some developed sensors are: Color sensors, X-ray transmission sensors, X-ray fluorescence (XRF) sensors, near infrared (NIR) sensors.
- Thermal properties. The separation is obtained exploiting two important material properties: the thermal capacity and the melting temperature.
- Biological and Chemical treatments

Biological processes can be both aerobical (in presence of oxygen) and anaero-bical (in absence of oxygen). Aerobic biological treatments are faster because of the high reaction velocity and are characterized by the production of water and carbon dioxide. Anaerobical processes are slower because of the slow reaction velocity and are characterized by the production of carbon dioxide and methane.

Generally aerobic biological processes and related technologies (aerobic biological reactors) are applied to wastes characterized by a high organic carbon content (and low amount of substances toxic for microorganisms such as: heavy metals, salts, halogenated organic compounds and so on) in order to obtain the stabilization and, eventually the fertilizer production (compost).

Anaerobical biological processes and related technologies are also applied to wastes characterized by a high organic carbon content (and low amount of substances toxic for microorganisms such as: heavy metals, salts, halogenated organic compounds and so on) mainly to obtain methane/energy production. The remaining solid residue after the anaerobic digestion, according to the final chemical and physical characteristics may be employed as a biological fertilizer.

Chemical processes are aimed to obtain a chemical reactions, eventually adding reactants and or air, thus achieving the transformation of some substances that are present in wastes.

The purposes of chemical processes applied to wastes may be different. Some examples are the inertization treatment, the leaching process and combustion/gasification reactions that will be illustrated in the following paragraphs.

The inertization treatment («process of solidification/stabilization») allows to achieve the change of the chemical/physical characteristics of the waste, in order to reduce its hardousness during the storage, transport, recovery and disposal phase. E.g. for stabilization is intended a process through which are created unsoluble compounds, able to catch one or more toxic elements in a stable cristalline structure; for solidification is intended a process that originates, starting from a liquid or pasty waste, a final solid product that can be easily transported, recovered or disposed.

Processes based on the cement or lime (hydraulic binders) have been developed essentially for the inertization of inorganic waste and are particulary suitable for those containing heavy metals. An other solutions could be the stabilization/solidification with asphalt binders. The asphalt, indeed, is a highly hydrofobic substance that creates an efficient immobilizing barrier in order to prevent the washout and the diffusion of pollutants from waste into the surrounding environment.

The leaching process may be aimed both to extract some substances from wastes that may be recovered by means of further treatments or to eliminate harzardous substances from wastes.

3 Energy Recovery and Landfill Disposal

(a) Incineration, gasification, pyrolysis

Thermal treatments are high—temperature chemical processes in which organic substances are broken down in order to produce other substances with a simpler chemical composition [4–6]. The primary goal of any thermal treatment is the transformation of the waste (with the production of substances that have less impact on the environment and on the human health) with the consequent reduction of the quantities and volumes to send in landfill and, at the same time, obtaining a recovery of energy [7].

In the waste sector, the following thermal treatments can be applied:

- Direct combustion in incineration plant;
- Gasification;
- Pyrolysis.

Among these, the incineration is the operation that has so far been most applied to solid waste, with experience on an industrial scale that is now very extensive; the other treatments have been developed as alternative technologies to incineration, which, however, have not yet given rise to significant experience on an industrial scale (from the viewpoint of the waste materials, a different discussion can be done from the viewpoint of other solid materials). The incineration process is based on the direct combustion with the use of the sensible heat of the flue gas to produce steam and from this steam to obtain electricity and/or thermal energy.

Alternative technologies, instead, essentially involve the production of a fuel gas (or of a gas and a liquid fraction), which can be burnt alternatively on site, to produce energy, or be used as feedstock for the production of potentially marketable fuels and/or feedstock for the chemical industry.

If we define R as the ratio between the used amount of oxidizing agent (air and/or oxygen) and the theoretical (stoichiometric) amount, the three main thermal treatment processes can be represented schematically as shown in Fig. 5.

The incineration of wastes is a thermal oxidation process, in which the fundamental elements that constitute the organic substances are oxidized, giving rise to simple molecules at gaseous state (flue gas). The organic carbon is oxidized to carbon dioxide (CO_2), hydrogen to water (H_2O), sulphur to sulphur oxide (SO_2) and so on; the inorganic part of the waste is oxidized too and leaves the process as a solid residue to be disposed or recovery. Since the process is oxidative, the presence of

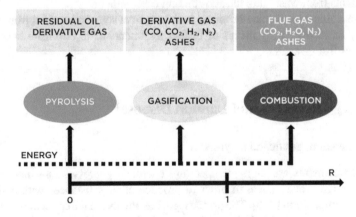

Fig. 5 Schematic representation of thermal treatments [9]

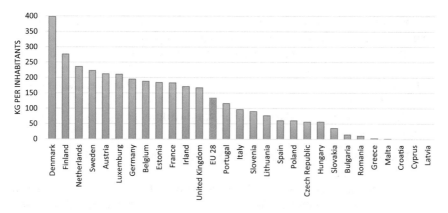

Fig. 6 Per capita incinerated wastes in Europe in the 2016. Modified from [10]

oxygen is necessary for the reactions. Air is normally used, supplied in excess to the stoichiometric amount in order to facilitate the chemical reactions [7].

The gasification is a partial oxidation of the solid materials (the amount of oxidant agent using during the process is lower than the stoichiometric amount) instead the pyrolysis is a thermal degradation process (without oxidant agent).

Figure 6 shows the situation concerning the amount of wastes (with a lower heating value higher than 11 MJ/kg) incinerated in the European countries in the year 2016.

Main sections of thermal treatments plants.

The main sections of a thermal treatment (incineration, gasification or pyrolysis) plant are the following [7]:

- Combustion chamber. The most common combustion technologies for the wastes treatment are the moving grate system and the fluidised bed system. The first one (moving grate system) is the most widely used technology in account to the flexibility of operation and reliability resulting from the numerous applications. The technology consists of a grid, inclined to the horizontal line, on which a bed of waste (several dozen centimetres thick) is placed. The grid consists of a set of elements called "fire bars", disposed in a way to allow the passage of the combustion air and its distribution inside the entire bed of waste. The combustion air is injected both under the grid and directly inside the combustion chamber. The residence time of the waste into the combustion chamber (and so on the grid) is generally between 30 and 60 min (in order to allow the completely oxidation of the organic substances). The residual bottom ash, generated from the process, is discharged from the final part of the grid with appropriate systems (water bath accumulation tanks). Temperature levels in the range of 950—1000 °C are considered sufficient, in correspondence with adequate oxygen contents (6—8%) and turbulence, to guarantee the complete oxidation of the organic components.

  The fluidized bed system consists of a combustion chamber inside which an amount of inert materials (the bed) is kept in suspension (fluid) by an upward air current air (which also acts as a comburent). The movement of the bed ensures the good comburent—fuel contact, as well as, considerable uniformity of temperature and mixing, which help to ensure constant and complete combustion. This equipment, initially develop in the petrochemical industry, was adapted to the combustion of rather homogeneous and small substances (such as for example wastewater sludge). Wastes must therefore undergo at least one shredding treatment in order to be reduced in size.

At the end of the thermal treatment process two different types of solid residues are generated:

- Bottom ashes, generally equal to 20% of the wastes in input to the plant. These kinds of residues are generally sent to recovery;
- Fly ashes, generally 4–7% of the wastes in input to the plant. These kinds of residues, removed through the flue gas depuration line, are hazardous wastes and generally are sent in a landfill for hazardous wastes.

- Flue gas depuration line. A thermal treatment plant generates three different output: Gaseous, liquid and solid emissions. Before being released into the atmosphere from the chimney, the flue gas must be treated in order to reduce the concentrations of the pollutants generated during the process. The flue gas treatment section is very articulated and complex, as a consequence of the increasingly strict limits imposed by the regulations and of a concrete technological progress, which has led in recent years to the development of sophisticated systems able to allow emission much lower than those that the legislation indicates as maximum.

The pollutants present in the flue gas can be grouped into two different groups:

- macro-pollutants, substances present into the flue gas in concentrations of the order of mg/Nm^3, such as PM (particulate matter), sulphur oxides (generally SO_2), nitrogen oxide (NO_x), carbon monoxide (CO) and halogen acids (generally HCl and HF);
- micro-pollutants, substances, present into the flue gas in much lower concentrations (in comparison with the concentrations of the macro-pollutants), which include both inorganic species, such as heavy metals (Cd, Cr, Hg, Pb, Ni, and so on) and organic species, such as PCDD, PCDF (Dioxins and Furans) and Polyciclic Aromatic Hydrocarbons (PAH).

Concerning the reduction of these pollutants (both macro and micro—pollutants), according to the current legislation it is necessary the use of the BAT (Best Available Techniques) defined and reported in the official document of the IPPC Bureau for the incineration plants [11].

- Energy recovery section. Energy recovery from incineration is commonly achieved from the steam derived through the cooling of the flue gases that is necessary for their subsequent treatment.

Energy recovery from gasification and pyrolysis, instead, are generally obtained from the combustion of the syngas generated during the process and so from the steam obtained from this combustion.

The possible configurations of the plant are two: Only electric configuration (with the production of only electric energy) or cogenerative configuration (with the production both of electric and thermal energy). Generally, in case of only electric configuration the gross yield is about to 30% and in case of cogenerative configuration the gross yield (from electric and thermal revenue) is about to 70%.

(b)   Landfill disposal

According to the EU's waste hierarchy previously reported, landfilling is the least preferable option and should be limited to the necessary minimum [12].

In the year 2018, 38,4% of all municipal waste generated in the EU countries was landfilled (as reported in Fig. 3). This can have dangerous effects on human health and on the environment. The generation of leachate can contaminate groundwater, also methane is produced, which is a greenhouse gas. In addition, where recyclable waste is landfilled, materials are unnecessarily lost from Europe's economy [12].

Landfills are divided into [12]:

- landfills for hazardous waste;
- landfills for non-hazardous waste;
- landfills for inert waste.

Solid Waste landfill refers to an entire disposal facility in a contiguous geographic space where waste is placed in or on land [13].

Landfills primarily use the "area fill" method which consists of waste placement on a liner, spreading the waste mass in layers, and compaction with heavy equipment. Daily cover is then applied to the waste mass to prevent odors, blowing litter, scavenging, and vectors (carriers capable of transmitting pathogens from one organism to another). Landfill liners may be comprised of compacted clay or synthetic materials to prevent off-site gas migration and to create an impermeable barrier for leachate. A final cover or cap is placed on top of the landfill, after an area or cell is completed, to prevent erosion, infiltration of precipitation, and for odor and gas control [13].

Modern engineered landfills are provided with nearly impermeable bottom liner and cover systems, gas collection systems, and groundwater monitoring systems to minimize the seepage of leachate and migration of gases into the atmosphere. As previously reported placement of waste in the landfill is performed in various stages, and subsequently different types of covers are applied (e.g., daily cover, intermediate cover, and final cover) to prevent exposure of waste to the surrounding environment at different stages of landfill operation. At the end of the day, a layer of soil (~ 150 to 300 mm thick) is placed over the daily placed and compacted waste as daily cover. Various alternative materials other than soil such as shredded tires, wood chips, removable textile cover or single use plastics are also used as daily cover materials as there are no regulations regarding hydraulic conductivity to such covers. Intermediate covers are applied at those sections of the landfills where another lift of waste will not be placed within 60–90 days of the waste placement. Like daily covers, there are no regulatory requirements governing hydraulic conductivity of the intermediate covers. The final cover is placed when the landfill reaches the designed waste capacity. The primary function of the final cover system is to prevent the inflow of precipitation into the waste and migration of gases from the landfill into the atmosphere. The minimum regulations require the landfill cover to have an infiltration layer and an erosion layer; however, landfill cover can have several layers depending on the site conditions, waste composition, and climatic conditions. The conventional final cover systems typically have one or more barrier layers to restrict the infiltration and gas migration [14].

By analyzing, in particular, the MSW landfill (landfill for non—hazardous waste) it is necessary to highlight that after placement in a landfill, a portion of organic waste (such as paper, food waste, and so on) decomposes [13].

Raw landfill gas (LFG) or source gas is produced by microorganism under anaerobic conditions and normally consisting of 50–60% methane and 40–50% carbon dioxide. In addition, LFG also contains hundreds of different compounds in trace amounts (combined ~ 1% in the total volume and referred to as 'trace gases'), originating either from disposed waste products and hazardous waste and/or from waste degradation processes occurring in the landfill body [15].

Landfill gas generation occurs under a four phase process [16]:

- First, CO_2 is produced under aerobic conditions (Phase I);
- After oxygen (O_2) is depleted, CO_2 and Hydrogen (H_2) are produced under anaerobic conditions (Phase II);
- Then CO_2 production depletes in proportion to the methane (CH_4) that is produced (Phase III);
- Finally, CH_4, CO_2 and nitrogen (N_2) production stabilize (Phase IV).

Significant landfill gas production typically begins one or two years after waste disposal in a landfill and can continue for 10 to 60 years or longer [13].

As indicated one of the gases generated into the landfill is the methane. Methane is twenty-three times more effective at trapping heat in the atmosphere than carbon dioxide, which is the most prevalent greenhouse gas. Since a lot of landfills contain organic waste like food and paper, the potential for methane emissions is very high.

Methane generation in landfills is a function of several factors, including [13]:

- the total amount of waste;
- the age of the waste, which is related to the amount of waste landfilled annually;
- the characteristics of the MSW, including the biodegradability of the waste;
- the climate where the landfill is located, especially the amount of rainfall.

Methane emissions from landfills are a function of methane generation, as discussed above, and:

- the amount of CH_4 that is recovered and either flared or used for energy purposes,
- the amount of CH_4 that leaks out of the landfill cover, some of which is oxidized.

4 A Reuse and Recycling Example: Recovery of Automotive Shredder Residues

The survey presented in this chapter analyzes the opportunity of assimilating several fractions extracted from the automotive shredder light part to a solid recovered fuel.

The suitable management of end-of-life cars and related residues that are produced from their shredding and metal separation plants is still a crucial issue. The target of 95% of reuse and recovery and 85% of reuse and recycling is reported by EU-Directives 2000/53/EC and 2018/849/EC. Reuse intends any technique by which components of end-of life cars are utilized for similar objective for which they were conceived; recycling intends the reprocessing in a fabrication procedure of the waste materials for the original objective or for other aims but excluding power recovery. Energy recovery means the use of combustible waste as a means to generate energy through direct incineration with or without other waste but with recovery of the heat. Actually end-of-life vehicles undergo to: dismantling, with separation and collection of materials at the dismantling stage and post-shredder treatments, with the separation of materials from automotive shredder residues after the shredding.

So according to Directive 2000/53/EC, power recovery is agreed to accept up to 10%. In Italian Republic the rules regarding solid recovered fuel (for example the fuel deriving from automotive shredder residues) are patched by DM 22/2013. These conditions include the compliance with the threshold values of three parameters, that are related to the domains of market (heating value), combustion method (chlorine content) and environment (mercury content). Furthermore, additionally the content of some heavy metals need to be lower than the set up threshold values.

This study analyzes the possibility of assimilating fractions coming from automotive shredder residues to a solid recovered fuel, to be used in plants such as cement factories, foundries or others with a high need of thermal energy [17]. The procedure for the assimilation has included a characterization of the automotive shredder residues collected from a dedicated shredding test in an industrial plant located in the north of Italy, tests for the separation of the most abundant and highest energy-content fractions, and finally the characterization of the separated products.

The flow sheet of the dismantling, shredding and sorting operations carried out in the industrial plant is shown in Fig. 7. Before entering the shredding and sorting plant, end-of-life vehicles undergo the treatments of dismantling and depollution. In the phase of dismantling, end-of-life vehicles are deprived of bumpers, tires and gas containers and all fluids are removed from the vehicles. After these operations end-of-life vehicles are introduced in a preliminary grinder and, subsequently, in the main shredder (a hammer mill).

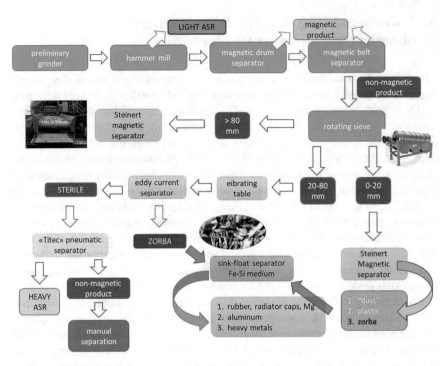

Fig. 7 Scheme of dry mechanical plus thermal reclamation treatment [17]

The final balance of the obtained products from the industrial tests performed on 435 end-of-life vehicles in the mentioned industrial plant is reported in the following table.

Fraction	Percentage [%]
Proler (steel)	69
Light shredder residue	23
Heavy shredder residue	2,7
Titec	1
Small zorba < 20 mm	2
Large zorba	2,3
Total	

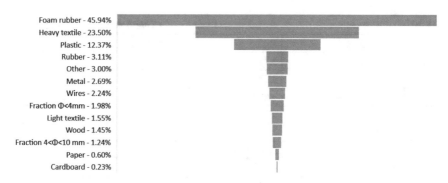

Fig. 8 Results of light automotive residues classification

The light automotive shredder residue accounted for 23% b.w. of the total weight involved in the shredding test. The results of the product composition analysis (Fig. 8) show that the sample was made up, for approx. 80%, of the three main products namely polyurethane foam (46%), heavy textile (24%) and plastic (12%). The results of the particle size analysis carried out on the four more abundant fractions are shown in the following Table.

Product	D_{10} (mm)	D_{50} (mm)	D_{90} (mm)
PUR	78	150	250
Heavy textile	105	220	500
Plastic	50	120	230
Metals	58	160	210

It was seen that amounts equal to 50% by weight (b.w.) of polyurethane foam and to 75% of heavy textile had sizes of more than 150 mm. That outcome suggested the possibility of separating the two most abundant fractions from the light shredder residue and subsequently, testing their suitability to be used as a solid recovered fuel.

According to Italian DM 22/2013, a waste product can assume the status of a solid recovered fuel if its characteristics satisfy the requirements for net heating value, chlorine, mercury and heavy metal content. The higher heating value found for polyurethane foam and heavy textile was of 28.04 ± 0.60 kJ/g and 22.64 ± 0.51 kJ/g respectively. On the basis of the net heating values, polyurethane foam could be placed in class I and the heavy textile in class II.

The following table reports the content of heavy metals of the two products. It can be seen that in both products the metal content was well below the threshold values fixed by DM 22/2013, with the exception of nickel. Nickel concentrations were 43.7% and 21.7% higher than the threshold values for heavy textile and foam rubber, respectively.

Metal	Heavy textile	Polyurethane foam	Threshold values 22/2013
	mg/kg	mg/kg	mg/kg
Cd	$1,40 \pm 0,17$	$1,75 \pm 0,18$	4
Cr	$51,3 \pm 2,6$	$33,3 \pm 0,6$	100
Cu	259 ± 30	270 ± 10	500
Pb	$24,7 \pm 6,0$	$29,8 \pm 6,2$	240
Fe	5668 ± 180	2549 ± 112	–
Mn	$92,9 \pm 4,3$	$58,6 \pm 0,8$	250
Zn	1974 ± 39	908 ± 15	–
Ni	$43,1 \pm 6,8$	$36,5 \pm 2,0$	30
Co	$6,76 \pm 0,60$	$3,54 \pm 1,64$	18

Post shredder treatments are at present utilized to divide the steel and the essential metallic parts from shredder residues. The valorization of the remaining waste products with the necessity of conforming with the statements of EU-Directives 2000/53/EC and 2018/849/EC is still a challenge. The objective of this survey was testing an operation capable to recover several waste products from automotive shredder residue that could be assimilated to a solid recovered fuel. The outcomes of the sieving test and characterization of the separated products illustrates that polyurethane foam and heavy textile had the characteristics to be assimilated to a opaque recovered fuel and consequently they were promising candidates to integrate the conventional fuel in cement factories, foundries or others.

References

1. Waste Framework Directive (WFD—Directive 2008/98/EC), available at EUR-Lex—32008L0098—EN—EUR-Lex (europa.eu), last access on 13th January 2022
2. TUA, Italian Legislative Decree 152/2006 available at Dlgs 152/2006—Norme in materia ambientale (camera.it), last access on 13th January 2022
3. Eurostat Statistics explained, available at Waste statistics—Statistics Explained (europa.eu), last access on 17th January 2022
4. CEWEP (Confederation of European Waste-to-Energy Plants), available at CEWEP—The Confederation of European Waste-to-Energy Plants, last access on 13th January 2022
5. Lindberg D, Molin C, Hupa M (2015) Thermal treatment of solid residues from WtE units: A review. Waste Manage 37:88–94
6. Lombardi L, Carnevale E, Corti A (2015) A review of technologies and performances of thermal treatment systems for energy recovery from waste. Waste Manage 37:26–44
7. Utilitalia (2020) White paper on Municipal Waste incineration, pp 96
8. De Stefanis P (2007) Sviluppi tecnologici dei trattamenti termici dei rifiuti, Convegno; Per una gestione sostenibile dei rifiuti: tecnologie a confronto, Bologna, 9 Luglio 2007
9. ENEA (2008) ENEA e le tecnologie per la gestione sostenibile dei rifiuti, Workshop, Roma, 18 Giugno 2008
10. ENI scuola energia e ambiente, available at Rifiuti urbani inceneriti pro capite in Europa, anno 2016—Eniscuola, last access on 13th January 2022

11. European Commission (2019) Best Available Techniques (BAT) reference document for waste incineration, available at Best Available Techniques (BAT) Reference Document for Waste Incineration (europa.eu) last access on 13th January 2022
12. European Commission, Landfill Waste, available at Landfill waste (europa.eu), last access on 17th January 2022
13. EPA—United States Environmental Protection Agency, Office of Air and Radiation (2011), Available and emerging technologies for reducing Greenhouse Gas Emissions from Municipal Solid Waste Landfills, June 2011, pp 28
14. ATSDR—Agency for Toxic Substances and Disease Registry (2001), Landfill Gas Primeer— An Overview for Environmental Health Professionals, Chapter 2: Landfill Gas Basics, available at http://www.atsdr.cdc.gov/hac/landfill/html/ch2.html
15. Chetri JK, Reddy KR (2021) Advancement in Municipal Solid Waste landfill cover system: A review. J Indian Inst Sci 101:557–588
16. Duan Z, Scheutz C, Kjeldsen P (2021) Trace gas emissions from municipal solid waste landfill: A review. Waste Manage 119:39–62
17. Ruffino B, Panepinto D, Zanetti MC (2021) A circular approach for recovery and recycling of automobiles shredder residues: material and thermal valorization. Waste Biomass Valorization 12:3109–3123

Thermodynamics and Kinetics of HT-Processes

Alessandro Pavese

1 Introduction

Thermodynamics is long being used as a powerful theoretical tool to describe, interpret and predict physical–chemical processes, both at micro-scale and macro-scale. Thermodynamics, in turn, develops at different degrees of complexity, which are reflective of the nature of the "systems" under study and their relationships with the "environment" around. Such an "environment" is composed of a "reservoir" and "walls": the former can be envisaged as an infinite supplier of energy in different forms, whereas the latter define the rules for the exchanges to take place. In this view, we categorize several distinct thermodynamic "*frames*", as a function of: (i) *how* and *what* system and reservoir exchange in terms of energy with one another (i.e. work-heat-matter); (ii) *homogeneity degree* of the system involved; (iii) occurrence and nature of *reactions* in the system, i.e.:

– equilibrium *versus* non-equilibrium thermodynamics (de Groot and [14],
– closed *versus* open systems [13],
– reversible *versus* irreversible transformations [31].

The exchanges between system and reservoir are controlled by the nature of the "*walls*" separating one from the other. In such a regard, walls are called "*restrictive*" with respect to a given observable, if they do not allow it to vary as a consequence of an interaction between reservoir and system. Therefore, the walls establish the boundary conditions that control how reservoir and system interact with one another.

Reservoir and walls define the environment we have mentioned above. In Fig. 1 system-reservoir-wall and the relationship of one with another are shown, according to a conventional representation that confines exchanges to heat and work.

A. Pavese (✉)
Earth Sciences Department, University of Turin, Valperga-Caluso Street, 35, 10125 Turin, Italy
e-mail: alessandro.pavese@unito.it

© The Author(s), under exclusive license to Springer Nature Switzerland AG 2023
M. Tribaudino et al. (eds.), *Minerals and Waste*, Earth and Environmental Sciences Library, https://doi.org/10.1007/978-3-031-16135-3_3

Fig. 1 The combination and relations between system (object of investigation), reservoir (thermal and work exchanger) and walls (boundary conditions)

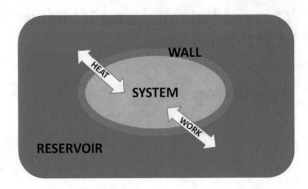

The aim of the present chapter is to provide a link between classic thermodynamics, in its simplest formulation, and "real" processes, the latter embodied by large scale transformations, characterized by relevant deviations from ideality,[1] like those occurring in solid waste incineration, for instance. The "system" is provided by the solid materials undergoing melting processes and solid–solid reactions; the "walls" are used to formally define the operating conditions under which the system transforms; the "reservoir" is the heat supplier that is supposed to provide matter too, in order to model a system whose chemical-physical complexity makes it difficult to reliably formulate a "compositional representation". In particular, fundamental notions of *ideal* thermodynamics have to be critically revised and adapted to conditions that, on the one hand, require the simple thermodynamic tools for comprehension of the involved phenomenology and, on the other hand, imply an "*unconventional*" use of classic thermodynamics, yet preserving its simplicity and efficacy, to practically treat problems difficult to be tackled otherwise.

2 Thermodynamic Fundamental Definitions

2.1 Equilibrium

From [13] we have that "there exist particular states (called equilibrium states) of simple systems that, macroscopically, are characterized completely by the internal energy U, the volume V, and the mole numbers N_1, ... of the chemical components", i.e. U-V-N sets of parameters. Such a statement may be complemented adding that (i) at equilibrium any system's observable is invariant with time and (ii) such a

---

[1] "Ideality" and "ideal" are here employed in association with conditions that are not found in real processes. In this light, the use we adopt in the present chapter is other than that commonly encountered in thermodynamics textbooks. There, "ideal" is often related to "in need of a correction that cannot be theoretically predicted" only on the basis of the classic thermodynamic principles. For instance, an "ideal mixing" indicates a mixing governed by pure configuration entropy, against a "real mixing", which requires to take also activity into account.

condition has been achieved with respect to a reservoir, once the boundary conditions have been defined. Let us consider the first equation of thermodynamics, i.e. the law of conservation of energy, formulated for a system that does conserve its chemical composition

$$dU = dQ + dW \tag{1}$$

in which "d" is used to address a differential, depending on the initial and final states only ("path independent"), whilst δ means infinitesimal change that is a function of the process causing it ("path dependent"). The nature of δW, the "work", allows us to fully describe how the reservoir acts upon the system for transferring energy, in a way other than exchanging heat. We mention the classical cases of the *mechanical work* at hydrostatic conditions, i.e.

$$dW = -PdV,$$

(P = pressure) the *electric* work, i.e.

$$dW = E_e dP_l,$$

(E_e = external electric field; P_l = polarization) and the *magnetic* work, i.e.

$$dW = m_0 H_e dI_H,$$

(μ_0 = permittivity of free space; H_e = external magnetic field; I_H = current generating H_e).

A special notion is that of "chemical work", that will be considered later, in the context of the irreversible processes.

Once the "work" has been defined in Eq. (1), there descends the notion of *equation of state*, i.e. the relationship that links to each other the thermodynamic variables, which suffice to univocally describe a system. In the present case, we shall restrict our attention to P-V-T-N, systems, i.e. such that they can be described by pressure, temperature, volume and chemical components. The latter are object of debate as to their intrinsic meaning [3], here we employ the terms "components" to address "substances" that allow us to describe the phases, reactants and products in a thermodynamic reaction, i.e. a basis vectors set of the phase vector space.

2.2 Non-Equilibrium

In this case, we say that *non*-equilibrium conditions occur when a system exhibits a U-V-N set of parameters that it cannot have at equilibrium, once a given environment is assumed. From elementary thermodynamics a system evolves towards a minimum of the Gibbs energy (G) at given P and T conditions so as to achieve a configuration of equilibrium with the environment ($dG < 0$). Therefore, we state, in short, that non-equilibrium conditions identify configurations of a system that are supposed to intrinsically change through transformations towards equilibrium. The reasons for which one observes a system at non-equilibrium, mimicking the seeming invariance of a proper equilibrium condition, are related to the occurrence either of a meta-equilibrium state (achievement of a local, and not absolute, minimum of the Gibbs energy) or an abrupt interruption of a reaction in progress, so that a "frozen state" of the system is induced as a barely "kinetic effect", often ascribable to a want of sufficient thermal energy to boost the required transformation into a definite equilibrium. Such a situation is commonly observed in industrial processes, in which thermal energy abruptly ceases to be provided for curbing production cost and the system undergoes quick cooling that leads to a frozen configuration, as a function of the thermal inertia of the furnace's walls.

The most general approach to treat a system at *non*-equilibrium conditions requires the use of classical field functions that describe matter, energy/heat flow, stress, dissipation and so on [14]. Although this approach accounts in principle for all the dynamic aspects involved in any process, and in this regard it is an exceedingly powerful tool, yet it needs nontrivial information that correlate observables to space and time. We shall not discuss these aspects, here, for the sake of brevity and simplicity.

2.3   Reversible and Irreversible Processes

Quasi-static processes are characterized by a succession of equilibrium states, defining an "equilibrium path", through which a system develops giving rise to an *ideal* transformation and taking an infinite time to pass from one state to another. Please, note that passing from one state to another along an equilibrium path cannot but require a virtual and "infinitesimal" destabilization of the system for it to change, otherwise the system would remain in an equilibrium state indefinitely. A system transforms quasi-statically from a state A to a final state B exchanging heat/work with a reservoir, according to the constraints/restraints posed by the wall. If the system, once in B, can be counter-transformed into A, thus realizing such a cyclic transformation that system and reservoir get back to their original states without any observable change (i.e. the system is characterized by the same state variables and the reservoir preserves its original energy/matter balance), then the transformation is said "*reversible*". Otherwise, a transformation is called "*irreversible*".

From elementary thermodynamics, reversible transformations are characterized by the following relationship between exchanged heat and entropy:

$$\oint_{\substack{reversible}}^{\substack{cyclic\,transformation}} \frac{\delta Q}{T} = 0 \tag{2.a}$$

$$dS_{rev} = \frac{\delta Q}{T}, \tag{2.b}$$

whereas the irreversible ones are such that

$$\oint_{\substack{\text{cyclic transformation} \\ \text{reversible}}} \frac{\delta Q}{T} < 0 \tag{3.a}$$

$$dS_{irrev} > \frac{\delta Q}{T}. \tag{3.b}$$

We take the liberty to remind here that entropy is a *state* function, which has been classically introduced by the need to have the differential of (2.b) because of Eq. (2.a). Any change in entropy due to an irreversible transformation implies fulfilment of the inequality (3.b). Let us see a practical and intuitive case. We start from a perfectly *adiabatic* system (i.e. thermally isolated) split into two sub-systems, a and b, equal to one another, at T. Their temperatures are changed in terms of $T_a = T + \delta T$ and $T_b = T - \delta T$, $\delta T > 0$. Therefore, the global system is shifted off equilibrium as proven by δT, which even if infinitesimal implies occurrence of a thermal gradient and break of homogeneity thereby. One of the possible statements, equivalent to each other, of the second principle of thermodynamics says that "*heat spontaneously flows from warmer bodies to colder*", which thing would lead to achievement of a uniform temperature in our split system. In this view, the a and b sub-portions exchange an amount of heat with one another, in terms of $\Delta Q_a = -\Delta Q$ and $\Delta Q_b = \Delta Q$, where $\Delta Q > 0$, and heat $a \to b$. This exchange of heat, occurring within an adiabatic system and tending to the achievement of $\nabla T = 0$, provides a trivial case of an "irreversible" transformation. One sub-system exchanges heat with the other, so that we write

$$\frac{\Delta Q_a}{T_a} + \frac{\Delta Q_b}{T_b} = -\frac{\Delta Q}{T + \delta T} + \frac{\Delta Q}{T - \delta T} = \frac{2\Delta Q \delta T}{T^2 - \delta T^2} \approx \frac{2\Delta Q \delta T}{T^2} > 0.$$

In other terms, although we do not have any heat exchange with a reservoir ($\delta Q = 0$), an irreversible flow of thermal energy occurs between sub-systems, unaccounted by Eqs. (2.a, 2.b, 3.a, 3.b) that see exchanges between system and reservoir. Such an "*internal*" heat exchange is the only phenomenon taking place in the system and related to its irreversible transformation towards equilibrium, and must produce an increase of entropy owing to Eq. (3.b). All this suggests that in the case of irreversible transformations

$$dS_{irrev} = \frac{\delta Q}{T} (exchange\ with\ reservoir)$$
$$+ \delta S (internal\ irreversible\ changes) \tag{3.c}$$

where the additional contribution to $\delta Q/T$ of Eq. (3.b) takes account of internal changes a system undergo, like reactions and any adjustment towards equilibration. This extension in the formulation of entropy provides an important step in approaching real processes.

2.4 Approximation of Real Processes

Real transformations are neither quasi-static nor reversible. Yet, the need of exploiting the power of thermodynamic tools motivates us to treat real transformations as "quasi-static and irreversible", i.e. a system passing through equilibrium-like states but under conditions of irreversibility.

3 Irreversible Transformations

3.1 Thermodynamics of Irreversible Transformations

How can we elementarily introduce the notion of "irreversibility" in classical thermodynamics? We use Eq. (3.b, 3.c). In fact, we can write

$$dS = \frac{\delta Q + \delta \Lambda}{T} \tag{4}$$

where $\delta \Lambda$ accounts for the "deviation" from the conditions of reversible transformation, other than those related to the bare heat exchange between system and reservoir, as suggested by Eq. (3.b, 3.c) *versus* Eq. (2.b). Our attention focusses now on $\delta \Lambda$ that, in turn, is formulated as a function of the nature of the irreversible process one aims to account for. In most transformations of interest to applications, the irreversible Λ-contribution is associated with the occurrence of "*reactions*", involving decomposition of reagents and formation of products, without constraints about a possible flow of matter between system and reservoir. So complex a frame was historically formalized using the notion of *chemical potential*, which originates from the need to describe how a change dn_j of the jth-component affects the entropy of a system. Other formulations are possible, which here we shall mention below, for the sake of completeness. Naturally, such an entropy change occurs even if the system transforms at adiabatic conditions. In this light, one introduces

$$\delta \Lambda = - \sum_{j=1,N} \mu_j dn_j \tag{5}$$

where N is the number of the components.

Using Eqs. (4) and (5), we set

$$dS = \frac{\delta Q - \sum_{j=1,N} \mu_j dn_j}{T} \tag{6}$$

dropping the subscript "*irrev*" for simplicity. Let us invoke the conservation of energy to replace δQ, as expressed by Eq. (1); then, Eq. (6) is readily recast into

$$dS = \frac{dU - \delta W - \sum_{j=1,N} \mu_j dn_j}{T}$$

From the equation above it descends

$$dU = \delta W + TdS + \sum_{j=1,N} \mu_j dn_j \tag{7.a}$$

and eventually, in the case of a mechanical work,

$$dU = -pdV + TdS + \sum_{j=1,N} \mu_j dn_j \tag{7.b}$$

It is trivial now to derive the other thermodynamic potentials using the Legendre transform that acts on the couples (p,V) and (T,S) of Eq. (7.b). In doing so, we obtain

$$dF = -pdV - SdT + \sum_{j=1,N} \mu_j dn_j \tag{7.c}$$

$$dH = Vdp + TdS + \sum_{j=1,N} \mu_j dn_j \tag{7.d}$$

$$dG = Vdp - SdT + \sum_{j=1,N} \mu_j dn_j \tag{7.e}$$

Equations (7.a, 7.b, 7.c, 7.d, 7.e) allow us to re-consider the meaning of the chemical potential, introduced here to account for the contribution to entropy that is ascribable to irreversibility. In fact, μ_j is the change in *energy* induced by a variation of the jth-component (dn_j), thus providing a more intuitive meaning for the chemical potential. It is a common approach to introduce the notion of chemical potential in terms of dU_{chim} caused by a reaction changing components into each other, so that Eq. (7.e) is often presented as the start point for any discussion that include transformations in a system.

Eventually, the inequality (3.b) leads to determining the "*direction*" along which an irreversible transformation evolves, that is

$$\delta Q = dU + pdV < TdS$$

and therefrom:
$dU < 0$ (irreversible at V and S constant);
$dF < 0$ (irreversible at V and T constant);
$dH < 0$ (irreversible at p and S constant);
$dG < 0$ (irreversible at p and T constant).
The inequalities above, in turn, provide further restraints on thermal–mechanical observables [37], among which we recall here:

$$K_T > 0$$

$$K_S > 0$$

$$C_V > 0$$

where K_T, K_S and C_V are isothermal bulk modulus, adiabatic bulk modulus and isochoric specific heat, respectively. Note that a violation of the conditions above is used to predict occurrence of instability, portending a change of state [27].

4 Open Systems

4.1 Open Versus Closed Systems

Let us consider a generic reaction formalized by the following equation

$$\sum_j v_j A_j = \sum_k v_k A_k \Rightarrow \sum_l v_l A_l \tag{8}$$

where A_j and A_k stand for left-hand side and right-hand side member "reactants", respectively; vs are the coefficients of the related reactions; subscript "l" is introduced for the sake of simplicity, to reckon components so that $v_k > 0$ and $v_j < 0$. Let us consider reaction (8) as a forward–backward transformation, for generality. If one is looking at a closed system, such as is not subject to any matter exchange with a reservoir, then the principle of conservation of mass constrains the change of each component to fulfil the equation beneath

$$\frac{dn_l}{v_l} = d\varepsilon \tag{9}$$

where $d\varepsilon$ is an infinitesimal fraction that is common for all the substances involved in the reaction and can be used to account for the degree of "reaction progress". Therefore, all of the substances involved in a given reaction fulfil the following equation

$$\frac{1}{v_l}\frac{dn_l}{dt} = \frac{d\varepsilon}{dt} = \dot{\varepsilon}(t) \tag{10}$$

Hence, each reaction is characterized by its own $\dot{\varepsilon}(t)$-function that describes the degree of progress of the transformation, and at equilibrium it expectedly satisfies the relationship shown below

$$\lim_{t \to \infty} \dot{\varepsilon}(t) = 0 \qquad (11.\text{a})$$

that implies

$$\lim_{t \to \infty} n_l(t) = N_{l,0} \qquad (11.\text{b})$$

$N_{l,0}$ in turn, is the amount of the l-component stabilized by the reaction, under the assumption of achieved equilibrium conditions.

Let now us assume the system under study to be *open*, in terms of having the possibility to exchange also matter with a reservoir, as schematized by Fig. 2. We represent the reservoir as composed of atoms, or molecules, that behave like isolated objects; in such a respect, they do not interact with each other in the reservoir. Hence, any interaction takes place in the system, only. The energy changes associated with an exchange of matter in "system + reservoir" occur because of the new interactions that rise in the system as a consequence of a change of composition. The evolution path is therefore provided by the driving inequality, $d\Delta G_{\text{irrev}} < 0$, where Δ means the difference with respect to the total energy of the isolated components' atoms, or molecules, and accounts for interactions in the system. Naturally, the way in which matter is exchanged between reservoir and system affects the nature of the "*openness*" of the involved transformations. We do not discuss further this point as it would lead us far from our focus.

In the case that matter is cross-border exchanged between system and reservoir via a flow, Eq. (10) becomes

$$\frac{dn_l}{dt} = \nu_l \frac{d\varepsilon}{dt} + \frac{d\Delta n_l}{dt} = \nu_l \dot{\varepsilon}(t) + \Delta nl \, \Delta n_l \qquad (11)$$

where one takes account of an in/out-flow related to the l-component by means of the second term of the right-hand side member of the equation above. If we split entropy into two parts, S_{ext} accounting for contributions due to an exchange with a reservoir (matter and heat), and S_{int}, related to the internal processes (reactions), then from Eqs. (6) and (11), we can write

Fig. 2 Layout of a thermodynamic system interacting with a reservoir in terms of matter, work and heat exchange

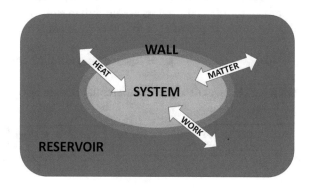

$$\frac{dS_{ext}}{dt} = \frac{1}{T}\frac{\delta Q}{dt} - \sum_l \frac{\mu_l}{T}\frac{d\Delta n_l}{dt} \qquad (12.a)$$

and

$$\frac{dS_{int}}{dt} = -\sum_l \frac{\nu_l \mu_l}{T}\frac{d\varepsilon}{dt}. \qquad (12.b)$$

We now introduce the notion of "*stationary state*", such that

$$\lim_{t\to\infty}(\nu_l\dot{\varepsilon}(t) + \Delta n_l(t)) = \lim_{t\to\infty} n_l(t) = 0 \qquad (12.c)$$

Therefore, in a stationary state no net change of the l-component takes place at time towards infinite, owing to a dynamic balance involving matter exchange and occurrence of internal reactions. Naturally, if one integrates over time the net exchange of matter between system and reservoir, then

$$\int_0^\infty \Delta \dot{n}_l(t')dt' = \Delta n_l(\infty) - \Delta n_l(0) = \Delta_l$$

The equation above shows that, in general, there is no particular restriction about Δ_l, meaning that the system does not necessarily preserve its composition with respect to the original one.

4.2 Representation of an Open System

The very key-question is how to formulate a strategy for treating open systems by elementary thermodynamics. Let us take the viewpoint of an "inner observer" (IO) who is looking at a stationary system in which reaction (8) is occurring, at $t \to \infty$, so that (12.c) holds.

IO does not perceive any matter exchange, as at t large enough n_ls are practically invariant, while P and T are set at equilibrium with a thermal-baristor. Taking account that the chemical potential of the lth-component is

$$\mu_l = \mu_{0l} + RT ln(x_l) + RT ln(\gamma_l),$$

the inner observer describes a differential change of G by

$$dG = \left[\sum_l \nu_l\mu_{0l} + RT ln\left(\prod_l x_l^{\nu_l}\right) + RT ln\left(\prod_l \gamma_{l,in}^{\nu_l}\right)\right]\delta\hat{\varepsilon} \qquad (13.a)$$

where $\delta\hat{\varepsilon}$ is the change in the degree of evolution of the reaction (8), as *seen* by IO; x_l is the molar proportion of the *l*th-substance involved; γs are activity-like coefficients; subscript *in* recalls that the equation is seen by IO. Moreover, IO considers the system as if it was at equilibrium, so that the fundamental equation reported below holds:

$$exp\left(-\frac{\sum_l \nu_l \mu_{0l}}{RT}\right) = \frac{\prod_k x_k^{\nu_k}}{\prod_j x_j^{\nu_j}} \times \frac{\prod_k \gamma_{k,in}^{\nu_k}}{\prod_j \gamma_{j,in}^{\nu_j}} = K(P,T)_{in} \qquad (13.b)$$

Let us take the case of an "external observer" (EO), who sees the occurrence of matter exchanges. Such an observer provides the following description of a Gibbs energy infinitesimal change:

$$dG = \left[\sum_l \nu_l \mu_{0l} + RT\ln\left(\prod_l x_l^{\nu_l}\right) + RT\ln\left(\prod_l \gamma_{l,ex}^{\nu_l}\right)\right]\delta\varepsilon$$
$$+ \left[\sum_l \delta\Delta n_l \mu_{0l} + RT\ln\left(\prod_l x_l^{\delta\Delta n_l}\right) + RT\ln\left(\prod_l \gamma_{l,ex}^{\delta\Delta n_l}\right)\right] \qquad (13.c)$$

where subscript *ex* indicates that an EO is involved.

In turn, $\delta\Delta n_l$ is associated with the fluctuations of the matter flow related to the *l*th-component. Owing to Eq. (12.c), it is possible to satisfy the following relationship

$$\delta\varepsilon \approx -\frac{1}{\nu_l}\delta\Delta n_l$$

at any arbitrary degree of precision, choosing *t* large enough. As a consequence, $dG = 0$ in Eq. (13.c), this leading EO to conclude that the system behaves as if it was at a minimum of its Gibbs energy. Therefore, if EO and IO look for a common description of the system under study, they converge on using an equation of the sort of Eq. (13.b) to formalize the reaction in progress. Therefore,

$$exp\left(-\frac{\sum_l \nu_l \mu_{0l}}{RT}\right) = \frac{\prod_k x_k^{\nu_k}}{\prod_j x_j^{\nu_j}} \times \frac{\prod_k \gamma_k^{\nu_k}}{\prod_j \gamma_j^{\nu_j}} = K(P,T) \qquad (14.a)$$

does provide a link between elementary thermodynamics and reactions occurring in a stationary regime. Note that we dropped the subscripts *in* and *ex*, given that the equation above is the descriptor shared by EO and IO.

Eventually,

$$\Phi(x_{1,...N-1}) = \frac{\prod_k \gamma_{k,}^{\nu_k}}{\prod_j \gamma_j^{\nu_j}}$$

where Φ is a function that depends on the components and the nature of the reactions that are involved.

Therefore, we conclude that in an open system, at stationary conditions, the following relationship holds

$$\frac{\prod_k x_k^{v_k}}{\prod_j x_j^{v_j}} = \frac{K(P,T)}{\Phi(x_{1,\ldots N-1})} = \hat{K}(P,T,X) \qquad (14.b)$$

which expresses the factorial ratio between right-hand side and left-hand side members of Eq. (8), thus decoupling the dependence of physical variables (P and T) from chemical variables (activity-like coefficients and component proportions), and introducing thereby a new equilibrium-like constant, i.e. $\hat{K}(P,T,X)$. The latter, in turn, can be either modelled or experimentally determined, as a function of one's targets, for a stationary system. The interest about Eq. (14.b) is that it provides a direct link between substance proportions in a reaction and a reference term that plays a role formally equivalent to the chemical equilibrium constant's. The reader interested in a formal approach to open systems is addressed to Appendix I.

4.3 Information Theory and Equilibrium

Let us assume that a system has a known average Gibbs energy value, i.e. $< G >$ at $P–T$, and that it may occupy given "states", each one with a probability $\{p_l\}$. Following a consolidated approach of statistical mechanics and information theory [22, 23], we state that the likeliest and least prejudicial $\{p_l\}$-set constrained to yield $< G >$ must correspond to an extreme of the expression beneath

$$\Xi = -R \sum_l p_l \ln(p_l) + \psi \left[\sum_l p_l G_l - \langle G \rangle \right] \qquad (15.a)$$

where G_l is the Gibbs energy of the lth-state; R is the gas constant and ψ is a lagrangian multiplier. If one requires that $\delta \Xi = 0$, then

$$p_l \propto \exp\left(-\frac{G_l}{RT}\right) \qquad (15.b)$$

taking $\psi = 1/T$. Let us "re-shape" our abstract system in terms of a multi-phase system, in which the "*states*" are represented by the occurring "*reactants*". It is as though one was reconstructing the system's energy by random samplings, each one giving one of the reactants with its Gibbs energy. Let the system undergo a reaction like (8) and be composed of the related reactants. We now analyse Eq. (8) in statistical terms. The occurrence of the left-hand or right-hand side members, can be modelled using the notion of "joint probability", thus obtaining

$$p_{left-hand\,side\,member} \propto \prod_j p_j^{v_j}$$

and

$$p_{right-hand\,side\,member} \propto \prod_k p_k^{v_k}$$

We take the ratio $p_{right-hand\,side\,member}/p_{left-hand\,side\,member}$, and observe that it can be either written as

$$\prod_l p_l^{v_l}$$

or formulated by Eq. (15.b) as

$$\prod_l \left[exp\left(-\frac{G_l}{RT} \right) \right]^{v_l} = exp\left(-\frac{\sum_l v_l G_l}{RT} \right)$$

Hence, the equations above lead to

$$\prod_l p_l^{v_l} = exp\left(-\frac{\sum_l v_l G_l}{RT} \right) \tag{15.c}$$

Equation (15.c) is readily likened to (14.a) by neglecting the activity coefficients and setting for each phase $G_l \equiv \mu_{0l}$. Altogether, a chemical equilibrium equation is thus re-formulated in a very simple and general fashion, which holds for stationary systems, too, once the occurrence of a reactant is unconventionally treated as a "state of the system". Therefore, Eqs. (14.a) and (15.c) coincide with one another, if one neglects the activity coefficient terms and likens the "lth-reactant proportion", x_l, to the "lth-reactant occurrence probability", p_l, i.e. $x_l \equiv p_l$.

4.4   Open/Closed Systems and Gibbs Rule

The constraint between phases, components and degrees of freedom is provided by the fundamental Gibbs rule, originally derived for closed systems. Passing to open systems, a new scenario discloses that depends on how the reservoir exchanges matter with the system. In this view, it is necessary to set boundary conditions about the way in which the reservoir "works". We see now how the Gibbs rule finds its natural formulation in closed systems and how it changes in the case of open systems that operate under the following assumption:

(1) one particular phase does exist, as a product of the reactions that are taking place;
(2) $d\Delta G_{irrev} < 0$ is the principle driving an exchange of matter between system and reservoir. ΔG is the Gibbs energy of the system with respect to the energy of the isolated atoms and reflects, in this respect, the interactions rising in the system;
(3) a stationary state exists and it is used as an equilibrium-like configuration, such that no decrease of ΔG due to an exchange of matter is possible any more.

The assumptions above provide a relevant simplification and make it possible to treat the question in a comparatively accessible way.

In fact, let us (i) indicate with $\mu_j{}^\alpha$ and $n_j{}^\alpha$ the chemical potential and number of moles of the jth-component with respect to the α-phase, respectively and (ii) assume one *independent* and *known* phase, "^{0}phase", to occur with a composition given by $n_j{}^0$; ξ is the number of moles of the ^{0}phase. The introduction of the constraints due to the ^{0}phase helps one reduce the complexity of a system, if this is multi-component and difficult to be treated. Phases other than ^{0}phase (they are here called *dependent* phases) are in total p; c indicates the number of components.

In this light, the Gibbs energy of the system turns out to be

$$G = \sum_{\alpha=1,p} \sum_{j=1,c} n_j^\alpha \mu_j^\alpha + \xi \sum_{j=1,c} n_j^0 \mu_j^0.$$

At given P–T conditions we have

$$dG = \sum_{\alpha=1,p} \sum_{j=1,c} \mu_j^\alpha dn_j^\alpha + \sum_{j=1,c} n_j^0 \mu_j^0 d\xi, \tag{16}$$

and at equilibrium $dG = 0$. If we assume the conservation of mass principle for components, i.e. closed system, then c-constraints are required to be fulfilled, i.e.

$$\sum_{\alpha=1,p} dn_j^\alpha + n_j^0 d\xi = 0. \tag{17}$$

Using the Lagrange multipliers (ψ_j) method, we combine (16) and (17), so that

$$\sum_{\alpha=1,p} \sum_{j=1,c} \mu_j^\alpha dn_j^\alpha + \sum_{j=1,c} n_j^0 \mu_j^0 d\xi + \sum_{j=1,c} \psi_j \left(\sum_{\alpha=1,p} dn_j^\alpha + n_j^0 d\xi \right) = 0.$$

Rearranging the terms of the equation above, we can write

$$\sum_{j=1,c} \left[\sum_{\alpha=1,p} (\mu_j^\alpha + \psi_j) dn_j^\alpha + (\mu_j^0 + \psi_j) n_j^0 d\xi \right] = 0. \tag{18}$$

Equation (18) requires that.

$$\mu_1^1 = \mu_1^2 = \ldots\ldots = -\psi_1 = \mu_1^0$$

$$\cdots\cdots\cdots\cdots\cdots\cdots\cdots$$

$$\mu_c^1 = \mu_c^2 = \ldots\ldots = -\psi_c = \mu_c^0$$

(19)

The set of Eq. (19) leads to the fully established equality between chemical potentials of the same component participating in the formation of different phases. Given that in the usual approach there is not any imposed ^{0}phase, we set it aside at the moment, so that we have p phases and c components. A set of $c \times p$ unknown $\{x_j^\alpha\}$-values (i.e. $x_j^\alpha = n_j^\alpha/\Sigma_{k=1,c}\, n_k^\alpha$) on which the chemical potentials depend, i.e. molar fractions of the components in each phase, must be determined. The $\{x_j^\alpha\}$-variables require more p-constraints of normalization, i.e. $\Sigma_{j=1,c}\, x_j^\alpha = 1$, in addition to the $c \times (p-1)$ constraints due to the equations of (19)-type. Hence, at given P and T, the classic relationship.

$$f_{\text{degrees of freedom}} = (p \times c) - c \times (p - 1) - p = c - p$$

is obtained and poses a severe constraint, given by $c - p \geq 0$.

Let us now focus on the case of the presence of one *independent* ^{0}phase and p *dependent* phases, which will be discussed as constituents of an open system. The constraining equations of (19)-type are $p \times c$, given that the lagrangian multipliers are determined by μ_j^0, i.e. $-\psi_j = \mu_j^0$. We now assume the system to be free of exchanging matter with a *reservoir* according to an irreversible transformation, so that $d\Delta G_{\text{irrev}} < 0$, until a stationary equilibrium-like state is achieved, where $d\Delta G = 0$. The p *dependent* phases behave as an "*open system*" in the equilibration-like process. We do know the system's composition at the start, but we do not know about its composition at the stationary equilibrium-like configuration, which is the result of a matter exchange with the reservoir. As $d\Delta G = 0$ and its composition changes no more, we treat it as if it was a virtual closed system. In this respect, the system is defined once the unknown $c \times p$ $\{n_j^\alpha\}$-values, $n_j^\alpha \geq 0$, have been determined. The $\{n_j^\alpha\}$-set is replaced by the $\{x_j^\alpha\}$-set and $\{\tau^\alpha\}$-set, the latter representing the number of moles associated with a α-phase and directly related to the ΔG minimization process and to the stationary equilibrium-like state. In full, one has $c \times p + p$ variables to describe the system, and $c \times p$ (like Eq. 19) + p (x-normalization) constraints. Therefore, the independent ^{0}phase pivots the equilibration-like process of the p *dependent* phases, regardless of p and c.

We now aim to introduce more than one *independent* phase: p_{indep}-phases. First, they must be at equilibrium with each other at P–T, namely they must satisfy the constraints of the Gibbs rule, i.e. $p_{\text{indep}} \leq c$ and the equality of the chemical potentials. The ψ_i's, determined by the chemical potentials of the *independent* phases accordingly, lead to the $c \times p$ constraining equations of (19)-type. The $c \times p$ $\{n_j^\alpha\}$-values that give the p *dependent* phases are determined, and an exchange of matter between *system* and *reservoir* takes place, as in the case of the ^{0}phase. It is worth noting that the obtained assemblage represents the least Δ-Gibbs energy system compatible with p_{indep} *independent* ($p_{\text{indep}} \leq c$) and *p dependent* phases.

5 Kinetics

5.1 Solid-State Reaction Kinetics and Its Phenomenological Description

A key to interpret *solid state transformations* occurring off equilibrium is the reaction kinetics which deeply affects the phenomenology under study. This viewpoint plays a fundamental role in many high temperature industrial process, in which the complexity of the reactions and the need to curb energy costs lead to an interruption of the transformations prior to their natural completion. Therefore, being able to measure the "*velocity*" of a reaction and choose when to interrupt it accordingly, allows us to design energy demanding processes in such a way that minimization of production costs and quality of the output are preserved.

Let us describe the evolution of a solid-state reaction by means of a "*coordinate*" (α) that may be associated with an observable normalized, for instance, so that

$$\lim_{t \to \infty} \alpha(t) = 1$$

and

$$\lim_{t \to 0} \alpha(t) = 0$$

Therefore, $\alpha(t)$, referred to as *conversion factor*, provides a description of the "*distance*" of a reaction from its completion, and in this view becomes a crucial marker of anthropic processes that are often interrupted prior to achievement of equilibrium. In some case, $t \to \infty$ can be replaced by $t \to t_c$, if the reaction is supposed to achieve completion at a given time t_c. By way of example, in the case of a mass measurement taking account of the mass change that the reactant undergoes, we have

$$\alpha(t) = \frac{m_0 - m(t)}{m_0 - m_\infty}$$

m_0 being the mass at $t = 0$.

The solid-state *reaction rate* is formulated as

$$\frac{d\alpha}{dt} = R(t, \alpha)$$

and, in practice, provides a formal description of the reaction progress velocity.

Experimental observations suggest to decouple t from α, so that the following factorization is commonly accepted

$$R(t, \alpha) = K(t) f(\alpha)$$

where $K(t)$ is the *rate constant* and $f(\alpha)$ accounts for the *reaction model*, the latter depending on the intrinsic mechanism that drives a transformation. In this light, the general reaction kinetics equation along an isotherm (*isothermal mode*) is

$$\frac{d\alpha}{dt} = Ae^{-\frac{E_a}{RT}} f(\alpha) \tag{20.a}$$

where

$$K(T) = Ae^{-\frac{E_a}{RT}} \tag{20.b}$$

A is the frequency (pre-exponential) factor, whereas E_a is the activation energy, which provides a measure of the "difficulty a reaction encounters" to proceed. The larger is E_a, the slower the reaction develops. Integrating Eq. (20.a) over t and α, one readily has

$$\int_{\alpha_0}^{\alpha} \frac{d\alpha'}{f(\alpha')} = g(\alpha) = Ae^{-\frac{E_a}{RT}} (t - t_0) \tag{20.c}$$

Passing to the logarithm of the right-hand side and left-hand side members of Eq. (20.c), we obtain

$$\ln(g(\alpha)) = ln\left(Ae^{-\frac{E_a}{RT}} \right) + \ln(t - t_0) \tag{20.d}$$

Let us now see how kinetic equations are reformulated in the case of *non-isothermal mode*, i.e. when α changes as a function of time and temperature, because during the experiment the sample's temperature increases regularly. Equation (20.a) is recast as follows

$$\frac{d\alpha}{dT} \frac{dT}{dt} = \frac{d\alpha}{dT} \beta = Ae^{-\frac{E_a}{RT}} f(\alpha) \tag{21.a}$$

where β is the temperature ramp that in many practical applications is set to be constant. Integrating Eq. (21.a) over α and T, one obtains

$$g(\alpha) = \frac{A}{\beta} \int_{T_0}^{T} e^{-\frac{E_a}{RT}} dT \tag{21.b}$$

Table 1 Kinetic reaction model classification

Model	$d\alpha/dt$	Phenomenology
Acceleratory	>0	The conversion factor increases as the reaction proceeds
Decelerator	<0	The conversion factor decreases as the reaction proceeds
Linear	≈0	The conversion factor is constant
Sigmoidal	Bell-shape	The conversion factor exhibits two regions: one in which it increases, and another adjacent wherein it decreases, as the reaction proceeds

In Eq. (21.b) the explicit dependence on time has been removed and replaced by temperature, the latter changing as a function of t via the heating ramp. The reaction models are classified by the trends of α *versus* t, as shown in Table 1.

5.2 Reaction Models: The Case of the Avrami–Erofeyev Equation

Several reaction models were developed and are reported in literature, and account for the particular features of the transformation processes that are supposed to take place and drive the change a solid undergoes.

Let us follow the derivation of one among the most used and general kinetic equations, leading to the "Avrami model" [6, 7, 12]. The fundamental assumption is that the transformation process of "A-phase → B-phase" can be split into two steps: (1) nucleation of B particles and (2) their successive growth. As to 1, we state that in a dt' interval the number of nuclei appearing in a volume V, N_V, changes in terms of

$$dN_v = V \left(\frac{dN}{dt'} \right)_t dt' = V \dot{N} dt'$$

where N is referred to the number of nuclei in a unitary volume.

Each nucleus is supposed to grow into a particle, whose size can be associated with a linear dimension, r, that increases from t to t' according to the relationship below

$$r(t, t') = \dot{D}(t' - t)$$

$\dot{D}$ being the linear increasing coefficient with time. Therefore, the increase of the volume occupied by B at t' because of the nuclei appeared at t is

$$\Delta V_{t,t'} = \int_t^{t'} \varphi \dot{D}^n (t' - \tau)^n V \dot{N} d\tau \tag{22.a}$$

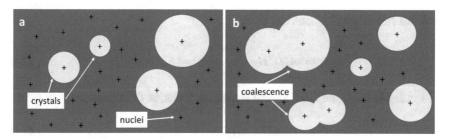

Fig. 3 a Spherical particle developing from nuclei, the latter represented by " + ", without coalescence; **b** as for (a), but assuming occurrence of coalescence

where φ is a form factor and n defines the actual dependence of a particle volume on the linear size r. Setting $t = 0$ and assuming $\dot{N} = Ct^m$, we replace $\Delta V_{t,t'}$ with $V_{0,t'}$ and have

$$V_{0,t'} = \varphi \, \dot{D}^n \, V \dot{N} \frac{1}{n+m+1} t'^{n+m+1} \tag{22.b}$$

However, $V_{0,t'}$ does not represent the real volume occupied by the B-phase as a consequence of the growth of particles from the nuclei at $t = 0$, as suggested by Fig. 3, in which the development of independent crystals vies with coalescence, the latter leading to a reduction of the expected B-volume.

Equation (22.b) provides hence an estimate of the *maximum* volume ascribable to the B-phase, under the assumption that coalescence does not take place.

Altogether, the occurrence of coalescence between particles that are increasing their size causes a decrease of the volume ascribable to the B-phase, and one must take this effect into account. We assume that an infinitesimal change of the "actual" volume of B, $V_{0,t}^{act}$, is proportional to the infinitesimal change of the *potential* volume of B, i.e. the one in which coalescence is excluded, scaled down by the fraction of volume still available to transforming A into B, i.e. $\frac{V - V_{0,t}^{act}}{V}$. All this leads to the following relationship

$$dV_{0,t}^{act} = dV_{0,t}\left(1 - \frac{V_{0,t}^{act}}{V}\right) \tag{23.a}$$

that can be easily integrated into

$$ln\left(1 - \frac{V_{0,t}^{act}}{V}\right) = -\frac{V_{0,t}}{V} \tag{23.b}$$

We remind that the actual fraction of A changed into B, i.e. α, is provided by

$$\alpha = \frac{V_{0,t}^{act}}{V}$$

and that from Eq. (22.b) one can write

$$\frac{V_{0,t}}{V} = \varphi C \dot{D}^n \dot{N} \frac{t^{n+m+1}}{n+m+1} = (K_{nucl}t)^{n+m+1} \tag{24}$$

Putting together the relationships above, i.e. Equation (23.b) and (24), we have

$$[-ln(1-\alpha)]^{\frac{1}{n+m+1}} = K_{nucl}t \tag{25.a}$$

the equation above corresponding to one of the possible formulations of the "Avrami–Erofeyev" equation [6, 7, 12], often reported in literature as

$$\frac{1}{n+m+1}ln(-ln(1-\alpha)) = ln(K_{nucl}) + ln(t) \tag{25.b}$$

Equation (25.a) is immediately related to Eq. (20.a), whereas Eq. (25.b) to Eq. (20.c), through the relation shown below

$$g(\alpha) = [-ln(1-\alpha)]^{\frac{1}{n+m+1}}$$

and taking $K_{nucl} = K$. In this view, we shall drop hereafter the subscript *nucl*.

The kinetic equation reported above has been derived following a model that describes the process of nuclei's formation and subsequent growth. Model other than Avrami's can be developed, and we address the reader to [25]. We mention here a very general representation provided by [32] for $g(\alpha)$, i.e.

$$g(\alpha) = \alpha^p (1-\alpha)^q (-\ln(1-\alpha))^r$$

that accounts for many of the most used models in literature.

5.3 Activation Energy Determination

There are two ways of treating kinetic data. One employs a reaction model, which is fitted to observations, using Eq. (25.a, 25.b) or their likes, and chosen as a function of the mechanism driving the transformation. Data are supposed to have been collected in *isothermal mode* at different temperatures, so that one follows the reaction progress with time at a given T. Each isotherm provides the related $K(T)$, and using then Eq. (20.b) one is able to extract E_a and A, through a $ln(K)$ *versus* $1/T$ plot.

Another way is provided by the *isoconversional methods* [34]. They rely upon the general principle that the reaction rate at given α depends on T only, and the activation energy is independent of β. For the sake of completeness, we report here some examples of how these methods work. The classical isoconversional model starts from Eq. (20.c) that is turned into

$$-ln(t) = ln\left(\frac{A}{g(\alpha)}\right) - \frac{E_a}{RT}$$

A plot of $ln(t)$ versus $1/T$ at given α allows determination of the activation energy independently of the model being used, i.e. $g(\alpha)$. Another case is provided by the model that exploits Eq. (21.b), which is recast in the following form

$$g(\alpha) = \frac{AE_a}{\beta R}p(x)$$

where $x = E_a/RT$. Approximations of $p(x)$ have been developed and are available in literature. In particular, using the [16, 17] linear approximation, the relationship above becomes [18, 30]

$$log(\beta) = log\left(\frac{AE_a}{g(\alpha)R}\right) - 2.315 - 0.457\frac{E_a}{RT}$$

Also in the present case, a plot of $log(\beta)$ versus $1/T$ makes it possible to extract the activation energy that is the slope of the linear interpolation.

Isoconversional methods lend themselves to explore possible dependence of E_a on α, i.e. $E_a(\alpha)$. This aspect enables to investigate reaction mechanisms in detail [2, 35]. An increase of accuracy to measure E_a is provided by [36], imposing the minimization of the special function

$$\Phi(E_a(\alpha)) = \sum_{i=1}^{n}\sum_{j\neq i}^{n}\frac{J[E_a(\alpha), T_i(t(\alpha))]}{J[E_a(\alpha), T_j(t(\alpha))]}$$

where

$$J[E_a(\alpha), T(t(\alpha))] = \int_0^{t(\alpha)} e^{-\frac{E_a(\alpha)}{RT(t')}} dt'$$

and $t(\alpha)$ means time for achievement of the α-value of the conversion factor, at a temperature given by $T(t(\alpha))$. Such an approach is only mentioned but not discussed here further, for sake of the brevity.

5.4 HT Transformations in Complex Systems: Municipal Solid Waste Incineration and Ferrous Slag

The thermodynamic modelling discussed in the sections above, relies upon quantities that are independent of time (leaving kinetics aside) and space. In this light, we are supposing that a general homogeneity holds across the samples under investigation. We would like now to look into "real" processes, and stress how far they lie from the conditions aforementioned. Nonetheless, the elementary thermodynamic models still remain crucial to drive interpretation and, with the due care, to develop predictions.

Transformations taking place in an incinerator provide an example of large spatial and time scale reactions, at non-ideal and off-equilibrium conditions. First of all, the large spatial scale involved poses serious questions about the *representativeness of the samples* used for any investigation. Therefore, sampling procedures have been developed [19, 28] to account for the intrinsic heterogeneity of the incineration products and, in such a light, an extra care is required in drawing conclusions, which must always be supported by an estimate of the oscillations and confidence intervals associated with the observables under study. The large time scale, in turn, reflects variations of the waste delivered over a year [5, 8], which give an ash changing with time, as an obvious consequence of the seasonality of waste. The abovementioned aspects are critical, in particular if attention is paid to comparatively small effects that may be misunderstood and mistaken for expressions of a general phenomenology, instead of a local one. An additional difficulty is that the incineration process' conditions are highly non-ideal. Suffice it to mention the heterogeneous temperature field, which imply severe gradients impairing definitely any model relying upon a homogeneous firing of waste. The occurring reactions, then, develop in presence of possible exchange of fluids and sublimation/evaporation processes, i.e. in a context that is far from any chance of contouring a "closed system" and difficult to be modelled even in an "open system" thermodynamic frame. Worst of all is that even an "open system" accounting for waste incineration implies matter exchanges whose correct modelling is out of reach. The reproducibility itself of the incineration process is a further concern, which biases inferences from observations on products. Eventually, transformation of waste upon incineration are composed of a variety of reactions that undergo abrupt interruption, and in this view they are mainly governed by kinetics, whose complexity has been mentioned in the sections above and requires more than one model to describe the reactions' progress.

Ferrous slag is classified into a variety of typologies that can be summarized as "blast slag" and "steel slag" [9]. The possibility to exploit ferrous slag in applications, and the nature of the field into which it can be reused/recycled, depend on the mineralogical and chemical composition of the waste [4]. As a consequence, the cooling path of the industrial process from which slag derives affects heavily the nature of the resulting waste, in terms of both stability field of the melt/solid system and crystallization kinetics. Microstructure formation that has a relevant say in any management of the slag is well known to be a function, on the one hand, of composition and, on the other hand, of the thermal path experienced by the system [24]. Leaving aside any claim of completeness in treating a subject with which literature teems, we mention here some of the commonest destinations, influenced by the slag's microstructure features, of such waste as secondary raw materials, like supplementary cementitious materials (de Matos et al., 2020), geopolymers [29], alkali-activated materials [1, 26] and fillers for road sublayer [11].

6 Applications

A basic and fundamental question is: how can we model in practice complex H*T*-reactions, involving solids and fluids? The answer can be articulated in three points: (i) approach, (ii) additional information, (iii) databases.

As to the approach, a common strategy [10] is to exploit equilibrium thermodynamics, using conditions provided by Eqs. (8) and (9), which lead to the general relationship reported below

$$\sum_j \nu_j \mu_j = \sum_k \nu_k \mu_k \qquad (26)$$

where μs are the molar Gibbs energy functions of the phases involved. We do not discuss about the construction of the Gibbs energy functions, for the sake of brevity, although this topic is extremely relevant. Suffice it to say that a shared approach consists in developing calculations using parametrizations of the key thermodynamic functions, such as specific heat at constant pressure/volume, isothermal/adiabatic bulk modulus and so on. These functions are then employed to determine the thermodynamic potentials and, thereby, the *P–T locus* of equilibrium between reactions, where the Eq. (26) holds. The stability field of the assemblage represented by either the right-hand side or left-hand side member of the Eq. (26) follows from the inequality shown below

$$\sum_j \nu_j \mu_j < \sum_k \nu_k \mu_k$$

meaning that the reaction is shifted towards the *j*-phases, or the inequality beneath

$$\sum_j \nu_j \mu_j > \sum_k \nu_k \mu_k$$

indicating a shift towards the *k*-phases.

As to the additional information to start a modelling, the complexity of the involved systems benefits much from further either constraints or restraints (guesses on the possible occurring phases, once the starting composition is at least partially known). In this view, a modelling is successful insofar as it is judiciously developed, and requires a solid competence about the system under investigation of those who aim to predict its behaviour. In other terms, modelling cannot be reduced to a straightforward use of programs, which thing, in turn, could lead to unphysical results.

As to the databases, they play a crucial role, providing the "numerical values" for the abovementioned parametrizations, which are then used by software packages to make calculations (see for instance: Perplex, https://www.perplex.ethz.ch; Thermo-calc, https://thermocalc.com/). Databases have been developed following a principle of "internal self-consistency", so as to correctly model reference reactions, or stability fields. In this view, databases tend to be focussed on given confidence ranges and specific classes of materials, in which they are expected to provide results coherent with the actual phenomenology. Moreover, databases may have been developed on purely experimental data, or complemented with theoretical calculations. All this aims to make the reader aware of the centrality played by the databases and the special attention to be paid in choosing one for modelling. By way of example, we conclude mentioning two reference papers addressing to widely employed databases for minerals and fluids: [20], for multipurpose applications and much used for metamorphic reactions, [33], for very high-pressure modelling.

7 Conclusions

In the case of waste from high temperature large scale processes, thermodynamic modelling (here proposed in its elementary formulation of "equilibrium thermodynamics", for the sake of simplicity) is a powerful tool to help interpretation, design actions and make decisions. However, the complex state of affairs due to the deviations from ideal conditions requires a judicious usage of thermodynamic tools, which are to be tuned and employed foremost to interpret general trends. The high heterogeneity of the reaction conditions leads to the formation of "local environments", which allow one to look into specific micro-scale reactions, yet obtaining results reflective of sub-systems' behaviour rather than of a wholesale phenomenology. In this view, although thermodynamics is fundamental to rationalize what happens in waste incineration processes, care is to be used in extending interpretative tools beyond the frontiers that contour their natural applicability, for which they were developed. In particular, we stress three aspects:

(i)   the capacity of a system to exchange matter with the environment affects the way of modelling it. In the present case, the extended use of the equilibrium constant formalism is proposed, so as to describe systems that are able to exchange matter with a reservoir;

(ii) the thermal path that systems experience affects the resulting products. In this light, tuning the heating and cooling processes has an important say in driving the output, thus contouring the context in which ash/slag may be introduced as a secondary raw material;

(iii) the interruption of reactions before achievement of equilibrium implies resort to reaction kinetics, as a tool that, on the one hand, makes it possible to determine the "distance" from a completion of transformation and, on the other hand, allows designing reaction environments that boost the changes required to confer specific properties to the resulting ash/slag.

Appendix

I. *Open systems*

Let us consider system and matter reservoir as a "*total system*", which obeys usual thermodynamic laws. An infinitesimal change of its Gibbs energy is

$$dG = \sum_{j=1,N\,system} \mu_j dn_j + \sum_{k=1,N\,reservoir} \mu_k dn_k \qquad (I.1)$$

where

$$dn_j = v_j d\epsilon + \delta\Delta n_j \qquad (I.2)$$

The second term of (I.2) accounts for the fraction of matter that is being exchanged between system and reservoir. In this light, in the summation we have $+\delta\Delta n$ and $-\delta\Delta n$, i.e. the same amount of matter is considered input for reservoir $\rightarrow$ system and output for system $\rightarrow$ reservoir. The *total system* evolves according to the fundamental principle of the irreversible transformations at P–T, that is

$$dG < 0,$$

and achieves at equilibrium the following relationship:

$$dG = 0.$$

We now model the reservoir in such a way that *no interaction takes place between its components*. This means that

$$\mu_k = \mu_{k,0}$$

and implies that the k-components have activity equal to unity.

Let us rewrite (I.2) using (I.2):

$$dG = \sum_{j=1,N\,system} \mu_j \left(v_j d\epsilon + \delta\Delta n_j\right) + \sum_{k=1,N\,reservoir} \mu_{k,0}\delta\Delta n_k$$

$$= \left(\sum_{j=1,N\,system} \mu_j v_j\right) d\varepsilon + \left(\sum_{j=1,N\,system} \mu_j \delta\Delta n_j + \sum_{j=1,N\,reservoir} \mu_{k,0}\delta\Delta n_k\right)$$

For the sake of simplicity, let us assume that all the components of the system are exchanged with the reservoir so that

$$dG = \sum_{j=1,Nsystem} \mu_j v_j d\epsilon + \sum_{j=1,Nsystem} \left(\mu_j - \mu_{j,0}\right)\delta\Delta n_j$$

Setting $dG = 0$, i.e. equilibrium conditions, we conclude:

$$\sum_{j=1,Nsystem} \mu_j v_j = 0 \qquad (I.3.a)$$

and

$$\sum_{j=1,Nsystem} \left(\mu_j - \mu_{j,0}\right)\delta\Delta n_j = 0 \qquad (I.3.b)$$

Equation (I.3.a) represents the *standard condition of equilibrium* of a forward–backward reaction. Equation (I.3.b) includes the mixing contribution only to the Gibbs energy and is associated with *a minimum of the mixing contribution*. In fact, Eq. (I.3.b) can be written as follows:

$$\sum_{j=1,\,N\,system} \left(\mu_j - \mu_{j,0}\right)\delta\Delta n_j = \sum_{j=1,\,N\,system} (RT\ln x_j + RT\ln \gamma_j)\delta\Delta n_j$$

that is $\delta G_{mixing} = 0$, where

$$G_{mixing} = \sum_{j=1,Nsystem} \left(RT\ln x_j + RT\ln\gamma_j\right)n_j$$

References

1. Adediran A, Yliniemi J, Illikainen M (2021) Development of sustainable alkali-activated mortars using Fe-Rich fayalitic slag as the sole solid precursor. Front Built Environ 7:653466. https://doi.org/10.3389/fbuil.2021.653466
2. Alzina C, Sbirrazzuoli N, Mija A (2010) Hybrid nanocomposites: advanced nonlinear method for calculating key kinetic parameters of complex cure kinetics. J Phys Chem B 114:12480–12487
3. Anderson GN, Crerar DA (1993) Thermodynamics in geochemistry: the equilibrium model. Oxford University Press
4. Andrews A, Gikunoo E, Ofosu-Mensah L, Tofah H, Bansah S (2012) Chemical and mineralogical characterization of Ghanaian foundry slags. J Miner Mater Charact Eng 11(02):183
5. Arm M (2004) Variation in deformation properties of processed MSWI bottom ash: results from triaxial tests. Waste Manag 24(10):1035–1042. https://doi.org/10.1016/j.wasman.2004.07.013
6. Avrami MJ (1939) Kinetics of phase change. I General Theory Chem Phys 7:1103
7. Avrami MJ (1940) Kinetics of phase change. II transformation-time relations for random distribution of nuclei. Chem Phys 8:212

8. Bielowicz B, Chuchro M, Jędrusiak R, Wątor K (2021) Changes in leachability of selected elements and chemical compounds in residues from municipal waste incineration plants. Energies 14(3):771. https://doi.org/10.3390/en14030771

9. Cardoso C, Camões A, Eires R, Motab A, Araújo J, Castro F, Carvalho J (2018) Using foundry slag of ferrous metals as fine aggregate for concrete. Res Cons Recycl 138:130

10. Connolly JAD (2005) Computation of phase equilibria by linear programming: a tool for geodynamic modeling and its application to subduction zone decarbonation. Earth Planet Sci Lett 236:524–541

11. Dondi G, Mazzotta F, Lantieri C, Cuppi F, Vignali V, Sangiovanni C (2021) Use of steel slag as an alternative to aggregate and filler in road pavements. Materials 14:345. https://doi.org/10.3390/ma14020345

12. Erofeyev BV (1946) Dokl Akad Nauk SSSR 52:511

13. Callen HB (1960) Thermodynamics. Wiley, N.Y.

14. de Groot SR, Mazur P (1985) Non-equilibrium thermodynamics. Dover Publications Inc

15. de Matos PR, Oliveira JCP, Medina TM, Magalhães DC, Gleize PJP, Schankoski RA, Pilar R (2020) Use of air-cooled blast furnace slag as supplementary cementitious material for self-compacting concrete production. Constr Build Mat 262:120102

16. Doyle CD (1962) Estimating isothermal life from thermogravimetric data. J Appl Polym Sci 6:639

17. Doyle CD (1965) Series approximations to the equation of thermogravimetric data. Nature 207:290

18. Flynn JH, Wall LA (1966) A quick, direct method for the determination of activation energy from thermogravimetric data. Sci B:Polym Lett 4:323

19. Gy PM (1992) Sampling of heterogeneous and dynamic material systems: theories of heterogeneity, sampling and homogenizing. Elsevier

20. Holland TJB, Powell R (2011) An improved and extended internally consistent thermodynamic dataset for phases of petrological interest, involving a new equation of state for solids. J Metamorphic Geol 29:333–383

22. Jaynes ET (1957a) Information theory and statistical mechanics (PDF). Phys Rev Series II 106:620–630

23. Jaynes ET (1957b) Information theory and statistical mechanics (PDF). Phys Rev Series II 108:171–190

24. Khater GA (2011) Influence of Cr2O3, LiF, CaF2 and TiO2 nucleants on the crystallization behavior and microstructure of glass-ceramics based on blast-furnace slag. Ceram Inter 37:2193

25. Khawam A, Flanagan DR (2006) Solid-state kinetic models: basics and mathematical fundamentals. J Phys Chem B 110:17315–17328

26. Komnitsas K, Yurramendi L, Bartzas G, Karmali V, Petrakis E (2020) Factors affecting co-valorization of fayalitic and ferronickel slags for the production of alkali activated materials. Sci Total Environ 721:137753. https://doi.org/10.1016/j.scitotenv.2020.137753

27. Merli M, Pavese A (2021) Melting temperature prediction by thermoelastic instability: an ab initio modelling, for periclase (MgO). CALPHAD: Comput Coupling Phase Diagr Thermochem 73:102259

28. Møller H (2004) Sampling of heterogeneous bottom ash from municipal waste-incineration plants. Chemom Intell Lab Syst 74(1):171–176. https://doi.org/10.1016/j.chemolab.2004.03.016

29. Nath SK, Kumar S (2013) Influence of iron making slags on strength and microstructure of fly ash geopolymer. Constr Build Mat 38:924

30. Ozawa T (1965) A new method of analyzing thermogravimetric data. Bull Chem Soc Jpn 38:1881

31. Prigogine I (1968) Introduction to thermodynamics of irreversible processes, 3rd edn. Wiley, N.Y., p 9

32. Sestak J, Berggren G (1971) Study of the kinetics of the mechanism of solid-state reactions at increasing temperatures. Thermochim Acta 3:1

33. Stixrude L, Lithgow-Bertelloni C (2011) Thermodynamics of mantle minerals—II. Phase Equilibria Geophys J Int 184:1180–1213
34. Vyazovkin S, Burnhamb AK, Criadoc JM, Pérez-Maquedac LA, Popescud C, Sbirrazzuoli N (2011) ICTAC kinetics committee recommendations for performing kinetic computations on thermal analysis data. Thermochim Acta 520:1
35. Vyazovkin S, Dranca I (2006) Isoconversional analysis of combined melt and glass crystallization data. Macromol Chem Phys 207:20–25
36. Vyazovkin S, Dollimore D (1996) Linear and nonlinear procedures in isoconversional computations of the activation energy of thermally induced reactions in solids. J Chem Inf Comp Sci 36:42–45
37. Wallace DC (1972) Thermodynamics of crystals. Wiley

Bio-mineral Interactions and the Environment

Giovanni De Giudici, Daniela Medas, and Carlo Meneghini

1 Biominerals: A Continuously Growing Family

Biominerals are the product of organism's activity leading to mineral formation within the cellular space or in the space surrounding the organism. In the last decades biominerals have received growing interest from a large interdisciplinary scientific community. Actually, biominerals are known from the geological record to play a pivotal role in biogeochemical cycles of elements [1, 2]. Thus, understanding biomineralization processes in widely different environments helps us to understand environmental changes induced by anthropic activities, as well the as environment resiliency [3, 4]. Moreover, they offer diverse examples to devise useful biobased materials and allow the development of technologies for environmental sustainability [5, 6]. This work is aimed to summarize our understanding of biominerals, their classification, and their impact in our society. Recent investigations on bio-mineral interactions are presented focusing on processes, investigation techniques, impact on the environment and sustainable technologies. We do not attempt to provide a comprehensive overlook of the whole field, but we place emphasis on specific aspects where we have first-hand experience.

Based on the causative effect of cellular activity, biominerals were classified for the first time by Lowenstam [7] into two main classes, namely biologically controlled mineralization (BCM) and biologically induced mineralization (BIM). More recently, Skinner [8] provided a thorough definition "*The simple definition of biominerals is that they are a subset of the mineral kingdom, created through the actions and activity of a life form. The term as used herein is meant to be very*

G. De Giudici (✉) · D. Medas
Department of Chemical and Geological Sciences, University of Cagliari, Cagliari, Italy
e-mail: gbgiudic@unica.it

C. Meneghini
Department of Sciences, University Roma Tre, Rome, Italy
e-mail: carlo.meneghini@uniroma3.it

general, and inclusive of the entire range of living creatures and their products. The range extends from the primitive, not-so-well-classified forms at the very bottom of the tree of life, the Archaea, and bacteria, through the eukaryotes with well-defined morphology that may contain mineral materials in sub-cellular, or extracellular, compartments, or tissues, up to and including, vertebrates and plants".

Biominerals can be classified using the same framework as that for other minerals, by composition based on the anionic constituents, and there are representatives in almost all the 10 mineral classes listed by Strunz [9]. The already long list provided by Skinner [8] is still growing with hundreds of investigations on novel natural ecological niches around the world, or at sites created by human activities through industry and manufacturing. Interestingly, biominerals also include minerals that in geological processes form only at high temperature and pressure.

The knowledge of the diversity of biomineral composition is increasing beyond all expectations. In 1963, Lowenstam identified 10 different mineral types; this increased to 19 biominerals by 1974 [10], 30 by 1981 [7], 39 by 1983 [11], 56 by 2003 [12], and 96 by 2008 [13]. Table 1 shows a state-of-the-art list of 160 different biogenic minerals identified to date and distributed between the life kingdoms. Figure 1 shows the distribution of the 160 biominerals among the classes.

As previously mentioned, most biominerals, despite their peculiar characteristics and the intimate association of their structures with organic molecules, can be classified as common mineral species based on the anionic constituents [8]. Biomineral composition depends on both the control played by the specific (micro)organism and by the available chemical species in the environment. Then, factors such as pH, pO_2, Eh, etc. can play a significant role on the final product. This leads to substantial differences, for example, between the locations where carbonate biominerals might form and the locations where sulphides might be produced [8]. For instance, carbonate (e.g., calcite, hydrozincite) bioprecipitation can occur in neutral or slightly alkaline environments often driven by the photosynthetic activity of bacteria [14], whereas sulphides (e.g., pyrite or sphalerite) will form under the mediation of sulphur-reducing bacteria, in a reducing environment where metals are available [15]. Table 1 shows that the most commonly occurring cations in biominerals are Ca, Fe, Mg, Na and K. Calcium and Mg are mainly hosted in carbonates and phosphates, whereas Fe occurs in sulphides and oxides/hydroxides. Also, potentially harmful elements such as Cu, Mn, Zn, As, Pb, etc. can occur in biominerals belonging to several mineral classes.

2 Mineral Surfaces and Biological Interfaces

Bio-mineral interfaces are the place where different microscopic processes take place resulting in controlled or induced biomineralization. Bio-mineral interfaces are then the key to understand biomineralization processes and to develop sustainable technologies for industry and environment. These processes are intrinsically

Table 1 List of minerals produced by biological precipitation over Strunz classes. Expanded from [12, 13]

Name	Formula
Elements (01)	
α-sulfur	S
γ-sulfur (Rosickýite)	S
Sulphides (02)	
Acanthite	Ag_2S
Amorphous pyrrhotite	$Fe_{1-x}S$ (x = 0–0.17)
Galena	PbS
Greigite	Fe_3S_4
Hydrotroilite	$FeS \cdot nH_2O$
Löllingite [16]	$FeAs_2$
Mackinawite	$(Fe, Ni)_9S_8$
Marcasite	FeS_2
Microcrystalline millerite [17]	NiS
Pyrite	FeS_2
Pyrrhotite	Fe_7S_8
Orpiment	As_2S_3
Sphalerite	ZnS
Wurtzite	ZnS
Halides (03)	
Atacamite	$Cu_2Cl(OH)_3$
Halite	NaCl

(continued)

Table 1 (continued)

Name	Formula
Sylvite	KCl
Ammineite [18]	$CuCl_2 \cdot 2NH_3$
Amorphous fluorite [19]	CaF_2
Fluorite	CaF_2
Hieratite	K_2SiF_6
Oxides and Hydroxides (O4)	
Amorphous iron titanate	$Fe^{2+}TiO_3$
Amorphous iron oxide	Fe_2O_3
Amorphous manganese oxide	Mn_3O_4
Anatase	TiO_2
Desert Varnish	MnO and FeO
Ilmenite	$FeTiO_3$
Lime [18]	CaO
Maghemite	$Fe_{2.67}O_4$
Magnetite	Fe_3O_4
Periclase	MgO
Portlandite [18]	$Ca(OH)_2$
Ice	H_2O
Amorphous mixed As(III)/As(V)-Fe(III) oxy-hydroxides [20]	
Birnessite	$NaMn_4O_8 \cdot 3H_2O$
Brucite	$Mg(OH)_2$

(continued)

Table 1 (continued)

Name	Formula
Ferrihydrite	$Fe_{4-5}(OH,O)_{12}$
Goethite	$\alpha\text{-FeO(OH)}$
Lepidocrocite	$\gamma\text{-FeO(OH)}$
Todorokite	$(Na,Ca,K,Ba,Sr)_{1-x}(Mn,Mg,Al)_6O_{12}\cdot3\text{-}4H_2O$
Tooeleite [20]	$Fe_6(AsO_3)_4(SO_4)(OH)_4\cdot4H_2O$
Carbonates and Nitrates (05)	
Amorphous calcium carbonates	$CaCO_3\cdot H_2O$ or $CaCO_3$
	(at least 5 forms)
Aragonite	$CaCO_3$
Baylissite [18]	$K_2Mg(CO_3)_2\cdot4H_2O$
Bütschliite [18]	$K_2Ca(CO_3)_2$
Calcite	$CaCO_3$
Chukanovite [18]	$Fe_2(CO_3)(OH)_2$
Dypingite [21]	$Mg_5(CO_3)_4(OH)_2\cdot5H_2O$
Fairchildite [18]	$K_2Ca(CO_3)_2$
Hydrocerussite	$Pb_3(CO_3)_2(OH)_2$
Hydrozincite [14]	$Zn_5(CO_3)_2(OH)_6$
Kalicinite [18]	$KH(CO_3)$
Lansfordite	$MgCO_3\cdot5H_2O$
Magnesite	$MgCO_3$
Mg-calcite	$(Mg_xCa_{1-x})CO_3$

(continued)

Table 1 (continued)

Name	Formula
Monohydrocalcite	$CaCO_3 \cdot H_2O$
Nesquehonite	$Mg(CO_3) \cdot 3H_2O$
Protodolomite	$CaMg(CO_3)_2$
Rhodochrosite	$MnCO_3$
Siderite	$FeCO_3$
Strontianite [22]	$SrCO_3$
Teschemacherite [18]	$(NH_4)H(CO_3)$
Trona [18]	$Na_3(HCO_3)(CO_3) \cdot 2H_2O$
Vaterite	$CaCO_3$
Gwihabaite	$(NH_4)NO_3$
Shilovite [18]	$Cu(NH_3)_4(NO_3)_2$
Sulphates (07)	
Aphthitalite	$K_3Na(SO_4)_2$
Arcanite [18]	K^2SO^4
Ardealite	$Ca_2[PO_3(OH)](SO_4) \cdot 4H_2O$
Barite	$BaSO_4$
Bassanite [23]	$CaSO_4 \cdot 0.5H_2O$
Celestine	$SrSO_4$
Epsomite	$MgSO_4 \cdot 7H_2O$
Feynmanite [18]	$Na(UO_2)(SO_4)(OH) \cdot 3.5H_2O$
Green rust [24, 25]	$[Fe^{2+}_{(6-x)}Fe^{3+}_{(x)}(OH)_{12}]^{x+}[(A^{2-})_{x/2} \cdot yH_2O]^{x-}$

(continued)

Table 1 (continued)

Name	Formula
Gypsum	$CaSO_4 \cdot 2H_2O$
Hexahydrite	$MgSO_4 \cdot 6H_2O$
Jarosite	$KFe^{3+}_3(SO^4)_2(OH)_6$
Langbeinite [18]	$K_2Mg_2(SO_4)_3$
Lecontite [18]	$(NH_4)Na(SO_4) \cdot 2H_2O$
Melanterite	$Fe^{2+}SO_4 \cdot 7H_2O$
Möhnite [18]	$(NH_4)K_2Na(SO_4)_2$
Schwertmannite	$Fe^{3+}_{16}O_{16}(OH)_{9.6}(SO_4)_{3.2} \cdot 10H_2O$
Witzkeite [18]	$Na_4K_4Ca(NO_3)_2(SO_4)_4 \cdot 2H_2O$
Phosphates (08)	
Amorphous calcium phosphate (at least 6 forms)	Variable
Amorphous calcium pyrophosphate	$Ca_2P_2O \cdot 2H_2O$
Amorphous hydrous Fe(III) phosphate	Variable
Bakhchisaraitsevite	$Na_2Mg_5(PO_4)_4 \cdot 7H_2O$
Bobierrite [18]	$Mg_3(PO_4)_2 \cdot 8H_2O$
Brushite	$Ca[PO_3(OH)] \cdot 2H_2O$
Carbonate-hydroxylapatite (dahllite)	$Ca_5(PO_4,CO_3)_3(OH)$
Chlorapatite	$Ca_5(PO_4)_3Cl$
Dittmarite [18]	$(NH_4)Mg(PO_4) \cdot H_2O$
Fluorapatite	$Ca_5(PO_4)_3F$
Francolite	$Ca_{10}(PO_4)_6F_2$

(continued)

Table 1 (continued)

Name	Formula
Hannayite	$Mg_3(NH_4)_2H_4(PO_4)_4 \cdot 8H_2O$
Hopeite [18]	$Zn_3(PO_4)_2 \cdot 4H_2O$
Hazenite	$KNaMg_2(PO_4)_2 \cdot 14H_2O$
Hydroxylapatite	$Ca_5(PO_4)_3(OH)$
Monetite	$Ca[PO_3(OH)]$
Newberyite	$Mg[PO_3(OH)] \cdot 3H_2O$
Octacalcium phosphate	$Ca_8H_2(PO_4)_6$
Pyromorphite [26]	$Pb_5[PO_4]_3(Cl, OH)$
Redondite [18]	$Al(PO_4) \cdot 2H_2O$
Schertelite [18]	$(NH_4)_2MgH_2(PO_4)_2 \cdot 4H_2O$
Stercorite [18]	$(NH_4)Na(PO_3OH) \cdot 4H_2O$
Strengite [27, 28]	$FePO_4 \cdot 2H_2O$
Struvite	$Mg(NH_4)(PO_4) \cdot 6H_2O$
Struvite-(K) [18]	$KMg(PO_4) \cdot 6H_2O$
Swaknoite [18]	$(NH_4)_2Ca(PO_3OH)_2 \cdot H_2O$
Variscite [29]	$AlPO_4 \cdot 2H_2O$
Vivianite	$Fe_3^{2+}(PO_4)_2 \cdot 8H_2O$
Whitlockite	$Ca_{18}H_2(Mg,Fe)_2(PO_4)_{14}$
Silicates (09)	
Amorphous silica	$SiO_2 \cdot nH_2O$
Amorphous hemimorphite [30, 31]	$Zn_4(Si_2O_7)(OH)_2 \cdot H_2O$

(continued)

Table 1 (continued)

Name	Formula
Buddingtonite	$(NH_4)(AlSi_3O_8)$
Chamosite [17]	$(Fe_5Al)(Si_3Al)O_{10}(OH)_8$
Forsterite [32]	Mg_2SiO_4
Kaolinite [33]	$Al_4(Si_4O_{10})(OH)_4$
Kerolite [34]	$Mg_3Si_4O_{10}(OH)_2 \cdot nH_2O$
Low-ordered (Fe,Al) silicate [35]	Variable composition
Nontronite [36]	$Na_{0.3}Fe^{3+}{}_2(Si,Al)_4O_{10}(OH)_2 \cdot nH_2O$
Quartz [18]	SiO_2
Stevensite [37]	$(Ca_{0.09}K_{0.01}Sr_{0.01})_{\Sigma=0.11}(Mg_{2.84}Fe_{0.02}Al_{0.03})_{\Sigma=2.89}(Si_{3.98}Al_{0.02}O_{10})(OH)_2 \cdot nH_2O$
Organic compounds (10)	
Abelsonite	$Ni^{2+}C_{31}H_{32}N_4$
Antipinite [18]	$KNa_3Cu_2(C_2O_4)_4$
Branchite [18]	$C_{20}H_{34}$
Calclacite [18]	$Ca(CH_3COO)Cl \cdot 5H_2O$
Ca malate	$C_4H_4CaO_5$
Ca tartrate	$C_4H_4CaO_6$
Carpathite	$C_{24}H_{12}$
Chanabayaite [18]	$CuCl(N_3C_2H_2)(NH_3) \cdot 0.25H_2O$
Coskrenite-(Ce) [18]	$(Ce,Nd,La)_2(C_2O_4)(SO_4)_2 \cdot 12H_2O$
Earlandite	$Ca_3(C_6H_5O_7)_2 \cdot 4H2O$
Ernstburkeite [18]	$Mg(CH_3SO_3)_2 \cdot 12H_2O$

(continued)

Table 1 (continued)

Name	Formula
Falottaite [18]	$MnC_2O_4 \cdot 3H_2O$
Flagstaffite [18]	$C_{10}H_{22}O_3$
Glushinskite	$MgC_2O_4 \cdot 2H_2O$
Guanine	$C_5H_3(NH_2)N_4O$
Hartite	$C_{20}H_{34}$
Hoganite	$Cu(CH_3COO)_2 \cdot H_2O$
Humboldtine [38]	$(FeC_2O_4 \cdot 2H_2O)$
Idrialite	$C_{22}H_{14}$
Joanneumite [18]	$Cu(C_3N_3O_3H_2)2(NH_3)_2$
Julienite [18]	$Na_2Co(SCN)_4 \cdot 8H_2O$
Kladnoite [18]	$C_6H_4(CO)_2NH$
Kratochvílite	$C_{13}H_{10}$
Levinsonite-(Y) [18]	$YAl(SO_4)_2(C_2O_4) \cdot 12H_2O$
Lindbergite	$MnC_2O_4 \cdot 2H_2O$
Mellite [18]	$Al_2C_6(COO)_6 \cdot 16H_2O$
Moolooite	$CuC_2O_4 \cdot nH_2O$
Oxammite [18]	$(NH_4)_2(C_2O_4) \cdot H_2O$
Paceite	$CaCu(CH_3COO)_2 \cdot 6H_2O$
Paraffin hydrocarbon	
Sodium urate	$C_5H_3N_4NaO_3$
Stepanovite [18]	$NaMgFe(C_2O_4)_3 \cdot 9H_2O$

(continued)

Table 1 (continued)

Name	Formula
Timnunculite [18]	$C_5H_4N_4O_3\cdot 2H_2O$
Triazolite [18]	$NaCu_2(N_3C_2H_2)_2(NH_3)_2Cl_3\cdot 4H_2O$
Urea	$CO(NH_2)_2$
Uricite	$C_5H_4N_4O_3$
Uroxite [18]	$[(UO_2)_2(C_2O_4)(OH)_2(H_2O)_2]\cdot H_2O$
Weddelite	$CaC_2O_4\cdot 2H_2O$
Whewellite	$CaC_2O_4\cdot H_2O$
Zhemchuzhnikovite [18]	$NaMgAl(C_2O_4)_3\cdot 9H_2O$
Zugshunstite-(Ce) [18]	$CeAl(SO_4)_2(C_2O_4)\cdot 12H_2O$

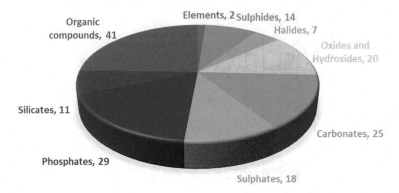

Fig. 1 Distribution of biominerals over Strunz classes

variable because they depend on many different factors and conditions that cellules can modulate to obtain the required biomineral structure and architecture.

2.1 Biologically Controlled and Induced Biomineralization

As previously mentioned, biominerals are classified based on the causative effect of cellular processes into two main classes [7, 12, 19]: biologically controlled mineralization (BCM) and biologically induced mineralization (BIM). In BCM organisms have extensive control over the mineral formation. BCM results in well-ordered mineral structures with minor size variations and species-specific crystal habits [39]. BCM can occur extracellularly or intracellularly. In the first case, the cellule fed ions to the mineralization area, generally the outermost cell wall, where a template, such as a biopolymer, drives the nucleation and subsequent particle attachment and biomineral formation (see e.g. so called "low magnesium" foraminifera, Fig. 2a–c). For intracellular BM, nucleation occurs within the cell, the nanocrystals are then fed to the area where the biomineral form (such as so called "high magnesium" or "porcelanaceous" foraminifera Fig. 2d–f). In BIM, organisms have no to minor control over the mineral formation. BIM generally results in heterogeneous mineral compositions with poor crystallinity, including large size variations, poorly defined crystal morphologies and the inclusion of impurities. Their formation is due to a change in bulk water chemistry related to cellular processes. Structures indistinguishable from BIM products can be also produced in the absence of organisms and highly organized minerals may form in the presence of organic abiotic substrates. Figure 2g shows an occurrence of "green rust" (a ferrous-ferric hydroxide) [40, 41], which is likely related to siderophore production and release, and results in the unexpected stabilization of ferrous iron during photosynthetic activity of bacteria. Red, muddy sediments in Fig. 2g are made of goethite, a ferric oxyhydroxide; the change from red to green sediments occurs at a length scale of tens of centimetres. This is indicative of the typical spatial variability of microorganism's communities and functions and their capacity of building and differentiating microenvironments.

Analysis of morphological and structural properties can lead to the attribution of the mineralization to BIM and BCM. Frankel and Bazylinski [24] list most of the Fe and Mn BIM and these mainly involve extracellular deposition of minerals. However, there are several reports of intracellular deposition of minerals that seem to blur the border between BIM and BCM. For example, many bacteria have iron storage proteins known as bacterioferritins that compartmentalize iron at concentrations far

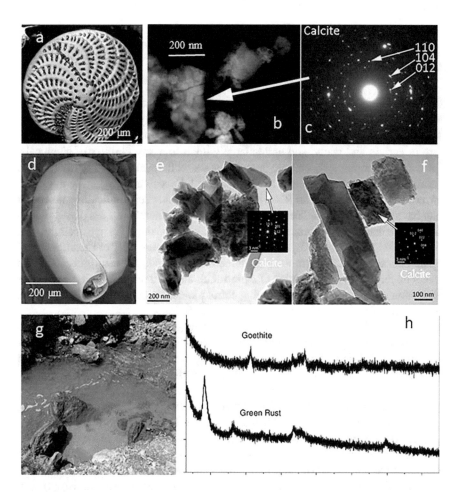

Fig. 2 a–f Images of Ca-carbonate biomineralizations: **a** SEM (scanning electron microscopy) image of extra-cellular BCM foraminifera *Elphidium*; **b** HRTEM image of a fragment and **c** selected area electron diffraction (SAED); **d** SEM image of intracellular BCM *Quinqueloculina* seminula; **e** and **f** HRTEM images with SAED of the calcite needles; **g** "green rust" occurring in a mine polluted river; **h** XRD patterns of the green sediments (bottom pattern) and the red sediments (top pattern) in (**g**)

above the solubility of Fe^{3+} [42]. Thus, attribution to BIM or BCM can be in some instances somewhat uncertain. For this reason, understanding biomineralization processes often requires the use of advanced investigation tools.

A strategy often adopted to understand the biomineralization process is to test the in vitro capability to drive mineralization of cultured cells, then the bio-mineralization effect of cell lysate and/or released polymers [43–45]. In this way, the role that in vivo cells, lysate of cell walls, and molecules released to the solution play on mineralization can be individually recognized allowing the attribution to BIM or BCM.

2.2 Mineral Surfaces and the Origin of Life

Since the original question about the physical origin of life [46] the scientific community has gained terrific insight into the complexity of life and its interaction with mineral surfaces interaction. Life functions through the specialized chemistry of carbon and water, and builds largely upon four key families of chemicals: lipids (cell membranes), carbohydrates (sugars, cellulose), amino acids (protein metabolism), and nucleic acids (DNA and RNA). Researchers generally think that current life descends from an RNA world, although other self-replicating molecules may have preceded RNA. There is a consensus that mineral surfaces play a pivotal role in biomineralization processes because they can favour selective adsorption of molecules and favour bio-polymerization processes. A large body of literature investigates how self-replicating molecules, or their components, came into existence. The role of minerals in polymerization of amino acids and nucleic acids, and the selective adsorption of organic species, including chiral molecules, onto mineral surfaces are two aspects of main interest for bio-mineral interactions.

A consistent body of experimental evidence suggests that vesicle-forming lipids and self-assembling lipid amphiphilic molecules [47] are at the base of the origin of life (see also [48], and references therein), in a step in which minerals may have played a useful, if not essential, role. However, besides self-assembling lipids, many of the key building blocks of life, including amino acids, sugars, and nucleic acid bases, are highly soluble in water and do not spontaneously self-organize. For these molecules, minerals may have provided a critical template for the formation of biopolymers.

Selective adsorption of organic molecules on mineral surfaces represents a viable mechanism for prebiotic molecular symmetry breaking. Chiral mineral surfaces abound in the natural world, and they have been shown to separate left- and right-handed molecules. Quartz is the only common chiral rock-forming mineral (i.e., it occurs naturally in both left- and right-handed crystals), but all centric crystals also have the potential to display chiral fracture or growth surfaces [49, 50], as well as chiral surface features such as steps and kink sites [51–53]. Common intrinsically chiral surfaces, in addition to those of quartz, include feldspar (110), clinopyroxene (110) and (111), olivine (111), clinoamphibole (110) and (011), calcite (214), and gypsum (110) and (111) faces.

Calcite is central to the origin of life in geochemically plausible origin scenarios because (1) calcite and other rhombohedral carbonate minerals were abundant in the Archean Era (e.g., [54, 55]); (2) calcite strongly adsorbs amino acids [19, 56–58]; (3) calcite's surface growth topology is dramatically affected by adsorbed L- versus D-amino acids [51, 53, 59–61]; (4) calcite scalenohedral (214) -type crystal faces demonstrated selective adsorption of D- and L- amino acids [52, 62, 63]. Through the co-evolution of life and minerals, deep modification occurred leading to the actual diversity of bio-mineral interactions.

The special role of clay minerals was advocated because these minerals have both high surface electrostatic charge and large reactive surface areas that facilitate the absorption of organic molecules. Clay minerals were found to concentrate and favour polymerization of amino acids to form small protein-like molecules [64] and can act as scaffolds in the formation of RNA, the polymer that carries the genetic message enabling protein synthesis [65–72]. Moreover, fine-grained clay particles may induce polymerization, though the molecular-scale mechanisms of the process are not yet fully understood [47].

DNA binding and conformation on mica surfaces as a function of solution compositions, has successfully been investigated by several means [73–75]. It is now well established that the DNA conformation on mica changes with cationic content and ionic strength, where a larger ionic potential (charge/density) favours adsorption. Valdrè et al. [76] pointed out that anisotropic surface properties of atomic-flat natural Mg–Al-hydroxysilicate substrate drives self-assembly and nanopatterning of nucleotides. RNA and DNA selectively adsorb on the surface of the Mg-hydroxide layer, with a higher concentration at the edge. No adsorption of RNA and DNA on surface of the TOT (tetrahedral–octahedral–tetrahedral sequence) layer was instead observed. Moro et al. [77] investigated interaction among di-glycine and clinochlore surfaces, showing preferential adsorption of di-glycine onto the hydrophobic brucite-like sheet, with the observed molecules organized as dot-like (single-molecules), agglomerates, filament-like and network structures by the surface, whereas only very few peptides were adsorbed onto the hydrophilic talc-like layer.

2.3 Biological Interfaces

Cellular membranes and cellular activity can modify physical properties of biological interfaces by effects on water dielectric properties, adding lipids and polysaccharides at the interface. In this way, cellular activity can exert a control on biomineralization kinetics, nucleation and growth regime, crystal size, crystal shape, and molecules aggregation [78]. These mechanisms explain the differences among mineral growth from aqueous solution, where growth regime dominates, and their biomineral analogue where nucleation regime generally dominates.

Regarding the role of water, it has become apparent that the interaction with water codetermines the molecular and supra-molecular organization of lipids and proteins within the cellular membrane ([79] and references therein). Membrane hydration drives the self-assembly of the bilayers, and studies of partially hydrated bilayers by X-ray scattering, neutron scattering and calorimetry indicated that the fluidity of the lipid phase—an essential parameter for membrane function—varies strongly with the degree of hydration [80]. Inversely, NMR (nuclear magnetic resonance) experiments have shown that the lipid head groups have a strong influence on the local water structure [81–86]. Biomembranes are bilayers of mainly amphiphilic phospholipids, which are composed of a glycerol unit with two hydrophobic fatty acid "tails" and one hydrophilic phosphate ester "head group". The self-assembled nature of the lipid bilayer membrane makes it a complex and dynamic system, whose behaviour and properties strongly depend on composition and temperature.

Interface between cellules and mineral surfaces is the place where organic molecules are released by living organisms to exert their physic-chemical control on the surrounding environment, namely solution pH and Eh, partial pressure of CO_2, ionic strength and so on.

Several critical biochemical phenomena involve electron transport. For instance, sulphur reducing bacteria and metal reducing bacteria entail unique biomolecular machinery optimized for long-range electron transport. Microorganisms have adapted multiheme c-type cytochromes to arrange heme cofactors into wires that cooperatively span the cellular envelope, transmitting electrons along distances greater than 10 nm. Recently, the first crystal structure of a representative decaheme protein was solved, but the mechanism of electron conduction remains difficult to probe experimentally. Therefore, at the molecular level, how these proteins shuttle electrons along their heme wires, navigating intraprotein intersections and interprotein interfaces efficiently, remains a mystery thus far inaccessible to experiment. Breuer et al. [87] reveal an evolutionary design principle significant to an entire class of heme proteins involved in mediating electron flow between bacterial cells and their environment. Insights into this phenomenon are of great importance for biomineral interactions and open the way to a multitude of potential biotechnological applications.

2.4 Nanocrystals and Mesocrystals in Biominerals

Understanding nanocrystal aggregation is a milestone for biomineralization studies. External morphology, microstructure, and texture provide important evidence of attachment-based growth, although they alone do not prove formation by a particle-based growth process. In fact, such features can be misleading. For example, irregular or branched morphologies can form through dendritic and spherulitic growth mechanisms from solution at high degrees of supersaturation [88].

Pioneer work of Penn and Banfield [89] shed light on mechanism of oriented aggregation of TiO nanocrystals. Banfield et al. [90] investigated biomineralization products of iron-oxidizing bacteria by high-resolution transmission electron microscopy (HRTEM); they revealed an alternative coarsening mechanism in which adjacent 2- to 3-nm particles aggregate and rotate so their structures adopt parallel orientations in three dimensions. Crystal growth is accomplished by eliminating water molecules at interfaces and forming iron-oxygen bonds. Self-assembly occurs at multiple sites, leading to a coarser, polycrystalline material. Point defects (from surface adsorbed impurities), dislocations, and slabs of structurally distinct material are created as a consequence of this growth mechanism and can dramatically impact subsequent reactivity. Meldrum and Cölfen [78] summarized the possible reaction paths for (bio)mineralization processes pointing out the many steps and activation energies needed to aggregate ions, form nanocrystals or droplets, perhaps from mesocrystals and eventually achieve macrocrystals. De Yoreo et al. [91] pointed out that pure oriented attachment rarely occurs, this implies very low (less than $10°$) or no misalignment among nanocrystals, while nearly oriented attachment often occurs with higher misalignment. This occurs in cyanobacterial biomineralization [92, 93]. Crystallization by particle attachment of amorphous precursors has been demonstrated in modern biominerals across a broad phylogenetic range of animals, including sea urchin spicules [94], spines [95, 96], and teeth [97]; the larval shells [98] and nacre [99] of mollusks; zebrafish bone [100] and mouse enamel [101]; and scleractinian coral skeletons [102].

3   Minerals and Life: A Co-evolution History

Biominerals are widespread in nature because they allow an organism to have a protective internal or external structure and to support physiological functions. The mutual impact of life evolution on mineral diversity is well documented in the geological record, and the scientific community considers that life and minerals coevolved through the geological time [13, 103]. Calcium carbonates probably forms the most known biomineralizations and it is the major constituent of the biogenic mineral reservoir. Since the earliest times, bacterial activity has been a driving force in Ca carbonate formation [104], as demonstrated by stromatolite formation dating at least 3.48 Ga, even though the greatest diversity of stromatolites in Earth's history was recorded between 2.25 and 2.06 Ga, in the aftermath of the Paleoproterozoic glaciations [13, 105]. Stromatolites can show remnants of fossilized microorganisms and are likely the most ancient evidence for BM. Unfortunately, cellularly preserved fossils and palimpsest microstructures are present only rarely in ancient stromatolites and the attribution of ancient stromatolites to BIM or BCM is still debated. Clearcut BCM examples appeared at a later stage of the coevolution history.

3.1   Ca Carbonate Biominerals

The history of BCM in trilobites, the dominant marine arthropods that lived during the Palaeozoic, is well documented. Across the time from life explosion in the Cambrian to mass extinction in the Permian, trilobites developed at least nine different taxa and three types of compound eyes, all with lenses supposed to consist mainly of primary calcite. The various kinds of trilobite eyes became highly diverse due to the demands provided by the various new environments.

Further examples of Ca-carbonate controlled biomineralization processes are offered by molluscs, clams, oysters, gastropods, foraminifera, coccolithophores, and corals (Fig. 3). These organisms developed the capability to select specific polymorphs, namely calcite, aragonite, vaterite, monohydrocalcite, and the amorphous calcium carbonate (ACC). These can occur, either singularly or together, in specific structures within an organism, or they may form sequentially during their development [8]. A typical case of phase transition was discovered in forming spicules in embryos of *Strongylocentrotus purpuratus* sea urchins. For instance, using X-ray absorption near-edge spectroscopy (XANES) and photoelectron emission microscopy (PEEM), [106] observed a sequence of three mineral phases: hydrated amorphous calcium carbonate (ACC·H_2O) $\rightarrow$ ACC $\rightarrow$ calcite. Interestingly, ACC·H_2O-rich nanoparticles can persist after dehydration and crystallization due to protein matrix components occluded within the mineral that inhibit ACC·H_2O dehydration. Weiss et al. [98] showed a similar function for ACC in the larval shells of the marine bivalves *Mercenaria mercenaria* and *Crassostreagigas*, where ACC transforms to aragonite.

In the last decades, it has been pointed out that the evolution of life implies adaptation to minerals, and this is genetically encoded in the organism's cellules. With life development across the geological record, biomineral numbers and type increased to support the different actual functions. Carbonate precipitation mediated by bacteria can occur as a by-product of metabolic activities (photosynthesis, ureolysis, denitrification, ammonification, sulphate reduction, and methane oxidation) that induce chemical variations at the microorganism-solution interface.

Other reported bacterial carbonate biominerals are rhodochrosite [113], hydrozincite [114], siderite [115], etc. More details on bacterial biomineralization can be found in the "Waters, metals and bacterial mineralization" paragraph.

3.2   Ca Phosphate Biominerals

The occurrence of Ca phosphate biominerals has been observed in living creatures from unicellular organisms to vertebrates [7, 116]. Bones are the most relevant example of phosphate biomineral in vertebrates. They are made up of a combination of inorganic calcium phosphate and an organic matrix. The inorganic phase comprises ~60–70% of the total bone mass [117] and consists of a nanocrystalline carbonate-hydroxyl-apatite (CO_3^{2-} at ~4.5 wt%). The remaining mass is formed by an organic matrix (20–30%, collagen fibres, glycoproteins and mucopolysaccharides) and water (10%). The combination and organization of the inorganic and organic components confers to the final material peculiar physical and mechanical properties providing toughness, the ability to withstand pressure, elasticity and resistance to stress, bending and fracture pressure [118].

Fig. 3 Images of Ca-carbonate biomineralizations: **a** mollusc shells (from Boettiger et al. [107], Copyright (2009) National Academy of Sciences); **b** Red Sea *Tridacna maxima* with its mantle exposed (from Lim et al. [108], Copyright (2020) Lim, Rossbach, Geraldi, Schmidt-Roach, Serrão and Duarte, under the Creative Commons CC BY license; **c** underwater photograph of coral reef in Indonesia with almost 100% cover of *Acropora* sp. (from Lesser [109], Copyright (2004), with permission from Elsevier); (d) coccolithophore species *Emiliania huxleyi* (from Triantaphyllou et al. [110], Copyright (2018) Triantaphyllou et al., under the Creative Commons Attribution License); **e** and **f** light micrographs of the foraminifera *Ammonia* sp. and *Haynesina germanica* from the Atlantic French coast intertidal mudflat and the Gullmar fjord (from LeKieffre et al. [111], Copyright (2018), with permission from Elsevier; **g** the pearl oyster *Pinctada fucata* (from Du et al. [112]), under the Creative Commons CC BY license

Bone constituents are highly hierarchically organised. Fibrillar collagen type I is the main component (~90%) of the organic fraction [117]. At the molecular level, the polarized triple helix of tropocollagen molecules is grouped in microfibres, and carbonate-hydroxyl-apatite crystals nucleate and grow within small cavities between their edges. Microfibres combine in larger fibres that represent the microscopic units of bone tissue. Finally, large fibres are arranged in different structural distributions to form the full bone [118]. Bone formation takes place by controlled nucleation and growth through an extracellular process. The inorganic mineral component and the organic matrix are linked to each other at the molecular level as Ca^{2+} is bonded to phosphoproteins along the collagen fibres at regular intervals, following the inorganic crystal structure of apatite [118]. The crystallographic axis c of bone crystals is not random oriented, but it is arranged in parallel to the collagen fibres and to the largest dimension of the platelet, while a- and b-axes are aligned along two other dimensions. Bone crystals are characterised by a hexagonal crystal structure, and they are 2–6 nm thick, 30–50 nm wide, and 60–100 nm long. The carbonate ions can substitute both phosphate (B-type substitution) and hydroxyl ions (A type substitution) in the lattice [116, 117]. The presence of carbonate ions and minor ions such as Mg^{2+}, Na^+, K^+, Cl^- and F^- leads to significant modifications in the lattice parameters with respect to purely inorganic apatites [118].

3.3 Fe Biominerals

Iron, together with Ca and Mg, is the most widespread metal in biominerals, due to Fe abundance on Earth, and to its important role in many metabolic processes. Bacteria represent the major mediators in the deposition of Fe biominerals in a host of different environments [8]. Iron biominerals allow organisms to accumulate the metal for future metabolic needs, avoiding high intracellular accumulation [119]. Other properties of Fe biominerals, potentially useful to organisms, include hardness, density and magnetism [119]. The ferric hydroxides or oxyhydroxides are an important class of Fe biominerals and can occur as amorphous or low-crystalline precipitates, such as ferrihydrite ($Fe_{4-5}(OH,O)_{12}$), or as crystalline phases such as lepidocrocite (g-FeO(OH)) or goethite (α-FeO(OH)) [33, 119, 120]. The Fe oxide magnetite, Fe_3O_4, is another important Fe biomineral. Magnetite is characterised by magnetic properties, high density (5.1 g/cm^3) and hardness. Uniformly sized particles of magnetite, often arranged in chains, are formed by magnetotactic bacteria allowing them to align according to the Earth's magnetic field [121]. Magnetite also occurs in the radular teeth of chitons, flattened molluscs, commonly found on hard substrata in intertidal regions of coastlines around the world ([122, 123] Fig. 4a–d). In chiton teeth, mineralization occurs within an organic matrix (α-chitin and protein) through several steps to form mature teeth, and chitons possess all stages of tooth development in one radula [124]. Specifically, four stages have been observed: (i) newly secreted unmineralized and transparent teeth, composed of α-chitin and proteins, and Fe transportation into the organic matrix; (ii) heterogenous nucleation of ferrihydrite on acidic proteins coating the alpha-chitin fibers; (iii) solid-state transformation ferri-hydrite → magnetite; (iv) fully mineralized teeth in both shell and core (the tooth core region is filled, [124]) (Fig. 4d, [125]). The deposition of these composite structures, refined by evolution over millions of years, confers unique properties to the chiton teeth, such as tensile strength, shock absorption and controlled wear and abrasion, resulting in highly efficient feeding tools [123, 126].

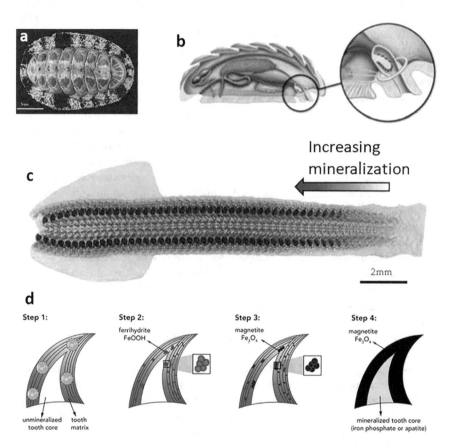

Fig. 4 Chiton, chiton teeth and their formation. **a** Chiton (*Acanthopleura gaimardi* species characterized by eight overlapping aragonite plates surrounded by a fleshy girdle covered by small aragonite spines) (from Brooker and Shaw [123], Licensee IntechOpen, under the terms of the Creative Commons Attribution 3.0 License); **b** internal anatomy of a representative chiton showing the location of the radula, a rasping, toothed conveyor belt-like structure used for feeding (from Weaver et a. [127]), Copyright (2010), with permission from Elsevier; **c** optical micrograph of the radula of *Acanthopleura gaimardi*; the arrow indicates the progressive stages of radular tooth development (from Brooker and Shaw [123], Licensee IntechOpen, under the terms of the Creative Commons Attribution 3.0 License; **d** hypothesized mechanism of Fe biomineral formation (ferrihydrite and magnetite) in chiton teeth (see text for explanation) (from Moura and Unterlass [125], Licensee MDPI, Basel, Switzerland, under the Creative Commons Attribution)

4   Geo-Bio Interactions and the Environment

4.1   Biofilm Composition and Structure

Most bacteria in the environment live associated with surfaces, in so called biofilms [128, 129]. Biofilms are the oldest form of life recorded on Earth [130] documented by petrified biofilms in Precambrian stromatolitic rocks [131]. They occur in nearly every moist environment where sufficient nutrient flow is available and surface attachment can be achieved [132, 133]. Biofilms can be formed by a single bacterial species, although they often consist of many species of bacteria, fungi, algae and protozoa that are attached to abiotic surfaces, such as minerals and rocks, or air–water interfaces, and to biotic surfaces, such as plants, roots, leaves and other microbes. Attachment is favoured on surfaces that are rough, hydrophobic and coated by surface conditioning films [128, 132], and are widespread both in subaerial and in subaquatic environments. Also, biofilms can grow in extreme environments characterised by extreme temperature, extreme pH values, high salinity, high pressure, poor nutrients, etc. Microorganisms that can survive in these extreme environments are called extremophiles [134], which include thermophiles, alkaliphiles, acidophiles, halophiles etc. according to the harsh condition that characterises the specific environment.

In this chapter, biofilm formation mechanisms and characteristics will be discussed briefly; for excellent comprehensive reviews on this topic see O'Toole et al. [135], Hall-Stoodley and Stoodley [136], Verstraeten et al. [137]. Biofilm matrix consists mainly of water (79–95%), which is held by the highly hydrated extracellular polymer substances (EPS), that represent 70–95% of the organic matter of the dry mass of the biofilm. The microorganisms are only a minor part of mass and volume but exert an important role by excreting the EPS and controlling the physical and chemical properties of biofilms [131, 138–140] (see Table 1 in Flemming and Wingender [141]). EPS are a complex mixture of highly hydrated biopolymers mainly consisting of polysaccharides, as well as proteins, nucleic acids, lipids and humic substances, which keeps the biofilm cells together. In Gram-negative cells, the EPS are made up of lipopolysaccharides, capsule polysaccharides, excreted polysaccharides and proteins; in Gram-positive cells, the main component of the EPS are lipoteichoic acids, polysaccharides and proteins [142]. EPS chemical composition and structure vary depending on the type of substrate upon which the cells are grown and are affected by environmental stress [128, 142, 143].

The development of a biofilm occurs, firstly, through cell attachment by physico-chemical interactions or extracellular matrix protein secretion to form a cell monolayer, this stage is followed by biofilm maturation and, finally, detachment of cells. Biofilms can be highly organized, and can form a single layer, a three-dimensional structure or even aggregates [144]. For subaquatic biofilms, the structure changes according to water flow conditions. In fastmoving waters, biofilms tend to form filamentous streamers (e.g., drainage run-off from acid mines, hydrothermal photosynthetic mats), whereas in quiescent waters, biofilms form mushroom or mound-like structures characterised by isotropic overall patterns, similar to those of stromatolites [128].

Biofilms actively interact with elements in waters (subaquatic biofilms) and with the attachment surface (subaerial and subaquatic biofilms). In the following paragraphs biofilm-metal interactions in aquatic environment and the influence of biofilms on the attachment surfaces will be presented.

4.2 Biofilm-Metal Interaction in Aquatic Environments

In aquatic environments, trace elements interact with biofilms through physical, chemical and/or biological processes [145–147]. Metal distribution, immobilisation and remobilisation will depend mainly on (i) the sorption properties of the biofilm, (ii) the type and concentration of the ligands within the biofilm matrix, (iii) the pH values and redox potential conditions at the cell/EPS surface, (iv) the physico-chemical characteristics of waters, and (v) the availability of reactive mineral surface sites (e.g. oxide, sulphide, phosphate, carbonate precipitates) ([145] and references therein). It has been demonstrated that low pH values favour the release of ions from a bound state, due to the competition with H^+ ions, while high pH values tend to favour their chelation. For example, Ferris et al. [148] grew microbial biofilms in acidic (pH 3.1) and near neutral water contaminated with metals from mine wastes, and they observed that biofilm metal uptake at a neutral pH level was enhanced by up to 12 orders of magnitude over acidic conditions. Moreover, adsorption strength values were usually higher at elevated pH levels.

Metal uptake occurs mainly through ion exchange, chelation, adsorption, and diffusion through cell walls and membranes [145], and it has been observed that EPS play an important role in the sorption of metals in biofilms due to the presence of anionic groups such as carboxyl, phosphoryl, and sulphate groups [142] that allow the formation of unidentate, bidentate and multidentate complexes of cations with anionic groups on the EPS molecules [145].

Both living and dead cells can accumulate metals through metabolism-independent association with cell walls and other external surfaces. Metabolism-dependent transport across the cell membrane and transport systems occurs only in living cells [149]. In the first mechanism, metal binding on cell walls is affected by their structure. In Gram-positive bacteria, the cell wall is composed of an intermingling of peptidoglycan and secondary polymers. Peptidoglycans consist of polysaccharide chains cross-linked by oligopeptides. Carboxylate groups at the carboxyl terminus of individual chains provide the bulk of the anionic character of peptidoglycans [146, 150]. Other secondary polymers include teichoic acids and teichuronic acids, which contain phosphate and carboxylate residues, respectively. Peptidoglycan and the cell wall acids are exposed to the external aqueous solution and form the surface of the bacterial cell. The anionic functional groups present in the peptidoglycan, teichoic acids and teichuronic acids of Gram-positive bacteria are the main components responsible for the anionic character and metal binding capacity of the cell wall [151]. In Gram-negative bacteria, cell wall is characterised by a more complex structure made up of (i) an outer porous and highly permeable membrane, rich in protein and lipopolysaccharide, (ii) a thin layer of peptidoglycan, enzymes and structural proteins in the periplasmic space. The peptidoglycan, phospholipids, and lipopolysaccharides are the primary components involved in the metal binding in Gram-negative bacteria. On the surface of bacteria, a well-ordered layer (S-layer), made up of protein and glycoprotein subunits, is frequently observed. Here, exposed anionic residues can react with dissolved metals [150, 151].

Metabolism-dependent transport of metals may be a slower process than biosorption on the cell wall, and can be affected by the temperature, the absence or presence of an energy source, the physiological state of cells and the nature and composition of the growth medium. Metals transported into the cell may be bound, precipitated, localized or translocated to specific intracellular structures or organelles depending on the metal and the bacteria species [152]. Passive biosorption and metabolic uptake can occur in the same bacteria species. Gourdon et al. [153] isolated Gram-positive and Gram-negative bacteria from activated sludge to evaluate Cd biosorption. At 30° and pH 6.6, Cd biosorption was higher (20%) in Gram-positive bacteria than Gram-negatives. This difference was attributed to the cell wall composition, rich in glycoproteins in the Gram-positive bacteria. These are characterised by a higher number of potential binding sites for Cd than phospholipids and lipopolysaccharides that characterised the external layer of the cell wall in Gram-negative bacteria. Biosorption was mainly attributed to the interaction of the metal with the bacteria surfaces, although metabolic uptake appeared to occur, especially in Gram-positive bacteria.

The sorption of pollutants on biofilms can be considered a dynamic process because biofilms are not chemically inert, and variations in environmental conditions can affect the microbiota and their physiology. Moreover, the detachment of biofilms can lead to microbial degradation processes of the biological binding sites, resulting in remobilization of metals, thus representing a secondary source of contaminants [154].

4.3 Waters, Metals and Bacterial Mineralization

In aquatic environments, bacteria are ubiquitous microorganisms that can precipitate a wide range of authigenic minerals, and drive both modern and ancient biogeochemical cycles from the microenvironment to global scales [155]. Indeed, bacteria have a remarkable potential to sequester and accumulate cations onto their surfaces mainly because (i) cell walls are ionized and naturally anionic (pH between 5 and 8), and because (ii) they are characterised by a large surface area to volume ratio due to their small size. Transition metals, due to their electronegativities, oxidation state, hydrated radii and hydration energies [156], are characterised by an extremely high affinity for the polymeric material present in the cell wall and outer membrane (carboxyl and phosphoryl groups), and in the surrounding capsules (carboxyl and hydroxyl groups) [157, 158]. During bacterial biomineralization, often, metals are electrostatically bound to the anionic surfaces of the cell wall and organic polymers, reducing the activation energy barriers to nucleation and providing sites for crystal growth [155, 159].

Due to its high concentration in natural waters compared to other trace metals [160], Fe is commonly bound to organic sites forming the greatest number of biominerals in waters (Fig. 5a, b). The microbial precipitation of Fe hydroxide is widespread in several aquatic systems, such as acid mine drainage environments, river sediments, deep subterranean groundwater, marine sediments, around deep-sea vents and in hydrothermal plumes [155] and references therein). In oxygenated waters, bio-Fe-hydroxides are commonly precipitated through different processes: (i) the binding of dissolved ferric species to negatively charged polymers, (ii) the reaction of soluble ferrous iron with dissolved oxygen and subsequent precipitation of ferric hydroxide on bacteria, and (iii) as a consequence of the metabolic activity of Fe^{2+} oxidizing bacteria that can induce ferric hydroxide precipitation as a secondary by-product [155]. *Acidithiobacillus ferrooxidans* and *Leptothrix* sp. are common microorganisms that produce Fe^{3+} sheaths and can be responsible for the production of copious amounts of Fe^{3+} mineral phases in the environment [161]. These bacteria utilize the oxidation of Fe^{2+} ions by O_2 as a source of energy [8, 119]. Ferric hydroxide can then evolve in more stable Fe oxides (e.g., hematite and goethite) via dehydration or dissolution–reprecipitation. Some peculiar bacteria, magnetotactic bacteria, can precipitate magnetite and greigite whose crystals are arranged in chains enclosed in membrane vesicles (magnetosomes) [119, 155]. The abundance and morphology of magnetosomes can reflect environmental conditions and they have been used as paleoenvironmental proxies as reported by He et al. [162].

Silica [163, 164], phosphate [165], carbonate [16], and sulphate/sulphur [166] available in solution may react with Fe bound to bacterial surfaces to form other authigenic mineral phases [161, 167]. The final product will depend on both the control of the bacteria on the precipitation process and on the pH values and redox conditions of the specific environment. For example, [165] investigated the depth variations of Fe and P speciation in Lake Pavin (Massif Central, France), by applying complementary research techniques such as X-ray diffraction (XRD), scanning electron microscopy (SEM), (HR)TEM (high resolution - transmission electron microscopy), synchrotron-based scanning transmission X-ray microscopy (STXM) and XANES at the Fe $L_{2,3}$-edges and the C K-edge. They found that Fe is hosted in different mineral phases: (i) in the shallower oxygenated water column (25 m), Fe is mainly hosted by Fe(III)-(oxyhydr)oxides and phyllosilicates, (ii) close to the chemocline (at 56 m depth), an additional amorphous Fe(II)–Fe(III)-phosphate phase was detected, (iii) in the deeper anoxic water (67 m and 86 m depths), vivianite ($Fe(II)_3(PO_4)_2 \cdot 8(H_2O)$) becomes dominant. A significant fraction of vivianite was observed at the surface of bacterial cells. Comparing field study with laboratory experiments, they proposed that Fe-oxidation may play a role in the precipitation of Fe-phosphates in the water column. Polyphosphate-accumulating microorganisms could also be involved in Fe-phosphate formation in the lake, by increasing dissolved phosphate concentrations in the monimolimnion. Both mechanisms play an important role in Fe and P cycling in

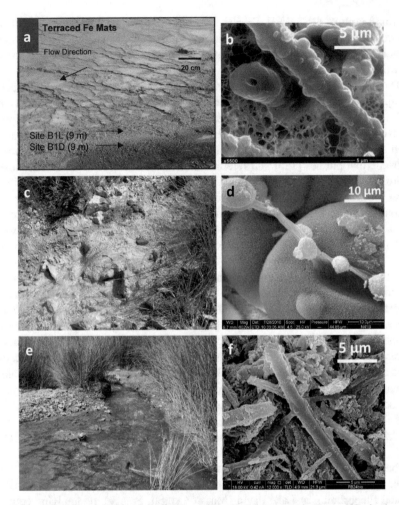

Fig. 5 Examples of bacterial biomineralizations: **a** Fe microbial mats at the Beowulf Spring located in Norris Geyser Basin, Yellowstone National Park; **b** SEM image of the filamentous As-rich Fe sheaths in the microbial mats; **a** and **b** from Inskeep et al. [161], modified, Copyright (2004), with permission from Elsevier; **c** and **e** microbial mats along the Naracauli stream responsible for formation of hydrozincite and amorphous Zn-silicate, respectively; **d** and **f** SEM images of hydrozincite and amorphous Zn-silicate, respectively

the investigated system. Along the Speed River (Ontario, Canada), Konhauser et al. [164] observed by TEM-EDS (energy dispersive spectroscopy) bacteria attached to different substrates and mineralized by Fe-rich capsular material to fine-grained (<1 μm) authigenic mineral precipitates. The authigenic grains are characterised by a wide range of morphologies, from amorphous gel-like phases to crystalline phases. The most abundant mineral is a complex (Fe, Al) silicate of variable composition. The gel-like phases are chemically similar to a chamositic clay, whereas the crystalline phases are more siliceous and have compositions between those of glauconite and kaolinite. The adsorption of dissolved constituents from the aqueous environment contribute significantly to the transfer of elements to the streambed sediments, considerably affecting the biogeochemical cycle of Fe, Si and Al. Ferric hydroxide and ferric hydroxysulfate precipitation by bacteria was observed by [166] in acid mine drainage lagoon sediments. TEM-EDS analysis revealed that bacteria are characterised by Fe-rich capsules, and Zn, Ti, Mn and K are incorporated into the mineralised matrix. In the subsurface, cells are associated with granular, fine-grained mineral precipitates, composed almost exclusively of Fe and S.

The type of bacterial biomineral is affected by the available ions in waters in which the microorganisms are growing [155, 167, 168], leading to a great variety of biomineralization patterns. In the Carnoulès AMD (acid mine drainage), Benzerara et al. [169] investigated As biomineralization by TEM, STXM, and near-edge X-ray absorption fine structure (NEXAFS) spectroscopy at the C K-edge, Fe $L_{2,3}$-edge, and As $L_{2,3}$-edge. Authors observed isolated spheres of Fe–As–S-rich precipitates (tooeleite and an amorphous phase) agglomerating outside the bacterial cell wall and forming, in some cases, thick continuous layers around the cells. Arsenic biomineralizations have been observed also in geothermal systems as reported by Tazaki et al. [16] and Inskeep et al. [161]. On the walls of the drainage systems of Masutomi Hot Springs (Yamanashi Prefecture, Japan), some bacteria accumulate Fe and As along with other trace elements to form various biominerals on the surface of the cell: hydrous iron oxides, calcite and lollingite ($FeAs_2$). Fourier-transform infrared spectroscopy (FT–IR) revealed the presence of organic components such as C—H, C = O, CNH, –COOH, and N—H, emphasizing the metal-binding potential of the bacteria [16]. Authors suggested that the polysaccharides of the bacteria may initially adsorb H_4SiO_4 and Ca ions from the spring water to form a mineral complex containing calcite. Then, Fe–As adsorption takes place through the cohesion of spherules, and finally lollingite precipitates over the calcite that encapsulated the bacterial cell surfaces.

Peculiar examples of bacterial biomineralizations occur along the Naracauli stream which drains the abandoned mining site of Ingurtosu (SW Sardinia, Italy). Specifically, at two distinct locations along the stream, bacterial activity drives the precipitation of two different Zn biominerals. At one location, in late spring-early summer, bioprecipitation results in the formation of hydrozincite ($Zn_5(CO_3)_2(OH)_6$), in association with a photosynthetic community including the cyanobacterium (*Scytonema* sp), and the microalga (*Chlorella* sp) [14] (Fig. 5c). Bio-hydrozincite precipitates on different substrates such as rocks, plant roots or stems on which the biofilm can adhere. SEM analysis (Fig. 5d) showed the association of the Zn hydroxycarbonate with the biological matrix such as bacterial sheaths, and extracellular polymeric substances. XRD and HR-TEM analysis indicated that, in comparison to abiotic hydrozincite, bio-hydrozincite is characterised by higher content in lattice defects (e.g., grain boundary, line defects), and by a higher a_o lattice parameter presumably reflecting differences in the stacking sequence of tetrahedral–octahedral–tetrahedral (TOT) units that are held together by distorted carbonate groups [114, 170] in the hydrozincite structure. Bio-hydrozincite is characterised by the presence of nanocrystals that aggregate according to an imperfect aggregation mechanism to form mesocrystals [93]. Further downstream, in summer, a colloidal Zn-silicate biomineralization occurs in association with the bacterium *Leptolyngbya frigida* [30] (Fig. 5e). The biomineral is made up of nanoparticles that precipitate on bacterial sheaths forming microtubules that are embedded in extracellular polymeric substances (Fig. 5f). ^{29}Si magic angle spinning and ^{29}Si/^{1}H cross polarization magic angle spinning analysis, FTIR and XAS analysis revealed a poorly crystalline phase closely resembling hemimorphite ($Zn_4Si_2O_7(OH)_2 \cdot H_2O$), a zinc sorosilicate [31, 171].

The reported examples suggest that bacteria can control biotransfer processes and biogeochemical cycles in different environmental conditions. They are able to concentrate heavy metals and metalloids through different mechanisms, such as adsorption, complexation, and active transport into the cell, that are influenced by external physicochemical parameters, such as the pH and ionic composition of the host water [16]. Bacterial biomineralizations in the aquatic environments contribute significantly to decrease metal dissolved in solution, offering clues to possible methods of bioremediation. Also, their metal-binding capacity makes these microorganisms potential candidates in biorecovery [172] of economically valuable metals [155].

4.4 Soils, Metals and Plant Activity

The advent of land plants, about 400 million years ago, contributed to alter Earth's surface appearance and to increase the rate of clay mineral production of at least an order of magnitude greater than the previous eras, favouring the production of soils [13, 103, 175–177]. Indeed, leaching through roots causes incongruent weathering of primary to secondary minerals, promoting the formation of clay-sized layer silicates and different oxides and hydroxides of Fe, Al, and Mn [178]. Fungi can cooperate with plants favouring mineral weathering [179, 180], although the precise mechanism by which mycorrhizae alter minerals is poorly understood. Recent ultramicroscopic and spectroscopic studies by Bonneville et al. [175] demonstrated that biotite weathering by fungi can occur through a biomechanical-chemical process, starting by physical distortion of the lattice structure of biotite and subsequent dissolution and oxidation reactions that lead to mineral neoformation (vermiculite and clusters of Fe(III) oxides). Furthermore, the action of plant activity has strongly influenced biogeochemical cycles of major and trace elements [1, 181]. Conley [182] reported that biogenic silica (phytoliths) that precipitates in living tissues of growing plants is characterised by an annual production of 60–200 Tmole Si yr^{-1}, a value comparable in magnitude to the oceanic production of biogenic silica by diatoms, silicoflagellates, and radiolarians (240 Tmole Si yr^{-1}, [183]). Phytoliths represent a sizable pool of Si [184–186] that remains in the soil after decomposition of organic material together with other biogenic detritus.

Another example of coevolution of life and minerals is offered by mycorrhizae, since their appearance, these symbiotic associations support plant roots in mobilizing phosphorous and other nutrients from soil minerals. As this process favours plant growth, it resulted in increasing the size of plants, forest development, formation of thick soil layers and, ultimately, increasing the number of clay minerals [13, 103, 173]. Nutrients mobilizing functions are also provided by soil microbial communities where bacteria and fungi act symbiotically [174].

The rhizosphere is the narrow region of soil that is directly influenced by root secretions and associated soil microorganisms [187], and it is characterised by processes that are dramatically different from those that occur in the bulk soil [188, 189]. Here plant roots exert a critical role to regulate nutrient availability and to detoxify undesirable metal pollutants by realising a large number of metabolites that change the pH or form metal–metabolite complexes [190]. The secreted compounds consist of a complex mixture of inorganic ions (H^+, HCO_3^-), gaseous molecules (CO_2, H_2), low-molecular-weight compounds (organic acids, amino acids, phenolics and sugar) and high-molecular-weight compounds (mucilage, polysaccharides, and ectoenzymes) [189–191]. Organic chelating anions have a significant influence on the nucleation reactions, transformations, morphology, and surface properties of soil precipitates. Violante and Caporale [189] reported that in the presence of organic ligands the specific surface and reactive sites of Al and Fe precipitates increase, and they can lead to the precipitation of short range ordered precipitates (ferrihydrites, noncrystalline Fe and Al oxides, poorly crystalline boehmite (AlOOH)) instead of well crystallized Fe or Al oxides (gibbsite, goethite, hematite, lepidocrocite) that precipitate in the absence of organic ligands.

To survive in metal-extreme environments, plants have developed different tolerance strategies. In metal-hypertolerant plants, exudation of organic acids has an important role in metal (e.g. Cd, Al, Ga, Cu, Mn, Zn and Pb) detoxification mechanisms, because chelators form stable complexes with metals limiting their absorption and/or translocation by plants, maintaining low levels of contaminants in the aerial parts [190, 192, 193]. The uptake of potentially toxic metals and/or metalloids may be also reduced by formation of (bio)mineral precipitates at the soil-root interface. Iron plaques (e.g. ferrihydrite, goethite, lepidocrocite, siderite, etc.), whose mineral composition is depends on the local biogeochemical factors at specific sites, have been observed on the roots of many aquatic and wetland plants (e.g. rice roots) [112, 194, 195] (Fig. 6a). Some researchers suggest that Fe plaques can act as a barrier because a significant amount of metals (e.g. Cr, Pb, Cu, Zn) bind to them by complex formation [196, 197] (and references therein) (Fig. 6b). A similar mechanism has been observed in *E. pithyusa* [198] and *P. lentiscus* [199] by SEM, STXM and Zn K-edge X-rays adsorption fine structure (XAFS) spectroscopy. These plant species uptake Zn and Si from soil minerals and precipitate an amorphous Zn-silicate at the soil-root interface. This rim acts as a physico-chemical barrier against metal stress, and its formation was interpreted as intrinsically biologically driven (Fig. 6c, d).

If uptake of metals occurs, the plants can manage excess of toxic elements mainly by the following tolerance mechanisms: (i) sequestration/compartmentalization, (ii)

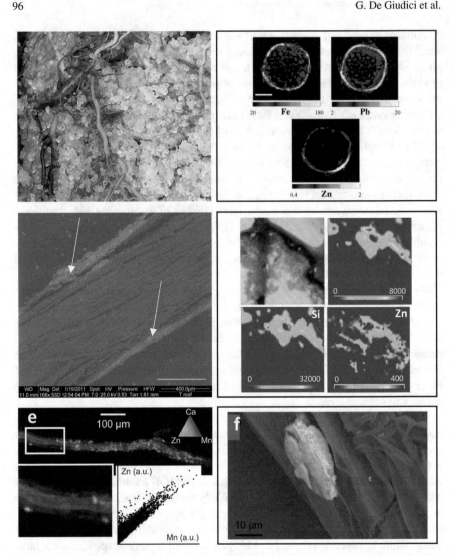

Fig. 6 Examples of plant biomineralizations. **a** Iron plaque formation at the rhizosphere-root interface of the wetland plant *Sparganium Americanum,* recovered from microcosm experiments (from
Chang et al. [195], Copyright (2014) American Chemical Society); **b** fluorescence microtomography showing Fe, Pb and Zn distributions on and within roots of *Phalaris arundinacea* (scale bar,
150 μm) (from Hansel et al. [197], Copyright (2001) American Chemical Society); **c** SEM image of
a *P. lentiscus* root (longitudinal section). The white arrows indicate the biomineral rim; **d** ordinary
light stereomicroscope image of a thin cross root section of *E. pithyusa*, and LEXRF (low-energy X
ray fluorescence) maps of Al, Si, and Zn (size of 80 μm × 80 μm, scan of 80 pixels × 80 pixels); **e**
tricolor (RGB) μ-XRF map of a root of *Festuca rubra* grown on a Zn-contaminated sediment with
Mn–Zn precipitates. Pixel size: 7×7 μm^2. The graph is a pixel-by-pixel scatterplot (Zn counts
vs. Mn counts) that shows the constant Zn:Mn ratio; **f** scanning electron microscope image with
backscattered electrons of Mn–Zn precipitates observed in the root epidermis of *Festuca rubra*. **e**
and **f** from Lanson et al. [207], Copyright (2008), with permission from Elsevier

binding/chelation, (iii) excretion from aerial plant parts, (iv) enzymatic and non-enzymatic antioxidants, (v) protection, stress recovery and repair of damaged proteins [192, 200] (and references therein), and (vi) biomineral precipitation [26, 201–203]. In the binding/chelation mechanism, metals are bound/chelated by several metal-binding molecules such as organic acids, amino acids, and phenolic compounds and/or by metal-binding peptides, such as metallothioneins and phytochelatins, resulting in low metal concentrations in the cytoplasm [192]. As stated above, metal stress can also lead to the excretion of harmful elements from aerial plant parts. For example, McNear et al. [204], by synchrotron-radiation based X-ray fluorescence and absorption-edge computed microtomographies, found that, in the Ni-hyperaccumulator *Alyssum murale*, Ni and Mn are colocalized at the trichome base throughout the entire leaf contributing significantly to metal detoxification and compartmentalization. Tobacco (*Nicotiana tabacum* L. cv Xanthi) plants [205], exposed to toxic levels of Zn, can precipitate Zn-containing biogenic calcite and other Zn-containing compounds on the head cells of trichomes and subsequently they are excreted to alleviate metal stress. Other plant species, such as birch and willow, shed their leaves in autumn together with the load of potentially toxic elements, thus tolerating the uptake of such elements [206].

Among the different functions of biominerals in the plant kingdom [201], detoxification of metals has been reported in different plant species. Several studies demonstrated the role of Ca oxalate crystals in the incorporation of metals [201, 208, 209]. For example, in *Eichhornia crassipes* plants, cultured in jars containing waters with different amounts of Pb and Cd, metal contents in Ca oxalate crystals increased progressively over time of exposure [209]. The presence of Cd in Ca oxalate crystals has been documented also in various tissues of stems of tomato plants [208].

Beside Ca oxalate, there are other biominerals that can contribute to metal detoxification in plants. Metal sequestration with phosphate has been reported in roots and needles of *P. sylvestris*, where pyromorphite ($Pb_5(PO_4)_3(Cl, OH)$) occurs as polycrystalline aggregates in bulges of the cell wall [26], and it was interpreted as a defence mechanism of the plant against Pb pollution. The graminaceous plant *Festuca rubra* (red fescue) precipitates Zn-rich phyllomanganate nanoparticles with constant Zn:Mn and Ca:Mn atomic ratios (0.46 and 0.38, respectively) [207] (Fig. 6e, f). Iron biominerals (mainly jarosite, ferrihydrite, hematite and spinel phases) were found at the cellular level in tissues of roots, stems and leaves of *Imperata cylindrica* from the mine-impacted Rio Tinto river (Iberian Pyritic Belt) [210, 211].

The beneficial role of silica in mitigating various abiotic stresses such as metal toxicity has been recognised by several authors for many instances [187, 201, 212]. Silicon can act through different mechanisms: (i) co-precipitation or complexation of toxic metals with Si, (ii) reducing active heavy metal ions in growth media, (iii) compartmentalization of metals within plants, (iv) reducing metal uptake and translocation, and (v) stimulation of antioxidant systems in plants. Silicon can decrease availability of phytotoxic metals affecting the soil properties [212] such as the pH value and metal speciation by formation of silicate complexes. Reduction of metal uptake and translocation in the presence of Si has been observed for Cd, Cu, Cr and Zn in many plant species such as rice, maize, cotton and wheat, and it can occur through several processes (e.g., stimulation of root exudates, reduction of the apoplasmic transport of metals, deposition of Si near the endoderm etc.). The effect of Si on metal uptake and translocation varies with plant species [187] (and references therein). Metal toxicity can be reduced also by the co-precipitation of Si with metals as observed by Neumann et al. [213] that reported the precipitation of a Zn-silicate in the epidermal cell walls of *Minuartia verna* ssp. *Hercynica*, and by Neumann and zur Nieden [214] that investigated Zn precipitation as silicate in the cytoplasm of *Cardaminopsis*.

4.5 Critical Zone

The Critical Zone (CZ) is defined as the thin outer skin of Earth's surface ranging from the top of the canopy layer down to the lower limits of groundwater [215]. The term CZ was first coined by the US National Research Council [216] as "*… the heterogeneous, near surface environment in which complex interactions involving rock, soil, water, air and living organisms regulate the natural habitat and determine availability of life sustaining resources*" [217, 218]. These complex biogeochemical-physical processes evolve in response to tectonic, climate, and anthropogenic forcing over vastly different timescales affecting the hydrosphere, lithosphere, pedosphere, atmosphere, and biosphere [215, 218–220]. In the CZ, the pore water can contain metals as free ions or complexed to inorganic or organic ligands, and both are subject mainly to the following processes: (i) diffusion in porous media and transport through the soil profile into groundwater, (ii) uptake by plants, (iii) sorption on mineral surfaces, natural organic matter, and microbes, (iv) precipitation as solid phases [221]. As a response to the variations in these processes, the CZ is characterised by a huge heterogeneity both vertically, recognizable in distinct layers of weathered rock, regolith, and soil (extending from organotrophic in the near surface to oligotrophic conditions at depth), and laterally, due to the diversity of landscapes and the distribution of soils across them [215, 222].

In the root and unsaturated zones, plants and their fungal symbionts physically open the regolith while seeking out water resources and mining for nutrients (e.g., P, K, Ca, Mg etc.) providing physical disturbance at the micro- and macroscales and inducing preferential flow paths. Also, fungal symbionts can increase fractures in regolith by fungal hyphae that are able to exploit microfractures inaccessible to plant roots alone [215]. The vadose zone is subject to fluctuations between wetting and drying conditions. During drying periods, evapotranspiration causes the increase of solute concentrations, leading to precipitation of soluble salts such as sulphates and carbonates. Subsequently, during wetting conditions, these salts are rapidly re-solubilized, and elements are re-mobilised into the pore water [223]. When pores are filled with aqueous solution, weathering reactions are enhanced by the production of organic acids, extracellular enzymes and complexing ligands (e.g. siderophores), pH modification due to CO_2 respiration, mineral mining by roots and fungal hyphae, and by microbial colonization of mineral surfaces (e.g., biotite, Fe oxides, pyrite)

[215]. Bioavailability of inorganic contaminants can change during weathering of primary phases and possible formation of authigenic minerals, depending on mineral transformation rates and the solubility of newly formed secondary minerals [178].

In the unsaturated zone, incongruent weathering of primary to secondary minerals promotes formation of clay-sized layer silicates and different oxides and/or hydroxides of Fe, Al, and Mn. These phases are effective sequestering agents for metals and metalloids (including radionuclides) due to their (i) high specific surface area (10–800 m^2/g), (ii) high surface charge and (iii) reactive surface functional groups [178, 224]. Although the saturated zone is often less altered than the root and unsaturated zones, due to the longer residence times of groundwaters that allow closer approach to equilibrium with soil minerals, (hydr)oxide precipitates can coat primary mineral surfaces, altering their surface reactivity and controlling metal mobility [178, 225]. Secondary clay minerals can interact with natural organic matter (complex mixture of biopolymers such as proteins, carbohydrates, aliphatic biopolymers, lignin), stabilizing it against microbial degradation. The association between minerals and organic compounds affects the behaviour of particle surfaces that can (i) retard the advective–diffusive transport and deposition of solutes and colloidal particles, or (ii) play a relevant role in the nucleation and growth of authigenic precipitates [178]. It is worth noting that the decomposition of natural organic matter by organisms leads to the formation of dissolved organic matter that form stable complexes with metals due to the presence of polar (e.g. carboxyl, hydroxyl) functional groups, potentially increasing pollutant mobility [178].

Davranche et al. [226] (and references therein) highlighted the role of electron transfer in the CZ, due to both biotic and abiotic mechanisms, that controls the fate of inorganic and organic contaminants, whether redox-sensitive or not. Usually, deep horizons and long-term waterlogged systems are characterised by lower amount of dissolved O_2 than upper layers, due to biomass consumption (for respiration), leading to the development of redox gradients. Manganese and Fe-bearing minerals can act as both electron donors and acceptors because Mn and Fe in these phases can have multiple redox states (e.g., magnetite, and green rusts or layered double Fe hydroxides), and they can precipitate or dissolve as a result of redox reactions providing or removing reactive sorption surfaces for chemicals, also they can activate co-precipitation processes. In addition, natural organic matter and microorganisms can promote abiotic electron transfer with mineral surfaces, catalyzing redox reactions (see [226] for a thorough review on natural organic matter and electron transfer).

Microbes, occurring mainly in biofilms spread within soils, actively contribute to the redox gradient of the CZ by forming microenvironments characterised by specific physico-chemical properties (e.g., pH, Eh, etc.). Metabolic processes allow to distinguish prokaryotes in phototrophs (conversion of the light energy into chemical energy when light is available), chemo-organotrophs (oxidation of organic compound), chemo-lithotrophs (oxidation of inorganic compounds, such as H_2, H_2S, NH_4^+, Fe^{2+}, Mn^{2+}, As^{3-}, etc.), anaerobes (anaerobic respiration using Mn^{4+}, Fe^{3+}, SO_4^{2-}, NO_3^-, etc.). Both chemo-lithotrophic and anaerobic microbes can influence the mobility and toxicity of inorganic pollutants [152, 226, 227]. For example, sulphur/sulphide-oxidizing bacteria cause chemo-lithotrophic leaching of particles, resulting in mobilization of pollutants from soil, while sulphate-reducing bacteria (SRB) can activate the precipitation of stable sulphides, removing chalcophile elements such as Zn, Cd, Cu, Co, Ni etc. [3, 152, 227].

Apart from being important because of their metabolism, microbes can influence metal and/or metalloid speciation, and thus their mobility, in a wide range of processes, including mineral bioweathering and biodeterioration, biosorption processes (e.g. cell wall and other structural biomolecules, metabolites, metal-binding peptides, EPS), intracellular accumulation by transport mechanisms, organellar localization, intracellular sequestration and bioprecipitation, extracellular biomineralization [228].

Several research approaches of varying complexity have been developed to quantify biogeochemical processes in the CZ. The hyporheic zone, the active ecotone between the surface stream and groundwater [229], is an important CZ whose evolution can be investigated by stream solute chemistry. Specifically, the tracer injection technique provides concentration-discharge data, a powerful tool to compare stream behaviour across catchments [230, 231]. In the hyporheic zone, solute release/uptake is ruled by dissolution, precipitation and sorption reaction rates, cation exchange capacity of clays, and biotic processes. In turn, concentration-discharge values in streams are mainly influenced by (i) bedrock lithology, (ii) geomorphic regime, (iii) organic matter and solute source mass distribution, (iv) water residence time, (v) subsurface flow paths, (vi) seasonality and storm events, (vii) cation exchange processes, and (viii) anthropogenic contributions [230, 232]. As a consequence of the balance among processes occurring in the hyporheic zone, this CZ can attenuate or contribute anthropogenic pollutants. Sewage discharges to surface waters can significantly increase pore water nutrient contents, altering hyporheic biogeochemical processes. Chemicals in agricultural runoff or contaminants from polluted areas can move from surface water into groundwater with little change in concentration or can be degraded and/or stabilised within the hyporheic zone. Also, metals and other pollutants can move from groundwater into surface water through the hyporheic interface [229].

The CZ can be considered as a biogeochemical reactor, open to fluxes of matter (gas, liquid and solid form) and energy (heat and reduced carbon compounds fixed through photosynthesis) and plays an essential role in natural and managed ecosystems. Indeed biogeochemical processes occurring at the atomic scale influence macroscopic to global scale processes [219, 222, 223]. The presence of several exposed solid surfaces (e.g., minerals, plants, microorganisms) affects transport, retention (i.e. sorption), and chemical transformation of solutes. At the nano and microscale, biogeochemical reactions drive the evolution of particle surfaces and their reactivity, whereas at the watershed scale, they control stream and groundwater quality [178].

5   Biomineral Processes and Sustainability

In this section we focus on the relevance of biomineral processes to the environment and environmental management. For this purpose, we present some examples of application of biomineral science to environmental management.

5.1   Biometallurgy and Circular Economy

Through geological time, microorganisms developed their capability to work as expert biometallurgists in many different microenvironments. To date, biometallurgy techniques are applied globally to recover Cu and many other metals [233, 234] (and references therein). As shown in Table 1, microorganisms are pioneers also in extreme environments and have the potential to extract or precipitate metals relevant to both industrial and environmental processes [235, 236]. For this reason, bacteria and fungi are used in the mining industry to increase recovery rates and offer unique tools for sustainable mining and reuse of large volumes of mine waste produced in the past activity.

Pyrite is the most common sulphur-bearing mineral in the Earth's crust, and ^{32}S enrichment in pyrite mineralisation driven by sulphate reducing microbes clearly provides evidence in the geological record [237]. Ohmoto and Lasaga [238] first pointed out that the rates of sulphate reduction by non-bacterial processes involving a variety of reductants are also dependent on T, pH, activities of S species in solution and appear to be fast enough to become geochemically important only at temperatures above about 200 °C. The process of sulphate reduction implies the transfer of eight electrons, is endothermic and is not likely to occur in nature at ambient temperature without microbial activity [239]. Sulphate reducing bacteria (SRB) play a key role in surface environments. SRB are part of the organism's microbiome and are ubiquitous in the environment. Framboidal morphology of minerals such as pyrite, greigite, magnetite, magnesioferrite, marcasite, as well of supposedly secondary minerals such as hematite, limonite, chalcocite, chalcopyrite, bornite, sphalerite, galena are known from the Archean [240, 241] and are often indicative of BIM. Though framboidal iron sulphides are ubiquitous in near-surface environments, SRB play a pivotal role in biomineral formation as they increase sulphide concentrations in water, and then induce formation of framboidal metal sulphides. Recently, zinc sulphide forming in SRB natural biofilms have been observed by mineralogists and environmental scientists [242].

Most often, biometallurgy involves heap leaching with use of non-autochthonous microbial strains that are commercially available. The challenge is now to use biometallurgy techniques to apply circular economy paradigm for many critical metals such as Sb, REE, Co, Ni and so on [243]. Staicu et al. [244] isolated *Bacillus sp. Abq*, belonging to *Bacillus cereus* sensu lato, finding its unusual property of precipitating Pb(II) by using cysteine, which is degraded intracellularly to produce hydrogen sulfide (H_2S). H_2S is then exported to the extracellular environment to react with Pb(II), yielding PbS (galena). Paganin et al. [245] investigated microbial diversity in core samples from different areas of the same abandoned mine district and obtained different inocula (Fig. 7a, b). Moreover, they were able to reprecipitate ZnS by using the selected inocula (Fig. 7c, d) from the mine polluted water with removal rates up to 100%. Summarizing, we are now accumulating a large body of knowledge that can be transferred soon to industrial scale applications that pave the way to sustainable technologies for recovering metals from wastes.

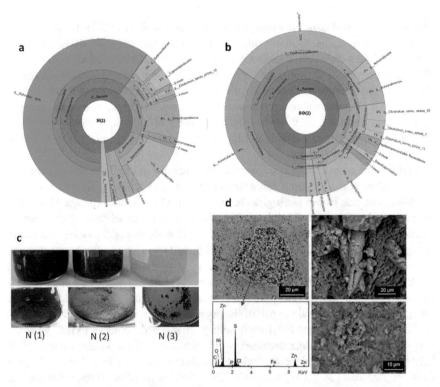

Fig. 7 a and **b** Krona plot at the genus level for two different areas indicating their microbial biodiversity; **a** inoculum N_2 cultured from the Naracauli mine area microbial community; **b** inoculum SG(2) cultured from the San Giorgio valley mine area microbial community; **c** growth of sulphate-reducing bacteria in the three different media inoculated with sediments of Rio Naracauli and details of Fe sulfide (black) and Zn sulfide (brownish) precipitates; **d** SEM–EDS analysis: BSE images of the bioprecipitates recovered from experiments performed with inoculum cultured from Rio Naracauli core sediments and mine polluted waters. Images from Paganin et al. [245], Copyright (2018) Paganin et al., under the Creative Commons Attribution License

5.2 Secondary Ores and Environmental Resilience

There is a growing body of literature that points out the interplay among biological activity and the response of environment to stress, also on a time scale of a few years. Figures 2g, 7a, b provide discrete evidence of the microbial biodiversity over a short spatial distance. Moreover, our research group found that cyanobacterial biofilm can lead to huge seasonal and spatial variability in biomineralization processes along the same riverbed without any apparent change in the geochemical properties of the investigated systems [30, 92, 93, 246].

Dore et al. [3] analysed biomineralization processes and the environmental status in four riverbeds affected by historical mine pollution. Growth of dense vegetation in a riverbed has a primary effect of favouring the sedimentation of fine sediments, and a secondary effect of favouring biogeochemical processes in the hyporheic zone, which is part of CZ. In fact, when the erosional regime is changed to a sedimentation regime due to the slow velocity of water retained from the stems like in a wetland-like system, many different biomineralization processes can become effective in trapping metals in the sediments that constitute the hyporheic zone of the riverbed. De Giudici et al. [247] and Dore et al. [3] show that the natural process of vegetation growth in historical-mine-activity-degraded riverbeds serves as an example for the development of effective industrial-scale biogeochemical processes previously described in the literature by, for instance, Labrenz et al. [242]. In the case of Rio San Giorgio in southwestern Sardinia, trapping of fine sediments within the stems of *Phragmites australis* and the riverbed allows the dominance of SRB communities, leading to the formation of abundant biogenically induced base metal sulphides. Plant roots also favour the formation of biominerals such as hemimorphite and hydrozincite, thus contributing to natural abatement of Zn and other metals. Notably, these sediments can be considered themselves as future secondary ores to be exploited with a new generation of biometallurgy techniques.

6 Remarks and Conclusion

In this chapter we summarized the impact of biomineralization on both Earth and environmental processes. Through the whole geological record, the co-evolution of life and minerals led to a significant diversity of biominerals, of which more than 160 are known today. About 120 biominerals are organic compounds, phosphates, carbonates, hydroxides, and oxides. This is partially related to the chemical composition of DNA and RNA and organic molecules in general, that are made of C, H, O, N and P. Diversity of biomineralization processes increased with evolution of life, and biominerals are made to assure many different physiological purposes such as detoxification. Actually, biominerals cover almost all the mineral classes comprising silicates, and biomineral processes involve toxic elements such as Pb and As.

Biomineralization processes co-exerted a profound impact on Earth by altering the composition of water, atmosphere, and soil. We stressed that biomineralization processes can have a profound impact on environmental systems even on short time scales (few years). Understanding the environment where biomineralization occur, such as biofilm and other water soil microenvironments, is then central to the understanding the biomineralization impact on our environment. In turn this shed light on the impact of biomineralization processes on sustainability and related technologies. As an example, we mentioned the role of authigenic minerals in the critical zone controlling mobility of metals.

As final remarks for the (bio)mineralogist, it is worth noting that biomineral studies are intrinsically interdisciplinary. The capacity building that the scientific community can develop in the area of mineral-biosphere interactions is proportional to our capability of selecting (micro)organisms able to build a specific microenvironment and to our skills in reproducing their biomineral processes. It should also be noted that the capability to reproduce biomineralization driven by microorganisms in industrial processes non in vivo industrial processes is still a visionary approach, that can reveal new and highly sustainable processes in future. Finally, characterization of crystal and mineral structure down to the nanoscale is central to understanding biominerals. Biominerals are often poorly crystalline, small, and dispersed in complex matrices. Their recognition and understanding of their role often need the use of advanced and multiscale/multiphysics approaches combining complementary techniques.

Acknowledgements We thank P. Lattanzi and R. B Wanty for their suggestions and comments on a preliminary version of this manuscript.

References

1. Van Cappellen P (2003) Biomineralization and global biogeochemical cycles. Rev Mineral Geochem 54:357–381. https://doi.org/10.2113/0540357
2. Trembath-Reichert E, Wilson JP, McGlynn SE, Fischer WW (2015) Four hundred million years of silica biomineralization in land plants. Proc Natl Acad Sci U S A 112:5449–5454. https://doi.org/10.1073/pnas.1500289112
3. Dore E, Fancello D, Rigonat N et al (2020) Natural attenuation can lead to environmental resilience in mine environment. Appl Geochem. https://doi.org/10.1016/j.apgeochem.2020.104597
4. Birarda G, Buosi C, Caridi F et al (2021) Plastics, (bio)polymers and their apparent biogeochemical cycle: an infrared spectroscopy study on foraminifera. Environ Pollut 279:116912. https://doi.org/10.1016/j.envpol.2021.116912
5. Fitzer SC, McGill RAR, Torres Gabarda S et al (2019) Selectively bred oysters can alter their biomineralization pathways, promoting resilience to environmental acidification. Glob Chang Biol 25:4105–4115. https://doi.org/10.1111/gcb.14818
6. Caraballo MA, Asta MP, Perez JPH, Hochella MF (2021) Past, present and future global influence and technological applications of iron-bearing metastable nanominerals. Gondwana Res. https://doi.org/10.1016/j.gr.2021.11.009
7. Lowenstam HA (1981) Minerals formed by organisms. Science 211(80):1126–1131

8. Skinner HCW (2005) Biominerals. Mineral Mag 69:621–641. https://doi.org/10.1180/002 6461056950275

9. Strunz H (1941) Mineralogische tabellen : im Auftrage der Deutschen Mineralogischen Gesellschaft/herausgegeben von Hugo Strunz. Becker & Erler

10. Lowenstam H (1974) Impact of life on chemical and physical processes. In: Goldberg ED (ed) The sea, marine chemistry. John Wiley and Sons, New York, pp 715–796

11. Lowenstam HA, Weiner S (1983) Mineralization by organisms and the evolution of biomineralization. In: Westbroek P, de Jong EW (eds) Biomineralization and biological metal accumulation. Springer, Netherlands, Dordrecht, pp 191–203

12. Weiner S, Dove PM (2003) An overview of biomineralization processes and the problem of the vital effect. Rev Mineral Geochem 54:1–29. https://doi.org/10.2113/0540001

13. Hazen RM, Papineau D, Bleeker W et al (2008) Mineral evolution. Am Mineral 93:1693–1720. https://doi.org/10.2138/am.2008.2955

14. Podda F, Zuddas P, Minacci A et al (2000) Heavy metal coprecipitation with hydrozincite [$Zn_5(CO_3)_2(OH)_6$] from mine waters caused by photosynthetic microorganisms. Appl Environ Microbiol 66:5092–5098

15. Wilkin RT, Barnes HL (1996) Pyrite formation by reactions of iron monosulfides with dissolved inorganic and organic sulfur species. Geochim Cosmochim Acta 60:4167–4179. https://doi.org/10.1016/S0016-7037(97)81466-4

16. Tazaki K, Rafiqul IA, Nagai K, Kurihara T (2003) $FeAs_2$ biomineralization on encrusted bacteria in hot springs: an ecological role of symbiotic bacteria. Can J Earth Sci 40:1725–1738. https://doi.org/10.1139/e03-081

17. Ferris FG, Fyfe WS, Beveridge TJ (1987) Bacteria as nucleation sites for authigenic minerals in a metal-contaminated lake sediment. Chem Geol 63:225–232. https://doi.org/10.1016/0009-2541(87)90165-3

18. International Mineralogical Association (2021) IMA database of mineral properties. https://rruff.info/ima/. Accessed 18 Nov 2021

19. Lowenstam HA, Weiner S (1989) On biomineralization. Oxford University Press, New York

20. Morin G, Juillot F, Casiot C et al (2003) Bacterial formation of tooeleite and mixed Arsenic(III) or Arsenic(V)—Iron(III) gels in the carnoulès acid mine drainage, France. A XANES, XRD, and SEM study. Environ Sci Technol 37:1705–1712. https://doi.org/10.1021/es025688p

21. Power IM, Wilson SA, Thom JM et al (2007) Biologically induced mineralization of dypingite by cyanobacteria from an alkaline wetland near Atlin, British Columbia, Canada. Geochem Trans 8:1–16. https://doi.org/10.1186/1467-4866-8-13

22. Schultze-Lam S, Beveridge TJ (1994) Nucleation of celestite and strontianite on a cyanobacterial S-layer. Appl Environ Microbiol 60:447–453. https://doi.org/10.1128/aem.60.2.447-453.1994

23. Dogan AU, Dogan M, Chan DCN, Wurster DE (2005) Bassanite from Salvadora persica: a new evaporitic biomineral. Carbonates Evaporites 20:2–7. https://doi.org/10.1007/BF03175444

24. Frankel RB, Bazylinski DA (2003) Biologically induced mineralization by bacteria. Rev Mineral Geochem 54:95–114. https://doi.org/10.2113/0540095

25. Zachara JM, Kukkadapu RK, James K et al (2002) Biomineralization of poorly crystalline Fe(III) oxides by dissimilatory metal reducing bacteria (DMRB) biomineralization of poorly crystalline Fe(III) oxides by dissimilatory metal reducing bacteria (DMRB). Geomicrobiol J 37–41

26. Bizo ML, Nietzsche S, Mansfeld U et al (2017) Response to lead pollution: mycorrhizal *Pinus sylvestris* forms the biomineral pyromorphite in roots and needles. Environ Sci Pollut Res 24:14455–14462. https://doi.org/10.1007/s11356-017-9020-7

27. Konhauser KO, Fyfe WS, Schultze-Lam S et al (1994) Iron phosphate precipitation by epilithic microbial biofilms in Arctic Canada. Can J Earth Sci 31:1320–1324. https://doi.org/10.1139/e94-114

28. Akai J, Akai K, Ito M et al (1999) Biologically induced iron ore at Gunma iron mine, Japan. Am Mineral 84:171–182. https://doi.org/10.2138/am-1999-1-219

29. Gadd GM, Rhee YJ, Stephenson K, Wei Z (2012) Geomycology: metals, actinides and biominerals. Environ Microbiol Rep 4:270–296. https://doi.org/10.1111/j.1758-2229.2011. 00283.x

30. Podda F, Medas D, De Giudici G et al (2014) Zn biomineralization processes and microbial biofilm in a metal-rich stream (Naracauli, Sardinia). Environ Sci Pollut Res. https://doi.org/10.1007/s11356-013-1987-0

31. Medas D, Lattanzi P, Podda F et al (2014) The amorphous Zn biomineralization at Naracauli stream, Sardinia: electron microscopy and X-ray absorption spectroscopy. Environ Sci Pollut Res 21:6775–6782. https://doi.org/10.1007/s11356-013-1886-4

32. Gorbushina AA, Boettcher M, Brumsack HJ et al (2001) Biogenic forsterite and opal as a product of biodeterioration and lichen stromatolite formation in table mountain systems (Tepuis) of Venezuela. Geomicrobiol J 18:117–132. https://doi.org/10.1080/014904501510 79851

33. Konhauser KO (1997) Bacterial iron biomineralisation in nature. FEMS Microbiol Rev 20:315–326. https://doi.org/10.1016/S0168-6445(97)00014-4

34. Zeyen N, Benzerara K, Li J et al (2015) Formation of low-T hydrated silicates in modern microbialites from Mexico and implications for microbial fossilization. Front Earth Sci 3:1–23. https://doi.org/10.3389/feart.2015.00064

35. Konhauser KO, Fyfe WS, Ferris FG, Beveridge TJ (1993) Metal sorption and mineral precipitation by bacteria in two Amazonian river systems: Rio Solimoes and Rio Negro, Brazil. Geology 21:1103–1106. https://doi.org/10.1130/0091-7613(1993)021%3c1103:MSA MPB%3e2.3.CO;2

36. Ta K, Peng X, Chen S et al (2017) Hydrothermal nontronite formation associated with microbes from low-temperature diffuse hydrothermal vents at the South Mid-Atlantic Ridge. J Geophys Res Biogeosciences 122:2375–2392. https://doi.org/10.1002/2017JG003852

37. Burne RV, Moore LS, Christy AG et al (2014) Stevensite in the modern thrombolites of Lake Clifton, Western Australia: a missing link in microbialite mineralization? Geology 42:575–578. https://doi.org/10.1130/G35484.1

38. Burford EP, Kierans M, Gadd GM (2003) Geomycology: fungi in mineral substrata. Mycologist 17:98–107. https://doi.org/10.1017/S0269915X03003112

39. Bazylinski D, Frankel R (2003) Biologically controlled mineralization in prokaryotes. Rev Mineral Geochem 54:217–247. https://doi.org/10.2113/0540217

40. De Giudici G, Medas D, Cidu R et al (2017) Application of hydrologic-tracer techniques to the Casargiu adit and Rio Irvi (SW-Sardinia, Italy): Using enhanced natural attenuation to reduce extreme metal loads. Appl Geochem 96:42–54. https://doi.org/10.1016/j.apgeochem. 2018.06.004

41. Frau F, Medas D, Da Pelo S et al (2015) Environmental effects on the aquatic system and metal discharge to the mediterranean sea from a near-neutral zinc-ferrous sulfate mine drainage. Water Air Soil Pollut 226:226–255. https://doi.org/10.1007/s11270-015-2339-0

42. Chasteen ND, Harrison PM (1999) Mineralization in ferritin: an efficient means of iron storage. J Struct Biol 126:182–194. https://doi.org/10.1006/jsbi.1999.4118

43. Woo KM, Jun J-H, Chen VJ et al (2007) Nano-fibrous scaffolding promotes osteoblast differentiation and biomineralization. Biomaterials 28:335–343. https://doi.org/10.1016/j.biomat erials.2006.06.013

44. Boraldi F, Burns JS, Bartolomeo A et al (2018) Mineralization by mesenchymal stromal cells is variously modulated depending on commercial platelet lysate preparations. Cytotherapy 20:335–342. https://doi.org/10.1016/j.jcyt.2017.11.011

45. Rui Y, Qian C (2021) The regulation mechanism of bacteria on the properties of biominerals. J Cryst Growth 570:126214. https://doi.org/10.1016/j.jcrysgro.2021.126214

46. Bernal JD (1951) The physical basis of life. Routledge and Paul, London

47. Hazen RM (2006) Mineral surfaces and the prebiotic selection and organization of biomolecules. Am Mineral 91:1715–1729. https://doi.org/10.2138/am.2006.2289

48. Antonietti M, Förster S (2003) Vesicles and liposomes: a self-assembly principle beyond lipids. Adv Mater 15:1323–1333. https://doi.org/10.1002/adma.200300010

49. Hazen R (2004) Chiral crystal faces of common rock-forming minerals. In: Palyi G, Zucchi C, Caglioti L (eds) Progress in biological chirality. Elsevier, Oxford UK, pp 137–151
50. Downs RT, Hazen RM (2004) Chiral indices of crystalline surfaces as a measure of enantios-elective potential. J Mol Catal A-Chem 216:273–285. https://doi.org/10.1016/J.MOLCATA. 2004.03.026
51. Orme CA, Noy A, Wierzbicki A et al (2001) Formation of chiral morphologies through selective binding of amino acids to calcite surface steps. Nature 411:775–779. https://doi.org/ 10.1038/35081034
52. Hazen RM, Sholl DS (2003) Chiral selection on inorganic crystalline surfaces. Nat Mater 2:367–374. https://doi.org/10.1038/nmat879
53. De Yoreo JJ, Dove PM (2004) Shaping crystals with biomolecules. Science 306(80):1301–1302. https://doi.org/10.1126/science.1100889
54. Bally AW, Palmer AR (1989) The geology of North America—an overview
55. Sumner DY (1997) Carbonate precipitation and oxygen stratification in late Archean seawater as deduced from facies and stratigraphy of the Gamohaan and Frisco formations, Transvaal Supergroup, South Africa. Am J Sci 297:455 LP–487. https://doi.org/10.2475/ajs.297.5.455
56. Carter PW, Mitterer RM (1978) Amino acid composition of organic matter associated with carbonate and non-carbonate sediments. Geochim Cosmochim Acta 42:1231–1238. https:// doi.org/10.1016/0016-7037(78)90116-3
57. Weiner S, Addadi L (1997) Design strategies in mineralized biological materials. J Mater Chem 7:689–702. https://doi.org/10.1039/A604512J
58. Aizenberg J, Tkachenko A, Weiner S et al (2001) Calcitic microlenses as part of the photoreceptor system in brittlestars. Nature 412:819–822. https://doi.org/10.1038/35090573
59. Teng HH, Dove PM (1997) Surface site-specific interactions of aspartate with calcite during dissolution; implications for biomineralization. Am Mineral 82:878–887. https://doi.org/10. 2138/am-1997-9-1005
60. Teng HH, Dove PM, Orme CA, De Yoreo JJ (1998) Thermodynamics of calcite growth: baseline for understanding biomineral formation. Science 282(80):724–727. https://doi.org/ 10.1126/science.282.5389.724
61. Teng HH, Dove P, Yoreo J (2000) Kinetics of calcite growth: surface processes and relation-ships to macroscopic rate laws. Geochim Cosmochim Acta 64:2255–2266. https://doi.org/10. 1016/S0016-7037(00)00341-0
62. Dana ED (1958) A textbook of mineralogy, 4th Editio. John Wiley and Sons, New York
63. Hazen RM, Filley TR, Goodfriend GA (2001) Selective adsorption of l- and d-amino acids on calcite: Implications for biochemical homochirality. Proc Natl Acad Sci 98:5487 LP–5490. https://doi.org/10.1073/pnas.101085998
64. Lahav N, White D, Chang S (1978) Peptide formation in the prebiotic era: thermal condensa-tion of glycine in fluctuating clay environments. Science 201(80):67–69. https://doi.org/10. 1126/science.663639
65. Ferris JP (1993) Catalysis and prebiotic RNA synthesis. Orig life Evol Biosph 23:307–315. https://doi.org/10.1007/BF01582081
66. Ferris JP (1999) Prebiotic synthesis on minerals: bridging the prebiotic and RNA worlds. Biol Bull 196:311–314. https://doi.org/10.2307/1542957
67. Ferris JP (2005) Mineral catalysis and prebiotic synthesis: montmorillonite-catalyzed formation of RNA. Elements 1:145–149. https://doi.org/10.2113/gselements.1.3.145
68. Holm NG, Ertem G, Ferris JP (1993) The binding and reactions of nucleotides and polynu-cleotides on iron oxide hydroxide polymorphs. Orig life Evol Biosph J Int Soc Study Orig Life 23:195–215. https://doi.org/10.1007/BF01581839
69. Ferris JP, Ertem G (1992) Oligomerization of ribonucleotides on montmorillonite: reaction of the 5′-phosphorimidazolide of adenosine. Science 257(80):1387–1389. https://doi.org/10. 1126/science.1529338
70. Ferris JP, Ertem G (1993) Montmorillonite catalysis of RNA oligomer formation in aqueous solution. A model for the prebiotic formation of RNA. J Am Chem Soc 115:12270–12275. https://doi.org/10.1021/ja00079a006

71. Ertem G, Ferris JP (1996) Synthesis of RNA oligomers on heterogeneous templates. Nature 379:238–240. https://doi.org/10.1038/379238a0

72. Ertem G, Ferris JP (1997) Template-directed synthesis using the heterogeneous templates produced by montmorillonite catalysis. A possible bridge between the prebiotic and RNA worlds. J Am Chem Soc 119:7197–7201. https://doi.org/10.1021/ja970422h

73. Zheng J, Li Z, Wu A, Zhou H (2003) AFM studies of DNA structures on mica in the presence of alkaline earth metal ions. Biophys Chem 104:37–43. https://doi.org/10.1016/s0301-462 2(02)00335-6

74. Cervantes NAG, Gutiérrez-Medina B (2014) Robust deposition of lambda DNA on mica for imaging by AFM in air. Scanning 36:561–569. https://doi.org/10.1002/sca.21155

75. Thomson NH, Kasas S, Smith et al (1996) Reversible binding of DNA to mica for AFM imaging. Langmuir 12:5905–5908. https://doi.org/10.1021/la960497j

76. Valdrè G, Moro D, Ulian G (2011) Nucleotides, RNA and DNA selective adsorption on atomic-flat Mg–Al-hydroxysilicate substrates. Micro Nano Lett 6:922–926(4). https://doi.org/10.1049/mnl.2011.0546

77. Moro D, Ulian G, Valdrè G (2020) Nano-atomic scale hydrophobic/philic confinement of peptides on mineral surfaces by cross-correlated SPM and quantum mechanical DFT analysis. J Microsc 280:204–221. https://doi.org/10.1111/jmi.12923

78. Meldrum FC, Cölfen H (2008) Controlling mineral morphologies and structures in biological and synthetic systems. Chem Rev 108:4332–4432. https://doi.org/10.1021/cr8002856

79. Bonn M, Bakker HJ, Tong Y, Backus EHG (2012) No ice-like water at aqueous biological interfaces. Biointerphases 7:20. https://doi.org/10.1007/s13758-012-0020-3

80. Singer SJ, Nicolson GL (1972) The fluid mosaic model of the structure of cell membranes. Science 175:720–731. https://doi.org/10.1126/science.175.4023.720

81. Steed JW, Atwood JL (2009) Supramolecular chemistry, 2nd edn. Wiley, Chichester

82. Lehn J-M (1988) Supramolecular chemistry—scope and perspectives molecules, super-molecules, and molecular devices (nobel lecture). Angew Chemie Int Ed English 27:89–112. https://doi.org/10.1002/anie.198800891

83. Lehn JM (1988) Supramolekulare Chemie – Moleküle, Übermoleküle und molekulare Funktionseinheiten (Nobel-Vortrag). Angew Chemie 100:91–116. https://doi.org/10.1002/ANGE.19881000110

84. Breslow R (2005) Artificial enzymes. Wiley-VCH, Weinheim

85. Pedersen C (1988) The discovery of crown ethers (noble lecture). Angew Chemie Int Ed English 27:1021–1027. https://doi.org/10.1002/anie.198810211

86. Pedersen CJ (1988) Die Entdeckung der Kronenether (Nobel-Vortrag). Angew Chemie 100:1053–1059. https://doi.org/10.1002/ange.19881000805

87. Breuer M, Rosso KM, Blumberger J (2014) Electron flow in multiheme bacterial cytochromes is a balancing act between heme electronic interaction and redox potentials. Proc Natl Acad Sci U S A 111:611–616. https://doi.org/10.1073/pnas.1316156111

88. Sand KK, Rodriguez-Blanco JD, Makovicky E et al (2012) Crystallization of $CaCO_3$ in water-alcohol mixtures: spherulitic growth, polymorph stabilization, and morphology change. Cryst Growth Des 12:842–853. https://doi.org/10.1021/cg2012342

89. Penn RL, Banfield JF (1998) Oriented attachment and growth, twinning, polytypism, and formation of metastable phases; insights from nanocrystalline TiO_2. Am Mineral 83:1077–1082. https://doi.org/10.2138/am-1998-9-1016

90. Banfield JF, Welch SA, Zhang H, et al (2000) Aggregation-based crystal growth and microstructure development in natural iron oxyhydroxide biomineralization products. Science 289(80):751–754. https://doi.org/10.1126/science.289.5480.751

91. De Yoreo JJ, Gilbert PUPA, Sommerdijk NAJM, et al (2015) Crystal growth. Crystallization by particle attachment in synthetic, biogenic, and geologic environments. Science 349(80):aaa6760. https://doi.org/10.1126/science.aaa6760

92. Medas D, De Giudici G, Podda F et al (2014) Apparent energy of hydrated biomineral surface and apparent solubility constant: an investigation of hydrozincite. Geochim Cosmochim Acta. https://doi.org/10.1016/j.gca.2014.05.019

93. De Giudici G, Podda F, Sanna R et al (2009) Structural properties of biologically controlled hydrozincite: an HRTEM and NMR spectroscopic study. Am Mineral 94:1698–1706. https://doi.org/10.2138/am.2009.3181

94. Beniash E, Aizenberg J, Addadi L, Weiner S (1997) Amorphous calcium carbonate transforms into calcite during sea urchin larval spicule growth. Proc R Soc London Ser B Biol Sci 264:461–465. https://doi.org/10.1098/rspb.1997.0066

95. Politi Y, Arad T, Klein E, et al (2004) Sea urchin spine calcite forms via a transient amorphous calcium carbonate phase. Science 306(80):1161–1164. https://doi.org/10.1126/science.1102289

96. Albéric M, Stifler CA, Zou Z et al (2019) Growth and regrowth of adult sea urchin spines involve hydrated and anhydrous amorphous calcium carbonate precursors. J Struct Biol X 1:100004. https://doi.org/10.1016/j.yjsbx.2019.100004

97. Killian CE, Metzler RA, Gong Y, et al (2011) Self-sharpening mechanism of the sea urchin tooth. Adv Funct Mater 21:682–690. https://doi.org/10.1002/adfm.201001546

98. Weiss IM, Tuross N, Addadi L, Weiner S (2002) Mollusc larval shell formation: amorphous calcium carbonate is a precursor phase for aragonite. J Exp Zool 293:478–491. https://doi.org/10.1002/jez.90004

99. DeVol RT, Sun C-Y, Marcus MA et al (2015) Nanoscale transforming mineral phases in fresh nacre. J Am Chem Soc 137:13325–13333. https://doi.org/10.1021/jacs.5b07931

100. Mahamid J, Sharir A, Addadi L, Weiner S (2008) Amorphous calcium phosphate is a major component of the forming fin bones of zebrafish: indications for an amorphous precursor phase. Proc Natl Acad Sci 105:12748 LP–12753. https://doi.org/10.1073/pnas.0803354105

101. Beniash E, Metzler RA, Lam RSK, Gilbert PUPA (2009) Transient amorphous calcium phosphate in forming enamel. J Struct Biol 166:133–143. https://doi.org/10.1016/j.jsb.2009.02.001

102. Mass T, Giuffre AJ, Sun C-Y, et al (2017) Amorphous calcium carbonate particles form coral skeletons. Proc Natl Acad Sci 114:E7670 LP-E7678. https://doi.org/10.1073/pnas.1707890114

103. Hazen RM, Ferry JM (2010) Mineral evolution: mineralogy in the fourth dimension. Elements 6:9–12. https://doi.org/10.2113/gselements.6.1.9

104. Zhu T, Dittrich M (2016) Carbonate precipitation through microbial activities in natural environment, and their potential in biotechnology: a review. Front Bioeng Biotechnol 4:4. https://doi.org/10.3389/fbioe.2016.00004

105. Schopf JW, Kudryavtsev AB, Czaja AD, Tripathi AB (2007) Evidence of Archean life: stromatolites and microfossils. Precambrian Res 158:141–155. https://doi.org/10.1016/j.precamres.2007.04.009

106. Gong YUT, Killian CE, Olson IC et al (2012) Phase transitions in biogenic amorphous calcium carbonate. Proc Natl Acad Sci 109:6088 LP–6093. https://doi.org/10.1073/pnas.1118085109

107. Boettiger A, Ermentrout B, Oster G (2009) The neural origins of shell structure and pattern in aquatic mollusks. Proc Natl Acad Sci 106:6837 LP–6842. https://doi.org/10.1073/pnas.0810311106

108. Lim KK, Rossbach S, Geraldi NR, et al (2020) The small giant clam, tridacna maxima exhibits minimal population genetic structure in the red sea and genetic differentiation from the Gulf of Aden. Front Mar Sci. https://doi.org/10.3389/fmars.2020.570361

109. Lesser MP (2004) Experimental biology of coral reef ecosystems. J Exp Mar Bio Ecol 300:217–252. https://doi.org/10.1016/j.jembe.2003.12.027

110. Triantaphyllou MV, Baumann K-H, Karatsolis B-T et al (2018) Coccolithophore community response along a natural CO_2 gradient off Methana (SW Saronikos Gulf, Greece, NE Mediterranean). PLoS ONE 13:e0200012. https://doi.org/10.1371/journal.pone.0200012

111. LeKieffre C, Bernhard JM, Mabilleau G, et al (2018) An overview of cellular ultrastructure in benthic foraminifera: new observations of rotalid species in the context of existing literature. Mar Micropaleontol 138:12–32. https://doi.org/10.1016/j.marmicro.2017.10.005

112. Du X, Fan G, Jiao Y et al (2017) The pearl oyster *Pinctada fucata martensii* genome and multi-omic analyses provide insights into biomineralization. Gigascience 6:gix059. https://doi.org/10.1093/gigascience/gix059

113. Rivadeneyra A, Gonzalez-Martinez A, Portela GR et al (2017) Biomineralisation of carbonate and sulphate by the halophilic bacterium Halomonas maura at different manganese concentrations. Extremophiles 21:1049–1056. https://doi.org/10.1007/s00792-017-0965-8

114. Medas D, De Giudici G, Podda F et al (2014) Apparent energy of hydrated biomineral surface and apparent solubility constant: an investigation of hydrozincite. Geochim Cosmochim Acta 140:349–364. https://doi.org/10.1016/j.gca.2014.05.019

115. Roh Y, Zhang C, Vali H et al (2003) Biogeochemical and environmental factors in Fe biomineralization: magnetite and siderite formation. Clays Clay Miner 51:83–95. https://doi.org/10.1346/CCMN.2003.510110

116. Combes C, Cazalbou S, Rey C (2016) Apatite biominerals. Miner 6

117. Beniash E (2011) Biominerals—hierarchical nanocomposites: the example of bone. Wiley Interdiscip Rev Nanomed Nanobiotechnol 3:47–69. https://doi.org/10.1002/wnan.105

118. Vallet-Regí M, Arcos Navarrete D (2016) Biological apatites in bone and teeth. In: nanoceramics in clinical use: from materials to applications (2). The Royal Society of Chemistry, pp 1–29

119. Frankel RB (1991) Iron biominerals: an overview. In: Blakemore R, Frankel R (eds) Iron biominerals. Springer, Boston, pp 1–6

120. Fleming EJ, Cetinić I, Chan CS et al (2014) Ecological succession among iron-oxidizing bacteria. ISME J 8:804–815. https://doi.org/10.1038/ismej.2013.197

121. Richard B (1975) Magnetotactic bacteria. Science 190(80):377–379. https://doi.org/10.1126/science.170679

122. Kaas P, Jones AM (1998) Class polyplacophora: morphology and physiology. In: Beesley PL, Ross GJB, Wells A (eds) Mollusca: the southern synthesis part A, Fauna of Australia. CSIRO, Melbourne, pp 163–174

123. Brooker L, Shaw J (2012) The Chiton Radula: a unique model for biomineralization studies. In: Seto J (ed) Advanced topics in biomineralization. pp 65–84

124. Kisailus D, Nemoto M (2018) Structural and proteomic analyses of iron oxide biomineralization in chiton teeth. In: Matsunaga T, Tanaka T, Kisailus D (eds) Biological magnetic materials and applications. Springer Singapore, Singapore, pp 53–73

125. Moura HM, Unterlass MM (2020) Biogenic metal oxides. Biomimetics (Basel, Switzerland) 5:29. https://doi.org/10.3390/biomimetics5020029

126. Wealthall RJ, Brooker LR, Macey DJ, Griffin BJ (2005) Fine structure of the mineralized teeth of the chiton *Acanthopleura echinata* (Mollusca: Polyplacophora). J Morphol 265:165–175. https://doi.org/10.1002/jmor.10348

127. Weaver JC, Wang Q, Miserez A et al (2010) Analysis of an ultra hard magnetic biomineral in chiton radular teeth. Mater Today 13:42–52. https://doi.org/10.1016/S1369-7021(10)70016-X

128. Ansari MI, Schiwon K, Malik A, Grohmann E (2012) Biofilm formation by environmental bacteria. In: Malik A, Grohmann E (eds) Environmental protection strategies for sustainable development. Strategies for sustainability. Springer, Dordrecht, pp 341–378

129. Karatan E, Watnick P (2009) Signals, regulatory networks, and materials that build and break bacterial biofilms. Microbiol Mol Biol Rev 73:310–347. https://doi.org/10.1128/mmbr.00041-08

130. Awramik SM, Schopf JW, Walter MR (1983) Filamentous fossil bacteria from the Archean of Western Australia. Precambrian Res 20:357–374. https://doi.org/10.1016/0301-9268(83)90081-5

131. Flemming HC (1993) Biofilms and environmental protection. Water Sci Technol 27:1–10. https://doi.org/10.2166/wst.1993.0528

132. Singh R, Paul D, Jain RK (2006) Biofilms: implications in bioremediation. Trends Microbiol 14:389–397. https://doi.org/10.1016/j.tim.2006.07.001

133. Simões M, Simões LC, Vieira MJ (2010) A review of current and emergent biofilm control strategies. LWT - Food Sci Technol 43:573–583. https://doi.org/10.1016/j.lwt.2009.12.008

134. Yin W, Wang Y, Liu L, He J (2019) Biofilms: the microbial "protective clothing" in extreme environments. Int J Mol Sci 20:3423. https://doi.org/10.3390/ijms20143423

135. O'Toole G, Kaplan HB, Kolter R (2000) Biofilm formation as microbial development. Annu Rev Microbiol 54:49–79

136. Hall-Stoodley L, Stoodley P (2002) Developmental regulation of microbial biofilms. Curr Opin Biotechnol 13:228–233. https://doi.org/10.1016/S0958-1669(02)00318-X

137. Verstraeten N, Braeken K, Debkumari B et al (2008) Living on a surface: swarming and biofilm formation. Trends Microbiol 16:496–506. https://doi.org/10.1016/j.tim.2008.07.004

138. Zhang X, Bishop PL, Kupferle MJ (1998) Measurement of polysaccharides and proteins in biofilm extracellular polymers. Water Sci Technol 37:345–348. https://doi.org/10.1016/S0273-1223(98)00127-9

139. Sutherland IW (2001) Biofilm exopolysaccharides: a strong and sticky framework. Microbiology 147:3–9. https://doi.org/10.1099/00221287-147-1-3

140. Sutherland IW (2001) The biofilm matrix—an immobilized but dynamic microbial environment. Trends Microbiol 9:222–227. https://doi.org/10.1016/S0966-842X(01)02012-1

141. Flemming HC, Wingender J (2010) The biofilm matrix. Nat Rev Microbiol 8:623–633. https://doi.org/10.1038/nrmicro2415

142. Flemming H-C (1995) Sorption sites in biofilms. Water Sci Technol 32:27–33. https://doi.org/10.1016/0273-1223(96)00004-2

143. Flemming H-C (2009) Why microorganisms live in biofilms and the problem of biofouling. In: Flemming HC, Murthy PS, Venkatesan RCK (ed) Marine and industrial biofouling. Springer Berlin Heidelberg

144. Banerjee S, Joshi SR (2013) Insights into cave architecture and the role of bacterial biofilm. Proc Natl Acad Sci India Sect B - Biol Sci 83:277–290. https://doi.org/10.1007/s40011-012-0149-3

145. van Hullebusch ED, Zandvoort MH, Lens PNL (2003) Metal immobilisation by biofilms: mechanisms and analytical tools. Rev Environ Sci Biotechnol 2:9–33. https://doi.org/10.1023/B:RESB.0000022995.48330.55

146. Beveridge TJ, Makin SA, Kadurugamuwa JL, Li Z (1997) Interactions between biofilms and the environment. FEMS Microbiol Rev 20:291–303. https://doi.org/10.1016/S0168-6445(97)00012-0

147. Tsezos M (2007) Biological removal of ions: principles and applications. Adv Mater Res 20–21:589–596. https://doi.org/10.4028/www.scientific.net/amr.20-21.589

148. Ferris FG, Schultze S, Witten TC et al (1989) Metal interactions with microbial biofilms in acidic and neutral pH environments. Appl Environ Microbiol 55:1249–1257. https://doi.org/10.1128/aem.55.5.1249-1257.1989

149. Gadd GM (2001) Phytoremediation of toxic metals; using plants to clean up the environment. John Wiley & Sons, Ltd

150. McLean RJC, Fortin D, Brown DA (1996) Microbial metal-binding mechanisms and their relation to nuclear waste disposal. Can J Microbiol 42:392–400. https://doi.org/10.1139/m96-055

151. Yee N, Fein J (2001) Cd adsorption onto bacterial surfaces: a universal adsorption edge? Geochim Cosmochim Acta 65:2037–2042. https://doi.org/10.1016/S0016-7037(01)00587-7

152. Gadd GM (2010) Metals, minerals and microbes: Geomicrobiology and bioremediation. Microbiology 156:609–643. https://doi.org/10.1099/mic.0.037143-0

153. Gourdon R, Bhende S, Rus E, Sofer SS (1990) Comparison of cadmium biosorption by Gram-positive and Gram-negative bacteria from activated sludge. Biotechnol Lett 12:839–842. https://doi.org/10.1007/BF01022606

154. Schorer M, Eisele M (1997) Accumulation of inorganic and organic pollutants by biofilms in the aquatic environment. Water, Air Soil Pollut 99:651–659. https://doi.org/10.1023/A:1018384616442

155. Konhauser KO (1998) Diversity of bacterial iron mineralization. Earth-Sci Rev 43:91–121. https://doi.org/10.1016/S0012-8252(97)00036-6

156. Ferris FG, Beveridge TJ (1986) Physiochemical roles of soluble metal cations in the outer membrane of *Escherichia coli* K-12. Can J Microbiol 32:594–601. https://doi.org/10.1139/m86-110

157. Beveridge TJ (1978) The response of cell walls of Bacillus subtilis to metals and to electron-microscopic stains. Can J Microbiol 24:89–104. https://doi.org/10.1139/m78-018

158. Violante A, Zhu J, Pigna M et al (2013) Role of biomolecules in influencing transformation mechanisms of metals and metalloids in soil environments. Mol Environ Soil Sci. https://doi.org/10.1007/978-94-007-4177-5_7

159. Mann S (1983) Mineralization in biological systems. Inorganic elements in biochemistry. Springer, Berlin Heidelberg, Berlin, Heidelberg, pp 125–174

160. Gaillardet J, Viers J, Dupré B (2003) Trace elements in river waters. Treatise Geochem 5–9:225–272. https://doi.org/10.1016/B0-08-043751-6/05165-3

161. Inskeep WP, Macur RE, Harrison G et al (2004) Biomineralization of As(V)-hydrous ferric oxyhydroxide in microbial mats of an acid-sulfate-chloride geothermal spring, Yellowstone National Park. Geochim Cosmochim Acta 68:3141–3155. https://doi.org/10.1016/j.gca.2003.09.020

162. He K, Roud SC, Gilder SA et al (2018) Seasonal variability of magnetotactic bacteria in a freshwater pond. Geophys Res Lett 45:2294–2302. https://doi.org/10.1002/2018GL077213

163. Peng XT, Zhou HY, Yao HQ et al (2007) Microbe-related precipitation of iron and silica in the Edmond deep-sea hydrothermal vent field on the Central Indian Ridge. Chin Sci Bull 52:3233–3238. https://doi.org/10.1007/s11434-007-0523-3

164. Konhauser KO, Schultze-Lam S, Ferris FG et al (1994) Mineral precipitation by epilithic biofilms in the speed river, Ontario, Canada. Appl Environ Microbiol 60:549–553. https://doi.org/10.1128/aem.60.2.549-553.1994

165. Cosmidis J, Benzerara K, Morin G et al (2014) Biomineralization of iron-phosphates in the water column of Lake Pavin (Massif Central, France). Geochim Cosmochim Acta 126:78–96. https://doi.org/10.1016/j.gca.2013.10.037

166. Clarke WA, Konhauser KO, Thomas JC, Bottrell SH (1997) Ferric hydroxide and ferric hydroxysulfate precipitation by bacteria in an acid mine drainage lagoon. FEMS Microbiol Rev 20:351–361. https://doi.org/10.1016/S0168-6445(97)00017-X

167. Konhauser KO, Fisher QJ, Fyfe WS et al (1998) Authigenic mineralization and detrital clay binding by freshwater biofilms: the brahmani river, India. Geomicrobiol J 15:209–222. https://doi.org/10.1080/01490459809378077

168. Konhauser KO, Ferris FG (1996) Diversity of iron and silica precipitation by microbial mats in hydrothermal waters, Iceland: implications for Precambrian iron formations. Geology 24:323–326. https://doi.org/10.1130/0091-7613(1996)024%3c0323:DOIASP%3e2.3.CO;2

169. Benzerara K, Morin G, Yoon TH et al (2008) Nanoscale study of As biomineralization in an acid mine drainage system. Geochim Cosmochim Acta 72:3949–3963. https://doi.org/10.1016/j.gca.2008.05.046

170. Sanna R, De Giudici G, Scorciapino AM et al (2013) Investigation of the hydrozincite structure by infrared and solid-state NMR spectroscopy. Am Mineral 98:1219–1226. https://doi.org/10.2138/am.2013.4158

171. Medas D, Meneghini C, Podda F et al (2018) Structure of low-order hemimorphite produced in a Zn-rich environment by cyanobacterium *Leptolingbya frigida*. Am Mineral. https://doi.org/10.2138/am-2018-6128

172. Garcia-Guinea J, Garrido F, Lopez-Arce P et al (2016) Analyzing materials in the microscopes: From the Sorby thin sections up to the non-destructive large chambers. AIP Conf Proc 1742:20002. https://doi.org/10.1063/1.4953121

173. Kochian LV (2012) Rooting for more phosphorus. Nature 488:466–467. https://doi.org/10.1038/488466a

174. Gianinazzi-Pearson V (1996) Plant cell responses to arbuscular mycorrhizal fungi: getting to the roots of the symbiosis. Plant Cell 8:1871–1883. https://doi.org/10.1105/tpc.8.10.1871

175. Bonneville S, Smits MM, Brown A et al (2009) Plant-driven fungal weathering: early stages of mineral alteration at the nanometer scale. Geology 37:615–618. https://doi.org/10.1130/G25699A.1

176. Bonneville S, Morgan DJ, Schmalenberger A et al (2011) Tree-mycorrhiza symbiosis accelerate mineral weathering: evidences from nanometer-scale elemental fluxes at the hypha-mineral interface. Geochim Cosmochim Acta 75:6988–7005. https://doi.org/10.1016/j.gca.2011.08.041

177. Moulton KL, West J, Berner RA (2000) Solute flux and mineral mass balance approaches to the quantification of plant effects on silicate weathering. Am J Sci 300:539–570. https://doi.org/10.2475/ajs.300.7.539

178. Chorover J, Kretzschmar R, Garica-Pichel F, Sparks DL (2007) Soil biogeochemicial processes within the critical zone. Elements 3:321–326. https://doi.org/10.2113/gselements.3.5.321

179. Landeweert R, Hoffland E, Finlay RD et al (2001) Linking plants to rocks: ectomycorrhizal fungi mobilize nutrients from minerals. Trends Ecol Evol 16:248–254. https://doi.org/10.1016/S0169-5347(01)02122-X

180. Lambers H, Mougel C, Jaillard B, Hinsinger P (2009) Plant-microbe-soil interactions in the rhizosphere: an evolutionary perspective. Plant Soil 321:83–115. https://doi.org/10.1007/s11104-009-0042-x

181. Song Z, Wang H, Strong PJ et al (2012) Plant impact on the coupled terrestrial biogeochemical cycles of silicon and carbon: implications for biogeochemical carbon sequestration. Earth-Sci Rev 115:319–331. https://doi.org/10.1016/j.earscirev.2012.09.006

182. Conley DJ (2002) Terrestrial ecosystems and the global biogeochemical silica cycle. Global Biogeochem Cycles 16:68. https://doi.org/10.1029/2002GB001894

183. Tréguer P, Nelson DM, Van Bennekom AJ et al (1995) The silica balance in the world ocean: A reestimate. Science 268(80):375–379. https://doi.org/10.1126/science.268.5209.375

184. Alexandre A, Colin F, Meunier J-D (1994) Phytoliths as indicators of the biogeochemical turnover of silicon in equatorial rainforest. Comptes Rendus - Acad des Sci Ser II Sci la Terre des Planetes 319:453–458

185. Cary L, Alexandre A, Meunier J-D et al (2005) Contribution of phytoliths to the suspended load of biogenic silica in the Nyong basin rivers (Cameroon). Biogeochemistry 74:101–114. https://doi.org/10.1007/s10533-004-2945-1

186. Fraysse F, Cantais F, Pokrovsky OS et al (2006) Aqueous reactivity of phytoliths and plant litter: physico-chemical constraints on terrestrial biogeochemical cycle of silicon. J Geochemical Explor 88:202–205. https://doi.org/10.1016/j.gexplo.2005.08.039

187. Adrees M, Ali S, Rizwan M et al (2015) Mechanisms of silicon-mediated alleviation of heavy metal toxicity in plants: a review. Ecotoxicol Environ Saf 119:186–197. https://doi.org/10.1016/j.ecoenv.2015.05.011

188. Hinsinger P (2011) Biogeochemical, biophysical, and biological processes in the rhizosphere. In: Huang PM, Li Y, Summer ME (eds) Handbook of soil science resource of management and environmental impacts. CRC Press, Taylor & Francis, pp 1–30

189. Violante A, Caporale AG (2015) Biogeochemical processes at soil-root interface. J Soil Sci Plant Nutr 15:422–448. https://doi.org/10.4067/s0718-95162015005000038

190. Chen YT, Wang Y, Yeh KC (2017) Role of root exudates in metal acquisition and tolerance. Curr Opin Plant Biol 39:66–72. https://doi.org/10.1016/j.pbi.2017.06.004

191. Zhu Y, Duan G, Chen B et al (2014) Mineral weathering and element cycling in soil-microorganism-plant system. Sci China Earth Sci 57:888–896. https://doi.org/10.1007/s11430-014-4861-0

192. Antoniadis V, Levizou E, Shaheen SM et al (2017) Trace elements in the soil-plant interface: phytoavailability, translocation, and phytoremediation–a review. Earth-Science Rev 171:621–645. https://doi.org/10.1016/j.earscirev.2017.06.005

193. Viehweger K (2014) How plants cope with salinity. Bot Stud 55:1–8

194. Xu B, Yu S (2013) Root iron plaque formation and characteristics under N_2 flushing and its effects on translocation of Zn and Cd in paddy rice seedlings (*Oryza sativa*). Ann Bot 111:1189–1195. https://doi.org/10.1093/aob/mct072

195. Chang H, Buettner SW, Seaman JC et al (2014) Uranium immobilization in an iron-rich rhizosphere of a native wetland plant from the Savannah River site under reducing conditions. Environ Sci Technol 48:9270–9278. https://doi.org/10.1021/es5015136

196. Tripathi RD, Tripathi P, Dwivedi S et al (2014) Roles for root iron plaque in sequestration and uptake of heavy metals and metalloids in aquatic and wetland plants. Metallomics 6:1789–1800. https://doi.org/10.1039/c4mt00111g

197. Hansel CM, Fendorf S, Sutton S, Newville M (2001) Characterization of Fe plaque and associated metals on the roots of mine-waste impacted aquatic plants. Environ Sci Technol 35:3863–3868. https://doi.org/10.1021/es0105459

198. Medas D, De Giudici G, Casu MA et al (2015) Microscopic processes ruling the bioavailability of Zn to roots of *Euphorbia Pithyusa* L. Pioneer plant. Environ Sci Technol 49:1400–1408. https://doi.org/10.1021/es503842w

199. De Giudici G, Medas D, Meneghini C et al (2015) Microscopic biomineralization processes and Zn bioavailability: a synchrotron-based investigation of *Pistacia lentiscus* L. roots. Environ Sci Pollut Res 22:19352–19361. https://doi.org/10.1007/s11356-015-4808-9

200. Mani D, Kumar C (2014) Biotechnological advances in bioremediation of heavy metals contaminated ecosystems: an overview with special reference to phytoremediation. Int J Environ Sci Technol 11:843–872. https://doi.org/10.1007/s13762-013-0299-8

201. He H, Veneklaas EJ, Kuo J, Lambers H (2014) Physiological and ecological significance of biomineralization in plants. Trends Plant Sci 19:166–174. https://doi.org/10.1016/j.tplants.2013.11.002

202. Medas D, De Giudici G, Pusceddu C et al (2019) Impact of Zn excess on biomineralization processes in *Juncus acutus* grown in mine polluted sites. J Hazard Mater 370:98–107. https://doi.org/10.1016/j.jhazmat.2017.08.031

203. Boi ME, Medas D, Aquilanti G, et al (2020) Mineralogy and Zn chemical speciation in a soil-plant system from a metal-extreme environment: a study on *Helichrysum microphyllum* subsp. tyrrhenicum (Campo Pisano Mine, SW Sardinia, Italy). Minerals 10:259. https://doi.org/10.3390/min10030259

204. McNear DH, Peltier E, Everhart J et al (2005) Application of quantitative fluorescence and absorption-edge computed microtomography to image metal compartmentalization in *Alyssum murale*. Environ Sci Technol 39:2210–2218. https://doi.org/10.1021/es0492034

205. Sarret G, Harada E, Choi YE et al (2006) Trichomes of tobacco excrete zinc as zinc-substituted calcium carbonate and other zinc-containing compounds. Plant Physiol 141:1021–1034. https://doi.org/10.1104/pp.106.082743

206. Reimann C, Englmaier P, Fabian K, et al (2015) Biogeochemical plant–soil interaction: variable element composition in leaves of four plant species collected along a south–north transect at the southern tip of Norway. Sci Total Environ 506–507:480–495. https://doi.org/10.1016/j.scitotenv.2014.10.079

207. Lanson B, Marcus MA, Fakra S et al (2008) Formation of Zn Ca phyllomanganate nanoparticles in grass roots. Geochim Cosmochim Acta 72:2478–2490. https://doi.org/10.1016/j.gca.2008.02.022

208. Van Balen E, Van De Geijn SC, Desmet GM (1980) Autoradiographic evidence for the incorporation of cadmium into calcium oxalate crystals. Z Pflanzenphysiol 97:123–133. https://doi.org/10.1016/s0044-328x(80)80026-2

209. Mazen AMA, El Maghraby OMO (1997) Accumulation of cadmium, lead and strontium, and a role of calcium oxalate in water hyacinth tolerance. Biol Plant 40:411–417. https://doi.org/10.1023/A:1001174132428

210. de la Fuente V, Rufo L, Juarez BH et al (2016) Formation of biomineral iron oxides compounds in a Fe hyperaccumulator plant: *Imperata cylindrica* (L.) P. Beauv. J Struct Biol 193:23–32. https://doi.org/10.1016/j.jsb.2015.11.005

211. Rodríguez N, Menéndez N, Tornero J et al (2005) Internal iron biomineralization in *Imperata cylindrica*, a perennial grass: chemical composition, speciation and plant localization. New Phytol 165:781–789. https://doi.org/10.1111/j.1469-8137.2004.01264.x

212. Liang Y, Sun W, Zhu Y-G, Christie P (2007) Mechanisms of silicon-mediated alleviation of abiotic stresses in higher plants: a review. Environ Pollut 147:422–428. https://doi.org/10.1016/j.envpol.2006.06.008

213. Neumann D, Nieden UZ, Schwieger W et al (1997) Heavy metal tolerance of *minuartia verna*. J Plant Physiol 151:101–108. https://doi.org/10.1016/S0176-1617(97)80044-2

214. Neumann D, zur Nieden U (2001) Silicon and heavy metal tolerance of higher plants. Phytochemistry 56:685–692. https://doi.org/10.1016/s0031-9422(00)00472-6

215. Moravec B, Chorover J (2020) Critical zone biogeochemistry. Biogeochem Cycles 131–149. https://doi.org/10.1002/9781119413332.ch6

216. NRC (2001) Basic research opportunities in the Earch sciences. National Academies Press, Washington

217. Giardino JR, Houser C (2015) Introduction to the critical zone. Dev Earth Surf Process 19:1–13. https://doi.org/10.1016/B978-0-444-63369-9.00001-X

218. White T, Brantley S, Banwart S et al (2015) The role of critical zone observatories in critical zone science. Elsevier B.V.

219. Brantley SL, Goldhaber MB, Vala Ragnarsdottir K (2007) Crossing disciplines and scales to understand the critical zone. Elements 3:307–314. https://doi.org/10.2113/gselements.3.5.307

220. Brantley SL, Eissenstat DM, Marshall JA et al (2017) Reviews and syntheses: on the roles trees play in building and plumbing the critical zone. Biogeosciences 14:5115–5142. https://doi.org/10.5194/bg-14-5115-2017

221. Sparks DL (2005) Toxic metals in the environment: the role of surfaces. Elements 1:193–197

222. Anderson SP, Von BF, White AF (2007) Physical and chemical controls on the critical zone. Elements 3:315–320. https://doi.org/10.2113/gselements.3.5.315

223. Perdrial J, Thompson A, Chorover J (2015) Soil geochemistry in the critical zone: influence on atmosphere, surface- and groundwater composition. In: Giardino JR (eds) Houser CBT-D in ESP. Elsevier, pp 173–201

224. Manceau A, Tamura N, Celestre RS et al (2003) Molecular-scale speciation of Zn and Ni in soil ferromanganese nodules from loess soils of the Mississippi basin. Environ Sci Technol 37:75–80. https://doi.org/10.1021/es025748r

225. Coston JA, Fuller CC, Davis JA (1995) Pb^{2+} and Zn^{2+} adsorption by a natural aluminum and iron-beariring surface coating on an aquifer sand. 59:3535–3547. https://doi.org/10.1016/0016-7037(95)00231-N

226. Davranche M, Gélabert A, Benedetti MF (2020) Electron transfer drives metal cycling in the critical zone. Elements 16:185–190. https://doi.org/10.2138/gselements.16.3.185

227. Gadd GM (2010) Microbial role in global biogeochemical cycling of metals and metalloids at the interfaces in the earth's critical zone. Mol Environ Soil Sci Interfaces Earth's Crit Zo. https://doi.org/10.1007/978-3-642-05297-2_2

228. Gadd GM (2013) Molecular environmental soil science. Springer, Dordrecht

229. Boulton AJ, Findlay S, Marmonier P et al (1998) The functional significance of the hyporheic zone in streams and rivers. Annu Rev Ecol Syst 29:59–81. https://doi.org/10.1146/annurev.ecolsys.29.1.59

230. Hoagland B, Russo TA, Brantley SL (2017) Relationships in a headwater sandstone stream. Water Resour Res Res 53:4643–4667. https://doi.org/10.1002/2016WR019717.Received

231. Mcintosh J, Schaumberg C, Perdrial J et al (2017) Geochemical evolution of the Critical Zone across variable time scales informs concentration-discharge relationships: Jemez River Basin Critical Zone Observatory. Water Resour Res 53:4169–4196. https://doi.org/10.1002/2016WR019712

232. Wondzell SM (2011) This file was created by scanning the printed publication. Text errors identified by the software have been corrected: however some errors may remain. The role of the hyporheic zone across stream networks t. Hydrol Process 25:3525–3532

233. Watling HR (2006) The bioleaching of sulphide minerals with emphasis on copper sulphides—a review. Hydrometallurgy 84:81–108. https://doi.org/10.1016/J.HYDROMET.2006.05.001

234. Brar KK, Magdouli S, Etteieb S, et al (2021) Integrated bioleaching-electrometallurgy for copper recovery—a critical review. J Clean Prod 291:125257. https://doi.org/10.1016/j.jclepro.2020.125257

235. Mishra D, Kim D-J, Ahn J-G, Rhee Y-H (2005) Bioleaching: a microbial process of metal recovery; A review. Met Mater Int 11:249–256. https://doi.org/10.1007/BF03027450

236. Cecchi G, Marescotti P, Di Piazza S, Zotti M (2017) Native fungi as metal remediators: silver myco-accumulation from metal contaminated waste-rock dumps (Libiola Mine, Italy). J Environ Sci Heal Part B 52:191–195. https://doi.org/10.1080/03601234.2017.1261549

237. Rickard D (2012) Sulfidic sediments and sedimentary rocks. Elsevier

238. Ohmoto H, Lasaga AC (1982) Kinetics of reactions between aqueous sulfates and sulfides in hydrothermal systems. Geochim Cosmochim Acta 46:1727–1745. https://doi.org/10.1016/0016-7037(82)90113-2

239. Sato M (1992) Persistency-field Eh-pH diagrams for sulfides and their application to supergene oxidation and enrichment of sulfide ore bodies. Geochim Cosmochim Acta 56:3133–3156. https://doi.org/10.1016/0016-7037(92)90294-S

240. Wilkin RT, Barnes HL (1997) Formation processes of framboidal pyrite. Geochim Cosmochim Acta 61:323–339. https://doi.org/10.1016/S0016-7037(96)00320-1

241. Wilkin RT, Barnes HL, Brantley SL (1996) The size distribution of framboidal pyrite in modern sediments: an indicator of redox conditions. Geochim Cosmochim Acta 60:3897–3912. https://doi.org/10.1016/0016-7037(96)00209-8

242. Labrenz M, Druschel GK, Thomsen-Ebert T et al (2000) Formation of sphalerite (ZnS) deposits in natural biofilms of sulfate-reducing bacteria. Science 290(80):1744–1747. https://doi.org/10.1126/science.290.5497.1744

243. Zhuang W-Q, Fitts JP, Ajo-Franklin CM et al (2015) Recovery of critical metals using biometallurgy. Curr Opin Biotechnol 33:327–335. https://doi.org/10.1016/j.copbio.2015.03.019

244. Staicu LC, Wojtowicz PJ, Pósfai M, et al (2020) PbS biomineralization using cysteine: *Bacillus cereus* and the sulfur rush. FEMS Microbiol Ecol 96:fiaa151. https://doi.org/10.1093/femsec/fiaa151

245. Paganin P, Alisi C, Dore E et al (2021) Microbial diversity of bacteria involved in biomineralization processes in mine-impacted freshwaters. Front Microbiol 12:778199. https://doi.org/10.3389/fmicb.2021.778199

246. De Giudici G, Wanty RB, Podda F et al (2014) Quantifying biomineralization of zinc in the Rio Naracauli (Sardinia, Italy), using a tracer injection and synoptic sampling. Chem Geol 384:110–119. https://doi.org/10.1016/j.chemgeo.2014.07.002

247. De Giudici G, Pusceddu C, Medas D et al (2017) The role of natural biogeochemical barriers in limiting metal loading to a stream affected by mine drainage. Appl Geochemistry 76:124–135. https://doi.org/10.1016/j.apgeochem.2016.11.020

Metals: Waste and Recovery

Gilberto Artioli

1 Introduction to Metals and Metallurgical Processes

Metals of different nature have been employed by mankind to produce ornaments and tools since prehistory (Fig. 1). In lay terms the crucial role of metals in providing technological power is well described by the common phrase that "metals make the world go round" ([26], Fig. 2).

It is not by chance that after the Neolithic Age the three age system of archaeological chronology is based on the introduction of metals (Copper Age, Bronze Age, Iron Age) and it is technologically similar in different places, although being shifted in terms of absolute chronology in culturally different areas.

The intrinsic properties of metals (thermal and electrical conductivity, ductility and malleability, toughness, etc.; see for example [37] are behind their fortune as basic materials. For example the unparalleled strength of metals compared to other types of materials (Fig. 3) exemplifies their importance in technology, although unfortunately their potential in the past as well as today has been mainly implemented through weapons ([8], Fig. 2). The other aspect of metals is that they are linked to the accumulation of richness: silver and gold stocks have dominated the measurement and quantification of the wealth of single individuals, nations, and empires [27].

The technological and socio-economical importance of metals is such that mankind has learned how to extract and to produce them since at least the 8th millennium BC, although important copper and silver metallurgical processes developed around the 5th millennium BC [23]. The increased circulation and use of metals is linked of course to the ever increasing exploitation of natural resources in terms of fuels and land use [39]. As an example, it is estimated that by using prehistoric technologies about 200 kg of wood/charcoal are necessary to produce 1 kg of copper [31]. The emergence of metallurgy is therefore linked to the extensive use of

G. Artioli (✉)
Dipartimento di Geoscienze and CIRCe Centre, Università di Padova, Padua, Italy
e-mail: gilberto.artioli@unipd.it

© The Author(s), under exclusive license to Springer Nature Switzerland AG 2023   117
M. Tribaudino et al. (eds.), *Minerals and Waste*, Earth and Environmental Sciences
Library, https://doi.org/10.1007/978-3-031-16135-3_5

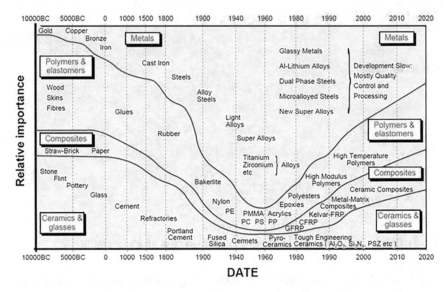

Fig. 1 Relative importance of metals in the history of mankind. Please note the non-linear time scale, and the overwhelming role of metals after the industrial revolution. Modified from Ashby [3]

Fig. 2 Textbook examples emphasizing the role of metals in history and prehistory

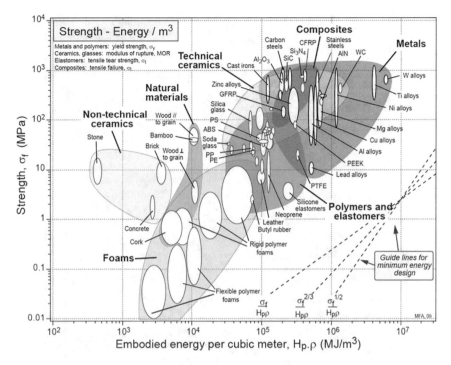

Fig. 3 One of the essential Ashby charts, showing the strength properties of metals compared to other materials. The chart illustrate well the almost linear relationship between strength and the amount of energy used to produce the material ([5], produced through the CES EduPack, Granta Design)

natural resources, to the extent that the industrial revolution and the massive production of steel marks the neat change between the sustainable and the unsustainable use of supplies of natural fuels and ores (Fig. 4). The exponential increase in good production is of course related to technological advances and the increase in global population, which is now reported at 7.9 billion people (source ourworldindata.org).

Extractive metallurgy is considered a branch of metallurgical engineering, which includes all methods and steps of extraction of metals from their natural mineral deposits. Depending on the types of ores, it covers all aspects such as washing, concentration, separation, chemical processes and extraction of pure metal and their alloying to suit various applications. Traditionally, we can distinguish the production of precious metals (gold, silver, platinum, etc.) from the production of base metals (i.e. common and inexpensive metals, such as iron, copper, lead, zinc, aluminum, etc.). Within the base metals, the processing of iron ores to produce iron, pig iron, and steel (ferrous extractive metallurgy) is distinct from the processing of non-ferrous ores, although many of the procedures and the by-products (slags, gangue separates) may be similar. From the engineering point of view the major steps involved in the metallurgical processes after ore excavation are: mineral processing, ore

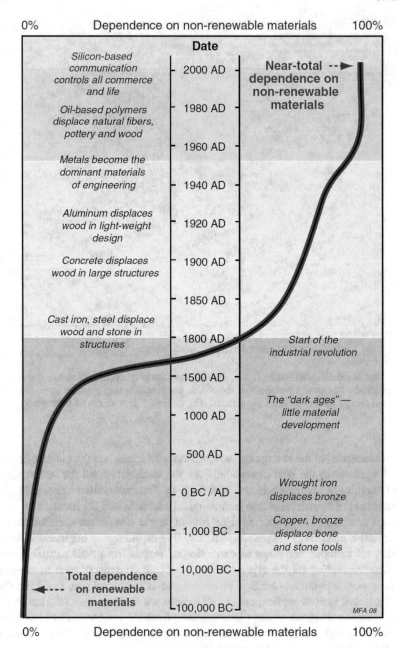

Fig. 4 The relative dependence of human products on renewable sources. Modified from Ashby [4]

concentration, metal extraction (by hydrometallurgy, pyrometallurgy, or electrometallurgy, depending on the nature of the ore and chemistry of the metal involved), and eventually metal purification.

If pyrometallurgy is adopted the process (called smelting) uses heat and a chemical reducing agent to decompose the ore, driving off other elements as gases or slag and leaving the metal base behind. The reducing agent is frequently a fossil fuel source of carbon, such as coke or charcoal. The minimum temperature is that is needed to process different ores is embedded in the well-known Ellingham diagrams (Fig. 5, [9]), which plot the standard free energy of a reaction as a function of temperature, so that the relative stability of a metal with respect to its oxide can be assessed as a function of temperature and oxygen fugacity.

2 Mine Wastes, Environmental Geochemistry, and Metal Recovery

During the whole metallurgical cycle there are emissions (gas, dust), and byproducts in form of fluids (percolating waters, acid mine drainage, flotation residues, etc.) and solids (overburden, gangue, slags, etc.) (Fig. 6).

Each of these material is a potential source of pollution and waste, and should be carefully controlled in terms of zero-waste objectives and environmental protection. However, each material is also a potential source for metal recovery, and this could help reducing the costs of the environmental treatment.

The field of low temperature geochemistry related to mining activities is a very important and active area of research, although the involved areas and quantity of materials are frequently huge and hard to tackle from practical and economic reasons. For example many mining areas in the US were inserted in the list of the superfund sites (i.e. superfund sites are polluted locations in the United States requiring a long-term response to clean up hazardous material contaminations, www.epa.gov/superfund/superfund-history). Some of the open-pit mines are possibly the largest holes created by man on the Earth's surface and are well visible from satellites. Figures 7 and 8 show the aerial and land aspects of two very large mines: the Rio Tinto area in Spain operated at least since Roman times [33], and the Bingham copper mine open pit, Utah [13].

The extraction of mineral resources requires access through underground workings, or open pit operations, or through drillholes for solution mining. Additionally, mineral processing can generate large quantities of waste, including mill tailings, waste rock and refinery wastes, heap leach pads, and slag. Thus, through mining and mineral processing activities, large surface areas of sulfide minerals can be exposed to oxygen, water, and microbes, resulting in accelerated oxidation of sulfide and other minerals and the potential for the generation of low-quality drainage [21, 22]. Water percolation through underground mining and interaction between surface waters and rain in open pit mining is the cause of the rapid oxidation of sulphide ore minerals,

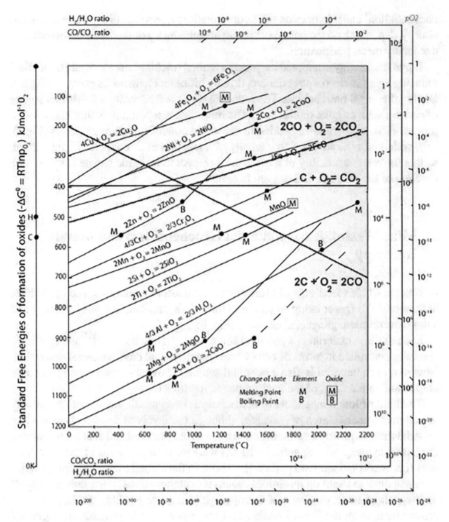

Fig. 5 Ellingham-type diagram showing the standard free energy of formation of several metal oxides [9]. At any given temperature the vertical difference between the ΔG values of two lines gives the value used in redox reactions. Inserted in the diagram are also the lines for the combustion reactions involving carbon, and the partial oxygen pressures (pO₂) needed to reduce the metal oxide or oxidize the element (scale at the right)

inducing profound alteration and transformation of the mineral–water system through a number of chemical and microbiological transformations. In most cases, the water leaching out of the mine is very acid and rich in metal and sulfate ions (Fig. 9), although depending on geological situation and the minerals extracted the water drained out of the mine can at times be near-neutral or even basic/saline [20]. The outflowing water from the mining areas is called "acid rock drainage" (ARD), "acid

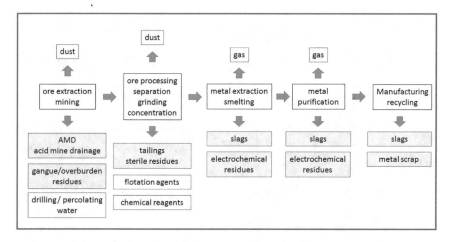

Fig. 6 Simplified diagram of metallurgical processing. In each step emissions and byproducts are indicated: they are invariably sources of potential pollution and waste. Major areas of possible intervention for environmental action and/or metal recovery are shaded

mine drainage" or "acid and metalliferous drainage" (AMD), "mining influenced water" (MIW), "saline drainage" (SD), or "neutral mine drainage" (NMD).

The increase in metal demand and the developments in extraction technologies has made possible to exploit in the last 100 years deposits with decreasing grade and concentration of metals (Fig. 10a). Of course this has stimulated a debatable alarm for the possible depletion of resources [32], but certainly from a sustainability perspective, these trends point to the scale of mines and the associated environmental footprint gradually increasing in the future. This is due to the increased solid wastes (tailings and waste rock) per unit mineral/metal production caused by declining ore grades and increased waste rock and open cut mining (Fig. 10b).

The exploitation patterns of the past, which had minimal or no concern for the environmental effects, have left huge mining areas where AMD processes are active, and largely out of control (Fig. 11).

The purely inorganic and chemical processes, essentially involving the oxidation of the sulphide minerals present in the mine wastes, may be substantially accelerated by the presence of the microbial community, which further promotes the oxidation of the iron and sulphide species in solution. In combination with evaporation, the concentration of the dissolved species in solution may become severe and lead to extremely acidic fluids. The process produces sulphuric acid and this is the trigger for the release of a number of metals, so that AMD essentially implies very acidic waters rich in dissolved metals. When these waters flow into the ground or spread into rivers and lakes, the whole ecosystem is affected, starting with aquatic organisms, then jeopardizing infrastructures and even putting freshwater supplies at stake. AMD critically affects human health, plant life, and aquatic life [36]. Given the complexity of the local geological, hydrological and geochemical systems, there is no general and clear cut limit of hazard and risk, though it is well known that many of the chemical

Fig. 7 Google Earth view of the Rio Tinto mines, in Spain. The open pit mine covers a large area, which is the result of continuous exploitation of copper, silver and gold ores from pre-Roman times till today

species transported in AMD are dangerous to microbial, plant, animal and human communities [12]. Metal toxicity is based on the fact that heavy metals persist for a long time in natural ecosystems. Further, they accumulate at different levels of the trophic chain, causing acute and chronic diseases because of metabolic disfunctions. In the organism of animals, heavy metal may accumulate in vital organs, affecting and disrupting important biochemical functions and substituting vital nutritional elements in proteins and other biological molecules. Malfunctions can be activated in animals

Fig. 8 Google Earth view of the Bingham mine open pit, Utah. The pit of the copper mine is possibly the largest hole created by man on Earth's surface

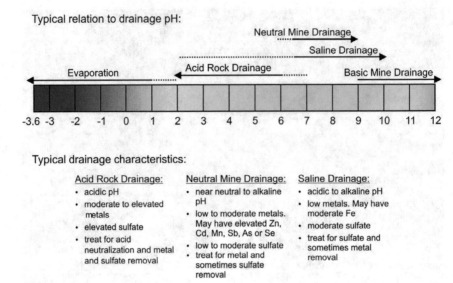

Typical relation to drainage pH:

Fig. 9   pH values of water drained from mines

and plants alike. Plant growth in fact can be severely affected, through cellular damage and disturbance of homeostasis. The whole physiology and morphology of the plant can be modified, so that concentration of metals in soils and absorption through roots has important ecological consequences.

Although geological sulphide-based mines are major sources of economical metals, and therefore AMD is the prevalent environmental concern associated with mining activities, different kind of mines (i.e. halides and other basic salts) may also generate neutral and alkaline waters containing high concentrations of dissolved metals or other toxic species, so that each mining ecosystem must be carefully evaluated.

Management of mining areas should imply prevention of AMD development, and eventually treatment of AMD dominated fluids [14, 15]. Prevention of AMD means to protect sulphide minerals from air, water, and bacteria. Since sulphide mining, extraction and processing is based on rock grinding and mineral concentration, this is of course an impossible task. It is a fact that by any kind of mining the geologically "passivated" surface layer of the deposit is disrupted and the alteration processes of the metal-loaded rocks are greatly accelerated. At best the oxidation process may be limited by appropriate processing and storage of rock wastes and tailings, usually by mixing with materials acting as neutralizers. Carbonates, hydroxides and silicate minerals may provide acid neutralization reactions, mainly through the formation of secondary phases that may incorporate toxic metals and thus decrease their concentration in water. Mine backfills therefore are often a mixture of mine tailings and other basic materials, such as soil, crushed limestone aggregate, sand, or even highly alkaline materials like Portland cement. Such treatments are usually referred to as

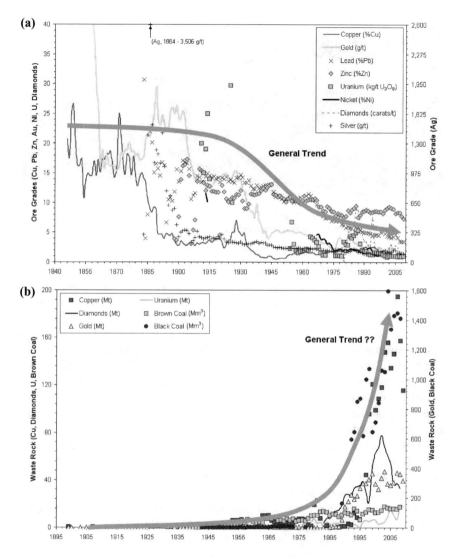

Fig. 10 a Decreasing grade of exploited ores for several metal elements in Australia; **b** increased volume of waste rocks [19]

passive treatments [36], they are generally of low cost and require long times for neutralization, being thus appropriate for abandoned mines. When the mines are in operation and large volumes of water need to be cleaned, the so-called active methods are adopted, in which the fluids used in mineral drilling and processing are continuously treated. They are more costly and require substantially more energy. Both passive and active remediation technologies can be enhanced by biological activities.

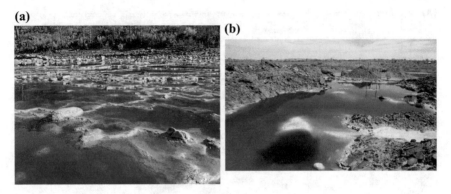

Fig. 11 **a** Precipitates from the "red river" of the Rio Tinto mines, Spain and **b** Macalder, Kenya: The site of a former copper mine that was closed in the 1960s

The sustainability of any remediation system of mine discharge is becoming critical. As discussed before, since deposits with lower metal concentrations are increasingly exploited, larger volumes of rocks are excavated, and wastes increase accordingly. The AMD treatment implies production of contaminated fluids (by metals or other chemicals used in mineral processing, such as surfactants or chelating agents), substantial amounts of metal-loaded sludges, and often also gas emissions. The residual sludges often contain toxic metals (i.e. As, Cd, Cr, Cu, Mn, Ni, Pb, etc.) well above the regulatory limits for surface or underground waters, so that they should be treated as hazardous waste. However, these materials may also be considered source of economically valuable elements, much the same way that urban wastes are starting to be looked at.

Based on the circular economy and zero-waste concepts, finally the management of residues is not only focused on the neutralization/remediation process, but also on the recovery of precious re-usable water and on the recovery of strategic metals. These by-products may help offset the cost of treatments. When the value of treated water and by-products exceeds the cost of treatment, it is economically advantageous to create enterprises that will provide economic benefits while dealing with the environmental problem, thus contributing to the sustainability of the process. The smart and sustainable processing of AMD requires novel concepts and advanced technologies. Innovative methods involving nano- and molecular-based technologies are increasingly applied and they are bound to greatly improve the long-term management of mining activities [7].

3 Metallurgical Slags: Mineralogy, Pozzolanic Behavior, Clinker-Free Concrete

Ore and metal processing produces a large amount of slags (Fig. 6), which are mostly treated as a waste or a by-product (Fig. 12). Depending on the details of the processing (i.e. ferrous or non-ferrous ores, blastfurnace or electrical furnace operations, air or water cooling, etc.) the slag may have very different chemical composition, phase mineralogy, and reactivity.

In most cases the mineral phases present in metallurgical slags are included in the C-A-S system (Fig. 13), and the overall composition of the slags overlaps substantially with the composition of important products such as cement binders. Portland-type cement [10, 40] is the most used product in terms of global volumes [4], and it can incorporate a substantial amount of GGBF (ground granulated blastfurnace slags, or BFS) as a SCM (supplementary cementitious material, see [38]. SCM's are materials that possess a certain degree of reactivity in alkaline environment (mostly due to the lime component) and therefore contribute to the long-term hydration of the cement composite and to the increase of mechanical strength. This behaviour is generally referred to as the pozzolanic property [16], that is a chemical reactivity-based behaviour that the Romans already knew very well two millennia ago [2, 18]. The continuous dissolution/precipitation processes intervening in the hardening of pozzolanic mortars and concretes [1] are the core of the durability and mechanical properties of the binders.

The fact that slags possess the correct composition and a substantial degree of reactivity makes them valuable materials and a resource for the construction industry. At present, virtually all the GGBF slags produced by the iron and steel industry are absorbed by the growing cement economy, and actually there are doubts whether the shrinking steel industry will supply enough material in the future for the global concrete needs. However on one hand GGBF slags are increasingly part of the circular economy targeted by EU countries (Fig. 14a), whereas many other types of steel

Fig. 12 Pouring of molten slags from the blastfurnace processing of ferrous ores

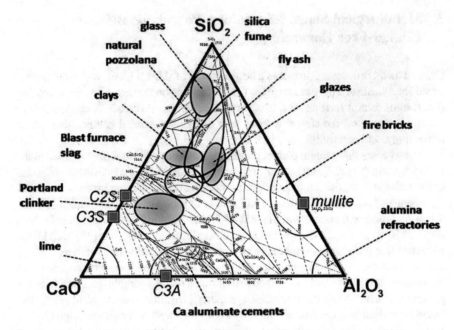

Fig. 13 Phase diagram of the CaO-SiO_2-Al_2O_3 system originally determined by Rankin and Wright [30], modified from Taylor [40], with overlapping compositional fields of several common industrial products and by-products.

furnace slags (SFS), such as those produced by the processing of scrap metal through electrical arc furnaces (EAF slags) or those produced by basic oxygen furnaces (BOF slags) are still disposed as wastes, or used inappropriately as inert materials, especially for road construction (Fig. 14b).

As a cautionary warning, there is no such thing as a totally inert slag. Therefore the use of metallurgical slags as large-sized inert components in road construction or even replacing gravel in concrete must be very carefully controlled. In a number of instances the swelling of the slag material due to long term hydration and carbonation caused substrate instability and even pop-up effect in masonry.

On the other hand, rather reactive slags such as those produced by specific smelters can even be used as the main reactive component in totally clinker-free concrete, such as those claimed for example by Murray and Roberts [6], Hoffmann Green Cements (www.ciments-hoffmann.com/low-carbon-cement/), or the Zeobond Group (www.zeobond.com/products-e-crete.html). The large amount of CO_2 emitted during clinker production (about 1 ton of CO_2 per ton of clinker, mainly due to limestone decarbonation and fuel combustion) is the main driving force behind the urgent request of decrease in clinker production. Despite a few promising large-scale applications (i.e. the City Deep Container Terminal, Johannesburg, the Brisbane Wellcamp Airport; etc., see a list of achievements in the "Zero Carbon Industry Plan: Rethinking Cement" document, available at https://apo.org.au/node/103031), most of the alternative to Portland cement fall into a fuzzy area of products employing to

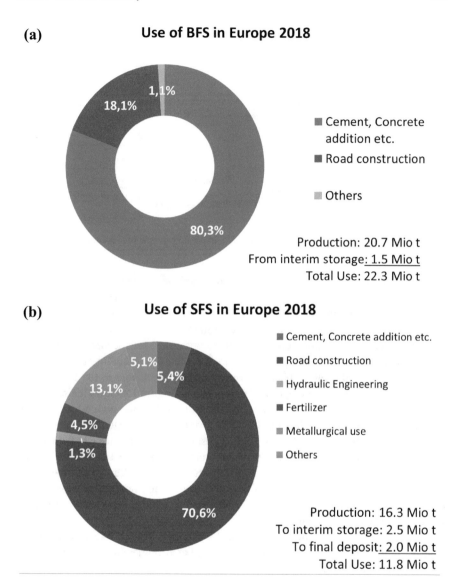

Fig. 14 **a** Use of blastfunace slags (BFS) and of steel furnace slags (SFS) in the EU. Statistics for year 2018 provided by Euroslag (www.euroslag.com/research-library-downloads/downloads/)

different degrees mixtures of reactive materials (slags, fly ash, calcined clays, waste glass, etc.), chemical activators (Na-metasilicate, alkaline solutions, etc.), buffering components (limestone, red mud, etc.), and even a portion of clinker. One of the most successful formulation at present is known as LC^3 cement, composed by almost equal parts of activated clay (mostly metakaolin), clinker, and limestone [11, 35]. All such

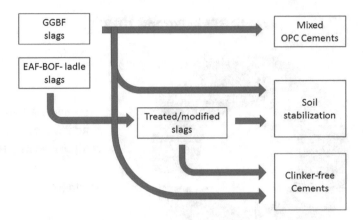

Fig. 15 Flow diagram of the incorporation of metallurgical slags into binder components

materials fall into the area of materials variously defined as geopolymers [28], alkali-activated materials [24, 29], or calcined-clays materials [17, 34]. Most companies are starting to offer products of this kind under the banner of eco-efficient (or green, or sustainable) concretes [25], though standardization is still lacking and control of short-time and long-time properties need to be optimized for specific applications.

Despite the complicated market situation, it is clear that metallurgical slags play today and may in the future play an even more important role in the formulation of more sustainable concrete alternative to Portland. Figure 15 shows the schematic pathways that slags can potentially follow in order to be incorporated into binder products. If appropriately characterized and formulated, the reactive slags provide sufficient resistance development at short times, superior resistance development at longer times, better resistance to corrosive attacks with respect to ordinary Portland cement (OPC) in construction applications. In case of the use of binders for the stabilization of contaminated soils (stabilization/solidification processes), the correct formulation may even help the formation of specific phases incorporating the hazardous species to be confined.

Overall, the transition of metallurgical by-products from waste to resource is clearly viable, and it depends more on social and environmental vision, rather than strict economical constraints.

References

1. Artioli G, Bullard JW (2013) Cement hydration: the role of adsorption and crystal growth. Cryst Res Technol 48:903–918
2. Artioli G, Secco M, Addis A (2019) The Vitruvian legacy: mortars and binders before and after the Roman world. In: Artioli G, Oberti R (eds) The contribution of mineralogy to cultural heritage. EMU Notes Mineral 20:151–202

3. Ashby MF (1988) Materials selection in mechanical design, 3rd edn. Elsevier/Butterworth-Heinemann
4. Ashby MF (2012) Materials and the environment: eco-informed material choice, 2nd edn. Elsevier
5. Ashby MF, Shercliff HR, Cebon D (2007) Materials: engineering, science, processing and design. Butterworth Heinemann
6. Attwell NC (2017) Hybrid concrete: advances in concrete activation. Murray and Roberts (Pty) Ltd. Unpublished
7. Brown GE, Hochella MF, Calas G (2017) Improving mitigation of the long-term legacy of mining activities: nano-and molecular-level concepts and methods. Elements 13:325–330
8. Diamond J (1997) Guns, germs, and steel. WW Norton & Company, New York, EUA
9. Ellingham HJT (1944) Reducibility of oxides and sulphides in metallurgical processes. J Soc Chem Ind London 63:125–133
10. Elsen J, Mertens G, Snellings R (2011) Portland cement and other calcareous hydraulic binders: history, production and mineralogy. In: Christidis GE (ed) Advances in the characterization of industrial minerals. EMU Notes Mineral 9:471–479
11. Ez-Zaki H, Marangu JM, Bellotto M, Dalconi MC, Artioli G, Valentini L (2021) A fresh view on limestone calcined clay cement (LC3) pastes. Materials 14:3037
12. Garland R (2011) Acid mine drainage—can it affect human health? Quest 7:46–47
13. James LP (1978) The Bingham copper deposits, Utah, as an exploration target; history and pre-excavation geology. Econ Geol 73:1218–1227
14. Johnson DB, Hallberg KB (2005) Acid mine drainage remediation options: a review. Sci Total Environ 338:3–14
15. Kefeni KK, Msagati TA, Mamba BB (2017) Acid mine drainage: prevention, treatment options, and resource recovery: a review. J Clean Prod 151:475–493
16. Malhotra VM, Mehta PK (2004) Pozzolanic and cementitious materials. CRC Press
17. Martirena F, Favier A, Scrivener K (eds) (2017) Calcined clays for sustainable concrete: proceedings of the 2nd international conference on calcined clays for sustainable concrete, vol 16. Springer
18. Mogetta M (2021) The origins of concrete construction in roman architecture: technology and society in Republican Italy. Cambridge University Press
19. Mudd GM (2009) The sustainability of mining in Australia: key production trends and their environmental implications for the future. Research report no RR5, Department of Civil Engineering, Monash University and Mineral Policy Institute, Revised—April 2009 (users.monash.edu.au/~gmudd/sustymining.html)
20. Nordstrom DK (2011) Mine waters: acidic to circumneutral. Elements 7:393–398
21. Nordstrom DK, Alpers CN (1999) Geochemistry of acid mine waters. In: Plumlee GS, Logsdon MJ (eds) The environmental geochemistry of mineral deposits, vol 6A. Society of Economic Geologists. Littleton, CO, pp 133–160
22. Nordstrom DK, Blowes DW, Ptacek CJ (2015) Hydrogeochemistry and microbiology of mine drainage: an update. Appl Geochem 57:3–16
23. Ottaway BS, Roberts B (2008) The emergence of metalworking. In: Jones A (ed) Prehistoric Europe: theory and practice. Wiley-Blackwell, pp 193–225
24. Pacheco-Torgal F, Labrincha J, Leonelli C, Palomo A, Chindaprasit P (eds) (2014) Handbook of alkali-activated cements, mortars and concretes. Elsevier
25. Pacheco-Torgal F, Jalali S, Labrincha J, John VM (eds) (2013) Eco-efficient concrete. Elsevier
26. Pare CF (ed) (2000) Metals make the world go round: the supply and circulation of metals in Bronze Age Europe: proceedings of a conference held at the University of Birmingham in June 1997. Oxbow Books Limited
27. Patterson CC (1972) Silver stocks and losses in ancient and medieval times. Econ Hist Rev 25:205–235
28. Provis JL, Van Deventer JSJ (eds) (2009) Geopolymers: structures, processing, properties and industrial applications. Elsevier

29. Provis JL, Van Deventer JS (eds) (2013) Alkali activated materials: state-of-the-art report, RILEM TC 224-AAM, vol 13. Springer Science & Business Media
30. Rankin GA, Wright FE (1915) The ternary system CaO-Al2O3-SiO2 with optical study. Am J Sci 4:1–79
31. Rehder JE (2000) Mastery and uses of fire in antiquity. McGill-Queen's Press-MQUP
32. Rötzer N, Schmidt M (2018) Decreasing metal ore grades—is the fear of resource depletion justified? Resources 7:88
33. Salkield LU (2012) A technical history of the Rio Tinto mines: some notes on exploitation from pre-Phoenician times to the 1950s. Springer Science & Business Media
34. Scrivener K, Favier A (eds) (2015) Calcined clays for sustainable concrete. RILEM
35. Sharma M, Bishnoi S, Martirena F, Scrivener K (2021) Limestone calcined clay cement and concrete: a state-of-the-art review. Cem Concr Res 149:106564
36. Simate GS, Ndlovu S (2014) Acid mine drainage: challenges and opportunities. J Environ Chem Eng 2:1785–1803
37. Smithells CJ (ed) (2013) Metals reference book. Elsevier
38. Snellings R, Mertens G, Elsen J (2012) Supplementary cementitious materials. In: Broekmans MATM, Pöllmann H (eds) Applied mineralogy of cement and concrete. Reviews in mineralogy & geochemistry 74. Mineralogical Society of America. Chapter 6, pp 211–278
39. Stephens L, the ArchaeoGlobe project (2019) Archaeological assessment reveals earth's early transformation through land use. Science 365:897–902
40. Taylor HFW (1997) Cement chemistry, 2nd edn. Thomas Telford Edition, London

Mineralogy of Metallurgical Slags

Daniel Vollprecht

1 Introduction

The term "slag" is used for different materials which result from solidification of a melt. Sometimes it refers to pyroclastic rocks [14], sometimes to ashes, e.g. from waste incineration, which have undergone sintering due to partial melting [93], but mostly for pyrometallurgical residues which result from the reaction of slag-forming agents with gangue or impurities in the metal in the molten state [14]. The utilization of the term "slag" for such a broad variety of materials of different chemical, physical and mineralogical properties as well as of different formation conditions may lead to confusion. Therefore, **it is suggested to restrict the term "slag" to pyrometallurgical residues only**, and to use other terms, e.g. bottom ash, for other materials. This chapter only deals with these metallurgical slags.

There is a comprehensive literature on slags ranging from slag metallurgy [66] via archaeology [34] to environmental sciences [69]. Recently, a full textbook on metallurgical slags [67] has been published which gives an excellent overview on the entire topic. In contrast, this chapter serves as scriptum of a lecture given within the **European Mineralogical Union (EMU) Summer School "Minerals and Waste: an Anthropocene Tale"**. Consequently, it is much shorter than the textbook, addresses especially students and provides some tasks to test their knowledge.

D. Vollprecht (✉)
Chair of Resource and Chemical Engineering, Augsburg University, Am Technologiezentrum 8, 86159 Augsburg, Germany
e-mail: daniel.vollprecht@uni-a.de

© The Author(s), under exclusive license to Springer Nature Switzerland AG 2023
M. Tribaudino et al. (eds.), *Minerals and Waste*, Earth and Environmental Sciences Library, https://doi.org/10.1007/978-3-031-16135-3_6

2   Slags in the Metallurgical Process

Slags are residues from pyrometallurgical processes which form in the molten state when a metallic and a silicate melt segregate. Slags are the silicate melt whose metallurgical function is to incorporate those chemical constituents which are undesired in the metal or alloy to be produced. A generalized illustration of slag formation is given in Fig. 1.

The formation of slags is inevitable in pyrometallurgy as every ore or scrap contains chemical elements which are undesired in the metal to be produced. This does not mean that these elements are of any environmental concern and the main chemical elements constituting metallurgical slags, i.e. calcium (Ca), silicon (Si), iron (Fe), magnesium (Mg) and aluminum (Al) are the same as in natural rocks. Chemical elements in slags occur mostly in an oxidized state whereas in the metallic phase they occur in the zero-valent state. Consequently, elements which are desired in the metal, such as chromium (Cr) in stainless steel, are intended to become or remain chemically reduced in the pyrometallurgical process. However, in certain pyrometallurgical processes such as the basic oxygen furnace (BOF) process (also called Linz-Donawitz (LD) process) oxidizing conditions prevail as carbon (C) shall be removed from the metal phase via oxidation. This also directs a fraction of the alloying elements into the slag. This fraction is lost for the metallurgical applications unless thermochemical treatment is applied in an additional converter to recover these alloying elements by chemical reduction [2].

Further metallurgical purposes of steel slags include prevention of re-oxidation of the metallic phase by chemical insolation as well as thermal insulation against heat losses [39].

Slags are separated from the metallic melt via tapping and solidify to form a "synthetic rock" composed of glassy and crystalline phases. Although the term "mineral" is restricted to phases which form by geological processes [62] the same phases may

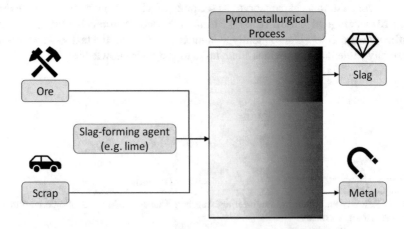

Fig. 1 Slag formation

also form in man-made environments. Therefore, the term "mineral phase" is used to cover crystalline inorganic phases from natural and synthetic origin. Thus, **slags are composed of glassy and mineral phases**.

Iron is by far the most used metal in the world [29]. Consequently, a classification of metallurgy into ferrous and non-ferrous metallurgy, and **ferrous and non-ferrous slags**, respectively is common. Since iron is mostly used for steel production and steel is a ferroalloy, for whose production also elements such as Cr, manganese (Mn) and nickel (Ni) are required, it has been suggested to consider slags resulting from the production of these elements as an own group, ferroalloy slags [68]. The past and present production routes of slags are presented in the subsequent chapters.

Test your knowledge
Which is the metallurgical function of slags?

(a) Uptake of elements undesired in the metal
(b) Reduction of the metals to their zero-valent state
(c) Provision of the heat for melting

Correct answer: (a)

3 Prehistorical and Historical Slags

Slags have been produced since the **Chalcolithic** 7000 years ago and are intensely studied interdisciplinarily by archaeologists and mineralogists to reconstruct prehistoric and historic technologies, trade routes and customs. Over the millennia, more and more metals have been utilized by mankind and correspondingly a broader and broader variety of slags has been produced which can be subdivided into slags from **nonferrous metallurgy** (e.g. copper, lead, zinc) and **ferrous metallurgy** (including both pig iron and steel production, the latter from either pig iron or scrap). Although utilization of slags as construction material has already been practiced by the Romans, tremendous amounts have been disposed of over the millennia.

Copper (Cu) was after gold (Au) the second metal used by mankind [20] and gave name to the prehistoric era Chalcolithic. The first metal objects were produced from native Cu, but already in the 6th millennial before Christ (BC) copper ores were smelted and the first Cu slags resulted, e.g. in Çatal Höyük [61]. In archaeological sites from the 4th and 3rd millennial BC slags are more abundant and are characterized by inclusions of mineral phases such as quartz (SiO_2) which have been interpreted as fluxes added by purpose [54], but are more likely to represent unmolten relicts of the batch [33]. This supported by the high metal content of 10–50% [7].

The main chemical components of chalcolithic and **Bronze Age** Cu slags are FeO/Fe_2O_3, SiO_2 and Al_2O_3 [35]. During the pyrometallurgical process, only at the grain boundaries melting occurred as temperatures only reached 1200 °C which is in the range of the eutectic point [33]. The share of melt depends on the proximity of the chemical composition to the eutectic, but was mostly too low to allow segregation of the slag and the melt in the liquid state which required mechanical separation in the solid state after cooling [33].

Since the middle of the 2nd millennium BC efficient fans were used which yielded higher temperatures at reducing atmosphere obtained from the use of charcoal. This, and the use of fluxes resulted in a higher, yet still incomplete melting degree. The partial melting at that time yielded a fayalitic (fayalite: Fe_2SiO_4) slag representing a low-melting composition and a restite which is enriched in SiO_2 and CaO or Fe-oxides [33].

Prehistorical and historical slags are investigated to deduce information regarding the technology used back then from the mineralogical composition found today. Later on (Chapter Eco-Design of Slags) it will be explained how this approach can be inverted to improve the environmental performance of present slags. One technological progress was the ability to create more and more reducing conditions. While in the 4th and 3rd millennium BC only Cu, lead (Pb) and tin (Sn) could be molten, in the 1st millennium BC also zinc (Zn) could be produced in some cases [33]. The valence of polyvalent elements such as iron, which is also associated with a change in mineralogy, can be used as a proxy to trace back the redox conditions [32]. The sensitivity of slag systems to redox conditions is especially valid for copper metallurgy of iron-bearing ores in which copper shall be reduced to its metallic state whereas iron must be present in divalent state to avoid both contamination of the metallic copper by metallic iron and freezing of the melt by crystallization of magnetite, $Fe^{3+[IV]}[Fe^{3+}Fe^{2+}]^{[VI]}O_4$ [33].

During the evolution of mankind more and more metals were used: at first, silver (Ag), lead (Pb) and tin (Sn) were used [20], forming together with gold the field of nonferrous metallurgy. In contrast, ferrous metallurgy dealing with **iron (Fe)** and its alloying elements started to flourish around 1200 before Christ (BC) with the beginning of the iron age [20].

In **Roman times**, bloomery furnaces were used to produce iron from ore [47]. The resulting iron contained impurities which were then removed by a blacksmith [17]. The iron had a various composition and some fragments had already the high hardness which is typical for steel. These fragments were than forged back together which yielded the first steel [92]. The first steel was probably produced by 1200 BC [6]. The slag is chemically composed of FeO, SiO_2 and Al_2O_3 with the composition depending on the production sites [36]. A certain share of FeO in the slag was necessary as at the applied temperature a low-Fe slag would be to viscous to be separated from the metal [86]. Mineralogically, bloomery slags consist of wuestite (FeO), fayalite (Fe_2SiO_4), solid solutions between magnetite (Fe_3O_4) and hercynite ($FeAl_2O_4$), and a calcium iron phosphate $Ca_{9-x}Fe_{1+x}(PO_4)_7$ [37]. Steelmaking from iron was not conducted via another melting process, but by carburization the diffusion of carbon into the structure of iron, which took already place by 800 BC in the Near East [6].

In the twelfth century, the blast furnace was developed in Sweden to remove the carbon from the iron [19]. The first blast furnace contained three arches, one for blowing, one for slag tapping and one for iron tapping [19]. Already the first blast furnace slags were granulated. But it was not applied as a binder with cement before 1862 [49]. In contrast to the bloomery furnace in which slag and metal were mixed and had to be separated by mechanical processing, in blast furnaces the iron incorporates 3–4% carbon which decreases the melting point and leads to the formation of a slag layer on top and a metal layer at the bottom [6]. The disadvantage of this high carbon content is the brittleness of the metal. Therefore, the metal was treated in a finery. There the metal, which is called pig iron, was melted in a bed of charcoal while air was blasted in. The iron is oxidized at the surface and resulting iron oxide reacts with the carbon from the pig iron to carbon monoxide [6]. However, it lasted until puddling process was invented in 1835 by which the pig iron is completely molten for steel melting [6], which lead to the formation of the first **steel slag**. Since then, the Bessemer process (1856) [8], Siemens Open Hearth process (1863) and the Basic steelmaking or Thomas process (1879) were further development which finally lead to the present electric arc furnace (1906) [6] and the Linz-Donawitz process (1949) [85]. Already these historical steel slags were used in unbound pavement base courses in road construction [44].

Pre-1900 BF slags are characterized by higher FeO_{total} (average 34%) and lower CaO (average 12%) compared to present BF slags (average 22% FeO_{total}, average 37% CaO) [68]. With respect to trace elements, they are poorer in Cr (9 compared to 328 mg/kg), but richer in Pb (74 compared to 20 mg/kg) [68].

In the last 200 years, the number of elements used by mankind increased tremendously, but many of these elements such as aluminium (Al) and uranium (U) are processed hydrometallurgically, not pyrometallurgically. However, **pyrometallurgical recycling** of electronic scrap [22] and lithium ion batteries [50] yielded new types of slags which are studied with respect to metal recovery [9] and leaching of contaminants limiting recyclability in road construction [64].

Test your knowledge
Order the slags according to their appearance in history:

(a) Copper slag, blast furnace slag, Thomas slag
(b) Thomas slag, copper slag, blast furnace slag
(c) Blast furnace slag, copper slag, Thomas slag

Correct answer: (a)

4   Recent Slags

In modern **ferrous metallurgy**, there are two main routes for steel production: the primary route converts iron ore via blast furnaces (BF) into pig iron, and pig iron via the BOF process into steel. The secondary route converts scrap via the electric arc furnace (EAF) into steel. The BF-BOF route contributes to 60% and the EAF route to 40% of the European steel production [25]. Subsequently, the liquid steel is further treated in various aggregates, and the resulting slags are referred to as secondary metallurgical slags. Consequently, there are three main types of present ferrous slags:

1. BF slag
2. BOF slag
3. EAF slag
4. Secondary metallurgical slags

BF slags are produced in the BF in which iron ore, partly after pre-processing such as sintering, slag-forming agents such as limestone and quartz react under reducing conditions [53]. The raw material mix ("burden") is chosen in a way that the basicity, i.e. the CaO to SiO_2 ratio, is about 1.1 [53]. Typically, BF slags consist of 40 wt% CaO, 36 wt% SiO_2, 10 wt% MgO and 10 wt% Al_2O_3 [28]. They are mostly granulated wetly and finely ground (ground granulated blast furnace slag, GGBFS) resulting in a glassy slag with chemical and hydraulic properties similar to ordinary Portland cement (OPC) [53]. A minor part of BF slags cools down slower into the crystalline blast furnace slag, which consists mainly of melilite, a solid solution between åkermanite ($Ca_2MgSi_2O_7$) and gehlenite ($Ca_2Al_2SiO_7$), and merwinite ($Ca_3MgSi_2O_8$) [77]. Main trace elements in BF slags are Ba (417 mg/kg in average), Cr (328 mg/kg in average) and V (205 mg/kg in average) [67].

BOF slags produced in the BOF or Linz-Donawitz (LD) process which aims to oxidize the dissolved carbon and phosphorous from the steel [41]. BOF slags have a basicity of >3 and are characterized by high FeO contents of 35–38 wt%, since the desired oxidation of carbon is accompanied with the partial oxidation of iron [94]. They are mostly used as industrial aggregates. They are crystalline and consist of calcium silicate and oxide phases. BOF slags are composed of >35 wt % CaO (thereof up to 12 wt% free CaO), up to 10–35 wt% FeO, 7–18 wt% SiO_2, 0.5–4 wt% Al_2O_3, and 0.4–14% MgO [53, 78, 79, 94]. The concentrations of trace elements in BOF slags are higher than in BF slags, e.g. for Cr ($3,172 \pm 5,954$ mg/kg average) and V ($5,094 \pm 8,292$ mg/kg average) [68].

EAF slags are also crystalline and are composed of silicates (larnite, melilite) and oxides (spinel, wuestite, brownmillerite). They are also used as industrial aggregate. They contain minor amounts of chromium and vanadium which are bound in spinel or brownmillerite. EAF slags consist chemically of 22–60 wt% CaO, 10–40% FeO, 6–34 wt% SiO_2, 3–14% Al_2O_3, and 3–13 wt% MgO [52, 53, 58, 94]. Trace elements in EAF slags are Cr ($16,873 \pm 38,920$ mg/kg), Ba ($1,195 \pm 1,169$ mg/kg) and V (873 ± 555 mg/kg) [68].

Secondary metallurgical slags are produced during the subsequent refining and alloying steps. They include a variety of slags, such as the ladle slag and the argon oxygen decarburization (AOD) slag, the latter accruing from the production of stainless steel [53]. In the AOD process, at first, oxygen is induced for decarburization creating a top slag rich in Cr_2O_3 (20–30%) and Al_2O_3 (18–22%), before ferrosilicon is used for chemical reduction yielding a composition rich in SiO_2 (30–40%) and

CaO (33–43%), but poor in Cr_2O_3 (1–3%) [53]. Ladle slags contain about 10 wt% SiO_2, 10 wt% MgO, 55 wt% CaO and 10 wt% Al_2O_3 and higher amounts of Cr, Ni, and Mn [53, 58, 74, 94]. With respect to trace elements, ladle slags are poorer in Cr than EAF slags ($2,997 \pm 4148$ mg/kg compared to $16,873 \pm 38,920$ mg/kg), but richer in V ($1,473 \pm 2,392$ mg/kg compared to 873 ± 555 mg/kg).

A general overview of slag production in ferrous metallurgy is given in Fig. 2.

In **non-ferrous metallurgy**, the yearly amount of slags produced is much lower than in ferrous metallurgy [67]. In Zn and Pb production, there are two main processes with specific types of slags: the Zn/Pb blast furnace (BF) process and the Imperial Smelting Process (ISP) [53, 90]. Pb BF slags contain up to 4% Pb and up to 12 wt% Zn, whereas ISP slags only up to 2 wt% Pb and up to 4 wt% Zn [53, 90]. To recover the metals from the slags, thermochemical cleaning in EAF and Waelz kilns is conducted [53]. Mineralogically, Pb–Zn slags consist of metal inclusions, oxides (e.g. PbO), sulfide (e.g. ZnS), spinels (e.g. $ZnFe_2O_4$) and silicates (e.g. willemite, $ZnSiO_4$) [24]. The production of Pb and Zn from secondary raw materials is often conducted via co-smelting with other residues in direct smelting furnaces, BFs, EAFs and other furnace types [38]. Copper slags are rich in fayalite, Fe_2SiO_4, and are subjected to a slag cleaning process in an EAF [53].

Although the pyrometallurgical processes were improved over the last millenia, there are still metal inclusions in recent slags. Therefore, magnetic separation and eddy current separation are applied to **recover** ferrous **metals** from ferrous slags and nonferrous metals from nonferrous slags, respectively. Additionally, accompanying elements may be incorporated in the slag and can be recovered by leaching methods [70].

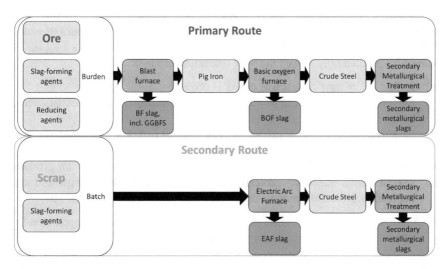

Fig. 2 Overview of iron and steel production in from ore (primary route) and scrap (secondary route), highlighting the slags (blue) produced along the process

Test your knowledge
Which mineral phases are typical for electric arc furnace slags?

(a) Wuestite-periclase solid solutions
(b) Spinel solid solutions
(c) Melilite solid solutions
(d) Plagioclase solid solutions

Correct answers: (a), (b), (c)

5 Slags: Waste and Secondary Raw Material

In this chapter, all kinds of slags have been addressed as "**residues**". This term from process engineering was chosen as it has no legal meaning and legislation on slags differs between different countries and slag type. The Cambridge Dictionary defines a residue as "the part that is left after the main part has gone or been taken away" [12]. However, as legal terms such as "waste" and "by-product" are often used in literature, it is worthwhile to explain and relate the different terminology in mining, process engineering and waste management (Fig. 2) [89].

In waste management, all materials are either a **waste** or a **product**, the latter including the "**by-product**" as a specific case. This disambiguation is based on the European Waste Framework Directive [26]. A waste is defined as "any substance or object which the holder discards or intends or is required to discard", whereas "a substance or object, resulting from a production process, the primary aim of which is not the production of that item, may be regarded as not being waste [...] but as being a by-product only if the following conditions are met:

(a) further use of the substance or object is certain;
(b) the substance or object can be used directly without any further processing other than normal industrial practice;
(c) the substance or object is produced as an integral part of a production process; and
(d) further use is lawful, i.e. the substance or object fulfils all relevant product, environmental and health protection requirements for the specific use and will not lead to overall adverse environmental or human health impacts" [26].

Consequently, the classification of a slag as a waste or a by-product depends on the specific case. Slag producers often try to classify a material as by-product to exempt it from waste legislation which requires additional administration and fulfilment of certain limit values for total and leachable contents of environmental pollutants for recycling. To avoid the term "waste" which has not only legal implications, but also negative connotations in public perception [72], terms from process engineering or mining are used to describe end-of-life materials. In this context it must be stressed that the terminologies from these three disciplines are not mutually exclusive, i.e. sentence such as "this is not a waste, it is a resource" are not correct because also resources may be waste.

Slags are formed in an industrial process, i.e. the pyrometallurgical process, and used in another process, e.g. in road construction. Consequently, terminology from process engineering can be used to describe slags. From the perspective of road construction or concrete production, slags are raw materials and the road or the concrete are the product. In this context it must be highlighted that the term "product" has a different meaning in waste management and process engineering! Considering slags as raw material, the specification "**secondary raw material**" may be used. This term has been defined "materials and articles which, after complete initial use (wear), may be used repeatedly in production as starting material" [73]. Slags are such secondary raw materials because they have been used as a metallurgical tool first and are then used repeatedly to produce a construction material [89].

Slags can be addressed from a mining perspective and terminology from mining economics [80] can be used. They compete with natural rocks for the utilization as aggregate in concrete and asphalt. As this utilization is currently economically feasible in most cases, slags can be classified as **anthropogenic reserve** following the definition of the Joint Ore Reserves Committee (JORC) (Joint Ore Reserves [42]. It is interesting to note that the term anthropogenic reserve is synonymous to the term secondary raw material [89]. In those cases, in which the utilization is not yet economically feasible, slags can only be considered as anthropogenic resources following the JORC definition. A slag pile whose mining is currently feasible can be called an anthropogenic deposit, whereas another slag pile might just be a geochemical anomaly (Fig. 3).

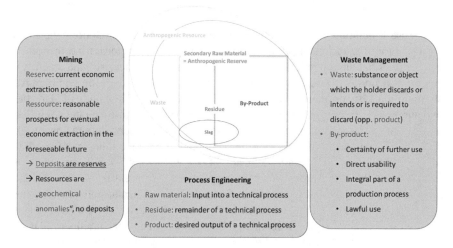

Fig. 3 Classification of slags into mining, process engineering and waste management terminology (translated and modified after [89])

In current European legislation, slags may either be a waste or a by-product, depending on individual case decisions of the authorities. Anyway, they represent not only residues of industrial processes, but valuable secondary resources as they can either be recycled (in case of a classification as a waste) or utilized (in case of a classification as a by-product) as aggregates for road construction or buildings or in case of ground granulated blast furnace slag as supplementary cementitious material substituting for clinker in cement production. Therefore, in the next chapter the different applications of slags as building materials are described.

Test your knowledge
Which sentences are mutually exclusive?

(a) Slags are wastes—Slags are resources
(b) Slags are wastes—Slags are by-products
(c) Slags are wastes—Slags are secondary raw materials

Correct answer: (b)

6   Slags as Building Materials

Slags can be used either as **industrial aggregate** which includes unbound applications in road construction as well as bound applications in asphalt and concrete, or in the production of **inorganic binders** which includes the use as substitute raw materials in clinker production as well as the use as supplementary cementitious material in cement production.

For use as building materials both technical and environmental requirements must be met. From the technical point of view, for the use as aggregate the resistance against physical and chemical weathering (e.g. frost resistance, volume stability), against abrasion (e.g. the Los Angeles index) and cracking (e.g. the compressive strength) must be considered. For the use in the production of inorganic binders, the two options, i.e. (i) prior to the firing process and (ii) after the firing process, must be distinguished. For the use prior to the firing process, i.e. as substitute raw material, the chemical composition must yield in combination with the other raw materials the desired clinker composition, i.e. $CaO/SiO_2 > 2$; $MgO < 5\%$ [10], whereas for the use after the firing process, i.e. as supplementary cementitious material, the hydraulic activity is the key parameter. From the environmental point of view the potential release of contaminants is the key concern. Details about the background are discussed in the Chapter "Mineralogy, Geochemistry and Leachability of Slags". However, some countries such as Austria do not only regulate the leachable content, but also the total content of possible environmental pollutants.

Besides these traditional applications, ferrous slags can also be used as filter materials in water treatment and environmental remediation, but also as feedstock material for carbon capture and utilization [31].

Slags from non-ferrous metallurgy are produced in lesser quantities and may also leach environmentally problematic elements, but are also applied as aggregate in construction materials [57, 75, 88].

Test your knowledge
How are slags called which are used instead of rocks in concrete or asphalt?

(a) Recycled aggregate
(b) Industrial aggregate
(c) Supplementary cementitious material

Correct answer: (b)

7 Mineralogy, Geochemistry and Leachability of Slags

The **phase composition** has a strong influence of the possible utilization of slags. On the one hand, for the utilization as inorganic binder, the hydraulic activity is a key parameter which can be increased by rapid cooling ("quenching") during wet or dry slag granulation yielding a high amorphous content. On the other hand, for the application as aggregate, the mineralogical bonding of potentially environmental relevant elements is important. Alloying elements such as chromium or vanadium as well as fluxes such as fluorine are partly also transferred to the slag in steelmaking using basic oxygen and electric arc furnaces. During utilization as building material, depending on the hydrogeochemical conditions at the interface between slag and surrounding aqueous solutions they might leach into soil and groundwater.

For nonferrous slags, weathering of sulfide phases, alloys and glassy phases, as well as precipitation of secondary phases play a key role for the leaching of metals and metalloids [67].

In contrast, for ferrous slags, silicate and oxide phases play the main role for the leaching of alloying elements such as Cr and V [67].

The **leaching** of slag constituents is either a desorption from the surface of the slag phases or a dissolution of the slag phases themselves. Anyway, it follows certain reaction kinetics which yield a certain thermodynamic equilibrium. Standardized leaching tests which run for 24 or 48 h are intended to reach this leaching equilibrium. Percolation tests, in which a column is filled with slag and water is pumped through it either upwards or downwards and sampled regularly, allow to study reaction kinetics. Reaction kinetics depend e.g. on particle size and mechanical processes (e.g. stirring or shaking of the reactor), which may vary for different standardized leaching tests. However, considering the typical applications of slags in engineering, the water flow rate through the material is very slow. Therefore, the concentrates of environmentally problematic elements obtained in the leachate after the leaching test in thermodynamic equilibrium are the ones which are regulated in legislation.

In thermodynamic equilibrium, the leaching of slag constituents is controlled by certain mechanisms. When slags are exposed to an aqueous solution, the mineral phases formed during the production of the slag in the pyrometallurgical process dissolve. In the easiest case, this leads directly to the final thermodynamic equilibrium when the aqueous solution gets saturated with respect to the corresponding mineral phase. However, this modified aqueous solution might be supersaturated with respect to another mineral phase. In this case, this so-called secondary mineral phase may precipitate and cover the surface of the primary mineral phase. This is called a couple dissolution/precipitation reaction. In most cases, the primary and the secondary phase differ with respect to chemical composition. This is called incongruent dissolution. In all cases of secondary mineral formation, it is not the solubility of the primary mineral phase, but that of the secondary mineral phase which controls the leachability of the respective chemical constituent. The incorporation of trace elements into crystal structures of other mineral phases is often observed in slag systems. Solid solutions in the olivine, spinel, brownmillerite and wuestite-periclase group occur and these structures allow the incorporation of several further chemical elements as well. If such a mineral phase dissolves the incorporated foreign ions are also released. Furthermore, the incorporation of foreign ions has an effect on the solubility of the phase. In this context it has to be mentioned that the leachability of a trace ion from a solid solution between two end members is lower than from a pure crystal of one end member coexisting with another pure crystal of the other end member. Another option is that the species produced by the dissolution of the primary phase are subsequently adsorbed to the surface a primary or secondary mineral phase. In this case, it is the adsorption/desorption equilibrium which controls the leachability.

Some elements present in slags such as Cr and V, may have negative impact on the environment [40]. However, for Cr only Cr(VI) is cancerogenic whereas Cr(III) is an essential trace element human beings [16], and also for V toxicity increases with increasing valency [21].

In steel slags **chromium** occurs mostly trivalent especially in spinel group phases [3] with the general formula AB_2O_4 or in solid solutions with periclase, MgO [82], or wüstite, FeO. Both phases are characterized by a low solubility. However, if leaching of chromium occurs, it may be associated with an oxidation into the hexavalent state depending on redox potential and pH [45]. This reaction is highly relevant for environmental issues, because trivalent chromium is an essential trace element for plants whereas hexavalent chromium is carcinogenic and phytotoxic [4]. For example, oxidation of chromium in contact with calcium phases and atmospheric oxygen can occur [71]. However, other sources state that only H_2O_2 as well as Mn(III)- and Mn(IV)-compounds can dissolve and subsequently oxidize the dissolved chromium to Cr(VI) [63]. The latter limitation seems reasonable because a comparing leaching study of numerous stainless steel slag—which are generally seen as more problematic than other steel slags—shows the concentration of Cr (VI) was always ≤0.01 mg/L and $\leq$0.1 mg/kg dry slag, respectively. However, experimental results and thermodynamic modelling indicate the formation of Cr(III) hydroxides at the **solid-water interface** as explanation for the observed dissolved Cr(III) concentrations [18]. This direct dissolution and precipitation of Cr(III) phases without intermediate oxidation is possible in blast furnace slags due to their low Ca content whereas at higher Ca concentrations prior oxidation of chromium spinels to Cr(VI)-containing **secondary phases** is required for leaching. In these cases it is rather not the pure chromatite ($CaCrO_4$) but the incorporation in $Ba(S,Cr)O_4$ solid solutions which governs chromium leachability [15]. E.g., it has been suggested that sulphate minerals are controlling factors for chromium leaching from EAF slags [27]. The neoformation of such phases leads to a decrease in chromium leachability during ageing of slags used in road construction [83]. Contrary, it was shown that wüstite-bound chromium from steel slags does not change its trivalent, octahedrally coordinated character during ageing [13]. Consequently, the formation of those phases which are resistant to weathering and oxidation should be favoured during slag production which can be done by decreasing the CaO/SiO_2 ratio and increasing the MgO content of the slag, whereas adverse conditions allow the formation of chromatite ($CaCrO_4$) and $Ca_5(Cr^{5+}O_4)_3F$ [11].

Molybdenum has a negative effect on plants due to its inhibition of copper uptake [46]. In steel slags molybdenum might occur as trivalent ion in Mo_2O_3, as this was suggested for copper slags [91]. After leaching from LD slags molybdenum occurs as MoO_4^{2-} anion which is preferentially adsorbed in the acidic pH range by iron hydroxides [51]. Furthermore, the distribution of molybdenum between the slag and the steel, but maybe also between different phases in the slag, might depend on the cooling rate, because it has been shown that rapid cooling of EAF slags decreases the Mo concentration in the slag [84].

Vanadium forms stable carbides in steel which increase strength and toughness at high temperatures while retaining ductility. In the slag vanadium is bound in trivalent form in spinel group phases [15]. Other authors [87] found out that larnite (=belite = Ca_2SiO_4 = C_2S) incorporates up to 1.7 wt% Vanadium and dissolves more rapidly with increasing carbonation of the slag. In EAF slags, it was found that Cr and V are incorporated into spinel and wuestite and minor amounts of olivine [60]. Release of these elements from olivine and subsequent adsorption onto hydrated phases formed during the leaching process were suggested to take place during leaching experiments [60]. It was found that the abundance of calcium silicates, spinel, and wuestite, which depends on the FeO/SiO_2 ratio is a key factor influencing the leaching of Cr and V [60]. Therefore, the leaching of Cr and V from EAF slags can be decreased by decreasing the FeO/SiO_2 ratio of the liquid slag [59]. This can be explained by the increase of the amount of spinel and the lower soluble calcium silicate phases by surpressing the formation of wuestite and higher soluble calcium silicates [59]. Further studies indicate the presence vanadium in brownmillerite ($Ca_2(Al,Fe)_2O_5$) both in trivalent form in octahedral and in pentavalent form in tetrahedral positions [13] and in spinels and in larnite mostly as V^{3+} [18]. Although especially dissolution of brownmillerite is very slow with a reaction rate constant of 10^{-9} mol m^{-2} s^{-2} it can convert vanadium into various species, ranging from zerovalent V^0 to pentavalent

VO_4^{3-} [15]. X-ray absorption near edge spectroscopy (XANES) measurements indicate the oxidation to the pentavalent state is common [13]. During hydration of steel slags larnite transforms into calcium silicate hydrate (CSH) phases. These phases can incorporate vanadium in their crystal/amorphous structure [30], although this is not always the case [18]. CSH phases are not the only secondary phases which incorporate vanadium: Incongruent dissolution of V-containing larnite leads to neoformation of vanadate phases like calciodelrioite ($Ca(VO_3)_2 \cdot 4 H_2O$) [15] whose solubility determines in this case the leachability of vanadium. Similarly, the formation of $Ca_2V_2O_7$ in LD slags stored outside has to be seen [1]. The effect of alteration of slags under special consideration of secondary phases on the leachability of vanadium is discussed controversially [83]. Leaching of vanadium from BOF slags can be interpreted as dissolution-controlled because the cumulative dissolved concentration increases linearly in a double log diagram in with increasing liquid/solid ratio. Dissolved vanadium in higher concentrations is phytotoxic [5]. Speciation of dissolved vanadium is important for environmental issues as pentavalent vanadium inhibits the formation of Na and K-ATPase stronger than tetravalent vanadium [65].

Fluorine in steel slags originates from the application of fluorite (CaF_2) as a flux. Unfortunately, high concentrations of fluorine in drinking water cause fluorosis in human beings [81]. Consequently, fluorine mineralogy in steel slags is a big issue. Fluorite has a very low solubility of 16 mg/L, whereas other fluorine-containing phase like fluoro-mayenite ($Ca_{12}Al_{14}(F,O)_{33}$) have a higher solubility. Consequently, the different formation conditions of both phases can be used to tailor the mineralogy of a slag with respect to its leachability: Fluorite forms at higher oxygen fugacities in a paragenesis with brownmillerite (($Ca_2(Al,Fe)_2O_5$) whereas mayenite forms at lower oxygen fugacities [48].

Slags from **nonferrous metallurgy** contain metals and metalloids as sulfides and intermetallic compounds whose dissolution may release larger amounts of these elements on the long term [23], which is especially relevant for copper slags used as construction material [76]. If coppers slag is used as a secondary raw material for tiles production, the firing process leads to immobilization of As and Pb [43].

Test your knowledge
What is meant by "leaching controlling mechanisms"?

(a) the chemical reactions whose equilibrium determine the concentration of a chemical element in the leachate
(b) the mineral phase in which a chemical element is bound to in a solidified slag
(c) the way a leaching test is conducted, e.g. shaking, stirring or percolation

Correct answer: (a)

8 Eco-design of Slags

Stable incorporation of environmentally problematic elements into distinct mineralogical phases during the production process of slags is a key process to allow the safe and environmentally friendly recyclability of slags as building material and represents an excellent example for the extension of the concept of eco-design from consumer products to industrial residues. The mineralogical bonding of environmentally relevant elements is determined during the cooling of the liquid slag when mineral phases crystallize from the melt. It is interesting to note that the relation between metallurgical process parameters and mineralogy of slags is often used in archaeometallurgy to reveal the details of (pre)historic techniques from the slag remainders found today in excavation sites. Recently, this approach has been reversed to tailor the mineralogy of steel slags by adjusting metallurgical process parameters [11, 59]. This inverse direction of using the same relationship is displayed in Fig. 4.

The addition of quartz to liquid EAF slag favours the formation of gehlenite and surpresses the formation of larnite, thereby decreasing the leaching of Ba, V and Cr [56]. The addition of magnesia favours the formation of spinels trapping Cr in their structure and decreasing their leaching [11]. On the other hand it increases the leaching of Ba, as Mg replaces Ba in calcium silicates [55]. Increasing the CaO content of the slag has a negative impact on Ba leaching from highly soluble calcium silicates, but a positive impact on V leaching, which can be explained by formation of calcium vanadates.

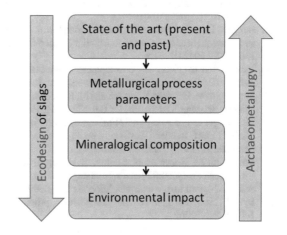

Fig. 4 Inverse approaches to link formation and properties of slags in environmental engineering (ecodesign of slags) and archaeometallurgy

Test your knowledge
What is Ecodesign of slags?

(a) Mixing solidified slag with other materials to dilute contaminants and
     keep limit values of environmental legislation
(b) Producing the slag in the pyrometallurgical process in a way that it can
     be utilized without negative environmental impact
(c) Claiming that slags are a by-product, not a waste

Correct answer: (b)

References

1. Aarabi-Karasgani M, Rashchi F, Mostoufi N, Vahidi E (2010) Leaching of vanadium from LD
    converter slag using sulfuric acid. Hydrometallurgy 102(1–4):14–21. https://doi.org/10.1016/
    j.hydromet.2010.01.006
2. Adamczyk B, Brenneis R, Adam C, Mudersbach D (2010) Recovery of chromium from AOD-
    converter slags. Steel Res Int 81(12):1078–1083. https://doi.org/10.1002/srin.201000193
3. Aldrian A, Raith J, Höllen D, Pomberger R (2015) Influence of chromium containing spinels in
    an electric arc furnace slag on the leaching behaviour. J Solid Waste Technol Manag 41(4):357–
    365. https://doi.org/10.5276/JSWTM.2015.357
4. Baceloux D, Barceloux D (1999) Chromium. J Toxicol Clin Toxicol 37(2):173–194. https://
    doi.org/10.1081/CLT-100102418
5. Baken S, Larsson M, Gustafsson J, Cubadda F, Smolders E (2012) Ageing of vanadium in soils
    and consequences for bioavailability. Eur J Soil Sci 63(6):839–847. https://doi.org/10.1111/j.
    1365-2389.2012.01491.x
6. Barraclough K (1981) The development of the early steelmaking processes. An essay in the
    history of technology. PhD thesis. University of Sheffield
7. Bartelheim M, Eckstein K, Huijsmans M, Krauss R, Pernicka E (2002) Kupferzeitliche Gewin-
    nung in Brixlegg, Österreich. In: Bartelheim M, Pernicka E, Kraus R (eds) Die Anfänge der
    Metallurgie in der Alten Welt. Forschungen zur Archäometrie und Altertumswissenschaft 1.
    Marie Leidorf, Rahden (Westfalen), pp 33–82
8. Bessemer H (1856) Patentnr. US16082A
9. Buchmann M, Borowski N, Leißner T, Heinig T, Reuter M, Friedrich B, Peuker U (2020)
    Evaluation of recyclability of a WEEE slag by means of integrative X-ray computer tomography
    and SEM-based image analysis. Minerals 10(4):309. https://doi.org/10.3390/min10040309
10. Bye G (1999) Portland cement: composition, production and properties (2 Ausg.). Thomas
    Telford Publishing
11. Cabrera-Real H, Romero-Serrano A, Zeifert B, Hernandez-Ramirez A, Hallen-Lopez M, Cruz-
    Ramirez A (2012) Effect of MgO and CaO/SiO$_2$ on the immobilization of chromium in synthetic
    slags. J Mater Cycles Waste Manag 14:317–324. https://doi.org/10.1007/s10163-012-0072-y
12. Cambridge University Press (2014) Cambridge Dictionary: residue. Abgerufen am 15. 3 2020
    von. https://dictionary.cambridge.org/de/worterbuch/englisch/residue
13. Chaurand P, Rose J, Briois V, Olivi L, Hazemann J-L, Proux O (2007) Environmental impacts
    of steel slag reused in road construction: a crystallographic and molecular (XANES) approach.
    J Hazard Mater 139(3):537–542. https://doi.org/10.1016/j.jhazmat.2006.02.060

14. Collins (2021) Collins dictionary. Abgerufen am 29. 12 2021 von. https://www.collinsdicti
onary.com/de/worterbuch/englisch/slag
15. Cornelis G, Johnson C, Van Gernven T, Vandecasteele C (2008) Leaching mechanisms of
oxyanionic metalloid and metal species in alkaline solid wastes: a review. Appl Geochem
23(5):955–976. https://doi.org/10.1016/j.apgeochem.2008.02.001
16. Costa M (1997) Toxicity and carcinogenicity of Cr(VI) in animal models and humans. Crit
Rev Toxicol 27(5):431–442. https://doi.org/10.3109/10408449709078442
17. Coustures M, Béziat (2003) The use of trace element analysis of entrapped slag inclusions to
establish ore-bar iron links: examples from two gallo-roman iron-making sites in France (Les
Martys, Montagne Noire, and Les Ferrys, Loiret). Archaeometry 45(4):599–613. https://doi.
org/10.1046/j.1475-4754.2003.00131.x
18. De Windt L, Chaurand P, Rose J (2011) Kinetics of steel slag leaching: batch tests and modeling.
Waste Manag 31(2):225–235. https://doi.org/10.1016/j.wasman.2010.05.018
19. den Ouden A (1985) The introduction and early spread of the blast furnace in Europe. Bull
Wealden Iron Res Group 5:21–35
20. Desjardins J (2014) The history of metals. Visual Captialist. Abgerufen am 21. 02 2022 von.
https://www.visualcapitalist.com/history-of-metals/
21. Domingo J (1996) Vanadium: a review of the reproductive and developmental toxicity. Reprod
Toxicol 10(3):175–182. https://doi.org/10.1016/0890-6238(96)00019-6
22. Ebin B, Isik M (2016) Chapter 5—Pyrometallurgical processes for the recovery of metals from
WEEE. WEEE Recycl: Res Dev Polic 197–137. https://doi.org/10.1016/B978-0-12-803363-
0.00005-5
23. Ettler V, Johan Z (2014) 12 years of leaching of contaminants from Pb smelter slags: geochem-
ical/mineralogical controls and slag recycling potential. Appl Geochem 40:97–103. https://doi.
org/10.1016/j.apgeochem.2013.11.001
24. Ettler V, Zdenek J, Kribek B, Sebek O, Mihaljevic M (2009) Mineralogy and environmental
stability of slags from the Tsumeb smelter, Namibia. Appl Geochem 24(1):1–15. https://doi.
org/10.1016/j.apgeochem.2008.10.003
25. Eurofer. The European Steel Association (2020) What is steel and how is steel made? Abgerufen
am 24. 2 2022 von. https://www.eurofer.eu/about-steel/learn-about-steel/what-is-steel-and-
how-is-steel-made/
26. European Parliament and Council (2008) Directive 2008/98/EC on waste and repealing certain
directives. Abgerufen am 5. 1 2022 von. https://eur-lex.europa.eu/legal-content/EN/TXT/?uri=
CELEX%3A02008L0098-20180705
27. Fällman A-M (2000) Leaching of chromium and barium from steel slag in laboratory and
field tests—a solubility controlled process? Waste Manag 20(2–3):149–154. https://doi.org/
10.1016/S0956-053X(99)00313-X
28. Geerdes M, Chaigneau R, Kurunov I (2015) Modern blast furnace ironmaking: an introduction.
IOS Press, Amsterdam
29. Global Chanbge Data Lab (2013) Our world in data. Abgerufen am 4. 1 2022 von. https://our
worldindata.org/grapher/metal-production-long-term
30. Gogar M, Scheetz B, Roy D (1996) Ettringite and C-S-H Portland cement phases for waste
ion immobilization: a review. Waste Manag 16(4):295–303. https://doi.org/10.1016/S0956-053
X(96)00072-4
31. Gomes H, Mayes W, Rogerson M, Stewart D, Burke I (2016) Alkaline residues and the
environment: a review of impacts, management practices and opportunities. J Clean Prod
112(4):3571–3582. https://doi.org/10.1016/j.jclepro.2015.09.111
32. Hauptmann A (2003) Rationales of liquefaction and metal separation in earliest copper
smelting: basics for reconstructing chalcolithic and Early Bronze Age smelting processes.
Archaeometallurgy in Europe. Milano, pp 459–486
33. Hauptmann A (2007) Alten Berg- und Hüttenleuten auf die Finger geschaut: Zur Entschlüs-
selung berg- und hüttenmännischer Techniken. In: Wagner G (ed) Einführung in die Archäome-
trie. Springer, Berlin, Heidelberg, New York, pp 115–137

34. Hauptmann A (2014) The investigation of archaeometallurgical slag. In: Roberts B, Thornton C (eds) Archaeometallurgy in global perspective. Springer, New York, pp 91–105. https://doi.org/10.1007/978-1-4614-9017-3_5

35. Hauptmann A, Rehren T, Schmitt-Strecker S (2003) Early Bronze Age copper metallurgy. In: Stöllner T, Körlin G, Steffens G, Cierny J (eds) Man and mining. Studies in honour of Gerd Weisgerber on occasion of his 65th birthday. Der Anschnitt, beiheft 16, pp 197–213

36. Hedges R, Salter C (1979) Source determination of iron currency bars through analysis of slag inclusions. Archaeometry 21(2):161–175

37. Heimann R, Kreher U, Spazier I, Wetzel G (2001) MIneralogical and chemical investigations of bloomery slags from prehistoric (8th century BC to 4th century AD) iron production sites in Upper and Lower Lusatia, Germany. Archaeometry 43(2):227–252

38. Hoang J, Reuter M, Matusewicz R, Hughes S, Piret N (2009) Top submerged lance direct zinc smelting. Miner Eng 742–751. https://doi.org/10.1016/j.mineng.2008.12.014

39. Holappa L (kein Datum) Treatise on process metallurgy. In: Seetharaman S (ed). Elsevier, Amsterdam

40. Jaishankar M, Tseten T, Anbalagan N, Mathew B, Beeregowda K (2014) Toxicity, mechanism and health effects of some heavy metals. Interdiscip Toxicol 7(2):60–72. https://doi.org/10.2478/intox-2014-0009

41. Jalkanen H, Holappa L (2014) Converter steelmaking. In: Seetharaman S (ed) Treatise on process metallurgy. Elsevier, Amsterdam, pp 223–270

42. Joint Ore Reserves Committee (2012) The JORC code. Abgerufen am 5. 1 2022 von. https://www.jorc.org/docs/jorc_code2012.pdf

43. Jordán M, Montero M, Pardo-Fabregat F (2021) Technological behaviour and leaching tests in ceramic tile bodies obtained by recycling of copper slag and MSW fly ash wastes. J Mater Cycles Waste Manag 23:707–716. https://doi.org/10.1007/s10163-020-01162-8

44. Josephson G, Sillers F, Runner D (1949) Iron blast-furnace slag: production, processing, properties and uses. United States Department of the Interior Buroeau of Mines, Washington, USA

45. Karaolu M, Zor Ş, Uğurlu M (2010) Biosorption of Cr(III) from solutions using vineyard pruning waste. Chem Eng J 159:98–106. https://doi.org/10.1016/j.cej.2010.02.047

46. Kubota J (1975) Areas of molybdenum toxicity to grazing animals in the western states. Rangel Ecol Manag/J Range Manag Arch 28(4):252–256

47. Lang J (2017) Roman iron and steel: a review. Mater Manuf Process 32(7–8):857–866. https://doi.org/10.1080/10426914.2017.1279326

48. Lee H, Kwon S, Jang S (2010) Effects of PO2 at flux state on the fluorine dissolution from synthetic steelmaking slag in aqueous solution. ISIJ Int 50(1):174–180. https://doi.org/10.2355/isijinternational.50.174

49. Lindqvist J, Balksten K, Fredrich B (2019) Blast furnace slag in historic mortars of Bergslagen, Sweden. HMC. University of Navarra, Pamplona

50. Makuza B, Tian Q, Guo X, Chattoadhyay K, Yu D (2021) Pyrometallurgical options for recycling spent lithium-ion batteries: a comprehensive review. J Power Sources (491):229622. https://doi.org/10.1016/j.jpowsour.2021.229622

51. Matern K, Rennert T, Mansfeldt T (2013) Molybdate adsorption from steel slag eluates by subsoils. Chemosphere 93(9):2108–2115. https://doi.org/10.1016/j.chemosphere.2013.07.055

52. Matinde E (2018) Mining and metallurgical wastes: a review of recycling and re-use practices. J South Afr Inst Min Metall 118:825–844. https://doi.org/10.17159/2411-9717/2018/V118N8A5

53. Matinde E, Steenkamp J (2021) Metallurgical overview and production of slags. In: Piatak N, Ettler V (eds) Metallurgical slags. Environmental geochemistry and resource potential. Royal Society of Chemistry, pp 14–58

54. Merkel J, Rothenberg B (1999) The earliest steps to copper metallurgy in the western Araba. In: Hauptmann A, Pernicka E, Rehren T, Yalcin Ü (eds) The beginnings of metallurgy. Der Anschnitt, Beiheft 9, pp 149–165

55. Mombelli D, Mapelli C, Barella S, Di Cecca C, Saout G, Garcia-Diaz E (2016) The effect of chemical composition on the leaching behaviour of electric arc furnace (EAF) carbon steel slag during a standard leaching test. J Environ Chem Eng 4(1):1050–1060. https://doi.org/10.1016/j.jece.2015.09.018

56. Mombelli D, Mapelli C, Barella S, Gruttadauria A, La Saout G, Garcia-Diaz E (2014) The efficiency of quartz addition on electric arc furnace (EAF) carbon steel slag stability. J Hazard Mater 279:586–596. https://doi.org/10.1016/j.jhazmat.2014.07.045

57. Murari K, Siddique R, Jain K (2015) Use of waste copper slag, a sustainable material. J Mater Cycles Waste Manag 17:13–26. https://doi.org/10.1007/s10163-014-0254-x

58. Ndlovu S, Simate G, Matinde E (2017) Waste production and utilization in the metal extraction industry. CRC Press, Boca Raton

59. Neuhold S, Algermissen D, Drissen P, Adamczyk B, Presoly P, Sedlazeck K, Vollprecht D (2020) Tailoring the FeO/SiO_2 ratio in electric arc furnace slags to minimize the leaching of vanadium and chromium. Appl Sci 10:2549. https://doi.org/10.3390/app10072549

60. Neuhold S, van Zomeren A, Dijkstra J, van der Sloot H, Drissen P, Algermissen D, Vollprecht D (2019) Investigation of possible leaching control mechanisms for chromium and vanadium in electric arc furnace (EAF) slags using combined experimental and modeling approaches. Minerals 9:525. https://doi.org/10.3390/min9090525

61. Neuniger H, Pittioni R, Siegl W (1964) Frühkeramikzeitliche Kupfergewinnung in Anatolien. Archaeologia Austriaca 35:98–110

62. Nickel E (1995) Definition of a mineral. Miner Mag 59(397):767–768. https://doi.org/10.1180/minmag.1995.059.397.20

63. Oze C, Bird D, Fendorf S (2007) Genesis of hexavalent chromium from natural sources in soil and groundwater. Proc Natl Acad Sci USA 104(16):6544–6549. https://doi.org/10.1073/pnas.0701085104

64. Pareuil P, Bordas F, Joussein E, Bollinger J-C (2010) Leaching properties of Mn-slag from the pyrometallurgical recycling of alkaline batteries: standardized leaching tests and influence of operational parameters. Environ Technol 31(14):1565–1576. https://doi.org/10.1080/09593331003801530

65. Patel B, Henderson G, Haswell S, Grzeskowiak R (1990) Speciation of vanadium present in a model yeast system. Analyst 115(8):1063–1066

66. Peng Z, Gregurek D, Wenzl C, White J (2016) Slag metallurgy and metallurgical waste recycling. J Miner Metals Mater Soc (JOM) 68:2313–2315. https://doi.org/10.1007/s11837-016-2047-2

67. Piatak N, Ettler V (2021) Metallurgical slags: environmental geochemistry and resource potential. R Soc Chem. https://doi.org/10.1039/9781839164576

68. Piatak N, Ettler V, Hoppe D (2021) Geochemistry and mineralogy of slags. In: Piatak N, Ettler V (eds) Metallurgical slags: environmental geochemistry and resource potential. The Royal Society of Chemistry, p 306. https://doi.org/10.1039/9781839164576

69. Piatak N, Parsons M, Seal R II (2015) Characteristics and environmental aspects of slag: a review. Appl Geochem 57:236–266. https://doi.org/10.1016/j.apgeochem.2014.04.009

70. Piatek N, Ettler V (2021) Introduction: metallurgical slags—environmental liability or valuable resource. In: Piatek N, Ettler V (eds) Metallurgical slags. Environmental geochemistry and resource potential. Royal Society of Chemistry, pp 1–13

71. Pillay K, von Blottnitz H, Petersen J (2003) Ageing of chromium(III)-bearing slag and its relation to the atmospheric oxidation of solid chromium(III)-oxide in the presence of calcium oxide. Chemosphere 52(10):1771–1779. https://doi.org/10.1016/S0045-6535(03)00453-3

72. Pongrácz E, Pohjola V (2004) Re-defining waste, the concept of ownership and the role of waste management. Resour Conserv Recycl 40(2):141–153. https://doi.org/10.1016/S0921-3449(03)00057-0

73. Prochorow A (1970–1979) The Great Soviet encyclopedia. Von. https://encyclopedia2.thefreedictionary.com/Secondary+Raw+Materialabgerufen

74. Radenovic A, Malina J, Sofilic T (2013) Characterization of ladle furnace slag from carbon steel production as a potential adsorbent. Adv Mater Sci Eng 198240. https://doi.org/10.1155/2013/198240

75. Saikia N, Cornelis G, Mertens G, Elsen J, Van Balen K, Van Gerven T, Vandecasteeele C (2008) Assessment of Pb-slag, MSWI bottom ash and boiler and fly ash for using as a fine aggregate in cement mortar. J Hazard Mater 766–777. https://doi.org/10.1016/j.jhazmat.2007.10.093

76. Schumkat A, Duester L, Ecker D, Schmid H, Heil C, Heininger P, Ternes T (2012) Leaching of metal(loid)s from a construction material: Influence of the particle size, specific surface area and ionic strength. J Hazard Mater 227–228:257–264. https://doi.org/10.1016/j.jhazmat.2012.05.045

77. Scott P, Critchley S, Wilkinson F (1986) The chemistry and mineralogy of some granulated and pelletized blastfurnace slags. Miner Mag 50(355):141–147. https://doi.org/10.1017/minmag.1986.050.355.19

78. Shen H, Forssberg E, Nordström U (2004) Physicochemical and mineralogical properties of stainless steel slags oriented to metal recovery. Resour Conserv Recycl 40(3):245–271. https://doi.org/10.1016/S0921-3449(03)00072-7

79. Shi C (2004) Steel slag—its production, processing, characteristics, and cementitious properties. J Mater Civ Eng 16(4):230. https://doi.org/10.1061/(ASCE)0899-1561(2004)16:3(230)

80. Slaby D, Wilke L (2005) Bergwirtschaftslehre. Teil 1: Wirtschafslehre der mineralischen Rohstoffe und der Lagerstätten. TU Bergakademie Freiberg

81. Spira L, Grimbleby F (1943) Fluorine in drinking water. Epidemiol Infect 43(2):142–145. https://doi.org/10.1017/S0022172400012730

82. Strandkvist I, Engström F, Pålsson K, Björkman B (2012) The influence of iron oxide on the chromium leachability of EAF slag: a full-scale study at Ovako Hofors. In: Scanmet IV: 4th International conference on process development in iron and steelmaking. Luleå, pp 329–338

83. Suer P, Lindqvist J-E, Arm M, Frogner-Kockum P (2009) Reproducing ten years of road ageing—accelerated carbonation and leaching of EAF steel slag. Sci Total Environ 407(18):5110–5118. https://doi.org/10.1016/j.scitotenv.2009.05.039

84. Tossavainen M, Engström F, Yang Q, Menad N, Lidstrom Larsson M, Bjorkman B (2007) Characteristics of steel slag under different cooling conditions. Waste Manag 1335–1344. https://doi.org/10.1016/j.wasman.2006.08.002

85. Tweraser K (2000) The Marshall plan and the reconstruction of the Austrian Steel Industry 1945–1953. In: Bischof G (ed) The Marshall plan in Austria. Transaction Publishers

86. Tylecote R, Austin J, Wraith A (1971) The mechanism of the bloomery process in shaft furnaces. J Iron Steel Inst 209:342–363

87. van Zomeren A, van der Laan S, Kobesen H, Huijgen W, Comans R (2011) Changes in mineralogical and leaching properties of converter steel slag resulting from accelerated carbonation at low CO_2 pressure. Waste Manag 31(11):2236–2244. https://doi.org/10.1016/j.wasman.2011.05.022

88. Vijayaraghavan J, Belin Jude A, Thivya J (2017) Effect of copper slag, iron slag and recycled concrete aggregate on the mechanical properties of concrete. Resour Policy 53:219–225. https://doi.org/10.1016/j.resourpol.2017.06.012

89. Vollprecht D, Pomberger R (2021) Ökodesign von Stahlwerksschlacken durch thermochemische Behandlung zur Erhöhung der Recyclingfähigkeit. BHM Berg- Huettenmaenn Monatsh 166(3):137–143. https://doi.org/10.1007/s00501-021-01090-6

90. Wang G (2016) The utilization of slag in civil infrastructure construction. Woodhead Publishing, Duxford

91. Westland A, Webster A (1990) Distribution of molybdenum between slag, copper matte and copper metal at 1300 °C. Can Metall Q 29(3):217–225. https://doi.org/10.1179/cmq.1990.29.3.217

92. Williams A (2003) The knight and the blast furnace: a history of the metallurgy of armour in the middle ages & the early modern period. Brill Academic Publishers, Leiden, Boston

93. Wong G, Gan M, Fan X, Ji Z, Chen X, Wang Z (2021) Co-disposal of municipal solid waste incineration fly ash and bottom slag: a novel method of low temperature melting treatment. J Hazard Mater 408:124438. https://doi.org/10.1016/j.jhazmat.2020.124438

94. Yildirim I, Prezzi M (2011) Chemical, mineralogical, and morphological properties of steel slag. Adv Civ Eng 463638. https://doi.org/10.1155/2011/463638

Bottom Ash: Production, Characterisation, and Potential for Recycling

Jacques Rémy Minane and Raffaele Vinai

1 Introduction

This chapter describes the main properties of the coarse fraction of combustion residues, called bottom ash, the standard industrial treatments related to its production, and possible reutilisation strategies. Typically, the term "incineration bottom ash" (IBA) is used for indicating the bottom ash produced in household waste incineration facilities, in order to make a distinction from the "furnace bottom ash" (FBA), i.e., the bottom ash produced by coal-fired power stations. Due to the increase in the number of waste incineration facilities worldwide, along with the current trends in decommissioning coal-fired power stations, the production and thus the availability of IBA (about 17.6 Mt/year from nearly 500 incinerator plants in Europe, Norway and Switzerland [1]) is significantly higher than that of FBA (about 3 Mt/year [2]), at least in Europe. For this reason, the discussion focusses mainly on the technological aspects of the production and treatments of IBA. A brief discussion on FBA is provided at the end of the chapter. Typical physical, chemical, and mineralogical properties of IBA and the main parameters influencing its final composition are recalled. In a second step, a review of the regulatory and legislative framework for the management of IBA worldwide, as well as the main recovery pathways for IBA in the construction sector are summarised. In a third part, current advanced separation technologies for processing IBA are presented. Eventually, some discussion

J. R. Minane
Department of Civil Engineering, National Advanced School of Engineering (Polytechnic), University of Yaounde I, Street 48, Melen, 3383 Yaounde, Cameroon
e-mail: remy.minane@univ-yaounde1.cm

R. Vinai (✉)
Department of Engineering, Faculty of Environment, Science and Economy, University of Exeter, Harrison Building, North Park Road, Exeter, EX4 4QF, UK
e-mail: r.vinai@exeter.ac.uk

on FBA production, environmental aspects of its handling and possible reutilisation pathways are provided.

2   Generalities on Non-hazardous IBA

2.1   Definition, Origin and Development of IBA

2.1.1   Household Waste Management and Incineration

The management of household and similar waste gathered momentum worldwide since the 90s. Household and similar wastes are generated by household and economic activities, and are collected by public waste disposal service [9].

Before 1990s, the main method of disposing of household waste was landfilling. Then, several EU countries started developing regulations for the recovery of valuable fraction and reduction of waste disposal volume and pollution potential. European countries, often under the impetus of directives delivered by the European Commission, started producing regulations and laws regarding the waste management practices and reduction, reuse, recycling and recovery options. In France for example, the Act No. 2020-105 dated 10 February 2020 (Act 2020–2015) set a specific target that by 1 January 2025 100% of plastic will be recycled (under Article L. 541-1 of the Environment Code). The law 13 July 1992 (Environmental Code) on the disposal of waste introduced the concept of "final waste", defined as any residue no longer able to be further treated under the current economic and technical conditions, either by recovery of the desirable fraction or by reducing its polluting or hazardous properties. Article L541-24 of the Environmental Code required that from 1 July 2002, only final waste may be landfilled [10]. In consideration to the increasing production of waste per capita [11], incineration is a very effective treatment as it can reduce the mass of waste on average by 70% and its volume by 90% [12].

2.1.2   Origin of IBA

IBA are the solid residues resulting from the combustion of household waste at the lower outlet of the furnace (Fig. 1). These residues, usually cooled with water, are rich in mineral fraction (silicon, aluminum, calcium, etc.) but also contain water (20–25%), 6–10% recoverable metal particles, and some non-valuable fraction [13].

In general, raw IBA is a heterogeneous material (Fig. 2) that consists of glass scrap, gravel-like particles, metal elements and water.

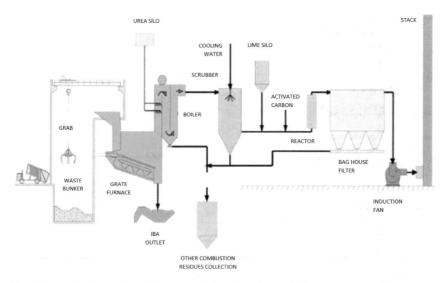

Fig. 1 Diagram of a non-hazardous waste incineration plant (grid furnace) [14, modified]

fraction 0/6 mm   fraction 6/12 mm   fraction 12/20 mm

Fig. 2 Different granular fractions of IBA [15]

2.1.3 Processing of IBA

Raw IBA (collected at the outlet of the furnace) usually needs to be treated before starting any recovery process. These treatments can be carried out either directly at the waste incineration plant (MWIP) or in a maturation and processing plant (MPP).

Treatment of IBA in MWIP

After an initial cooling phase, IBA undergoes manual or mechanical screening for removing oversized residues. Further to this operation, metals are removed using magnetic overhead separators. IBA is then stored on site for maturation, or, if necessary, transferred to a maturation and processing facility.

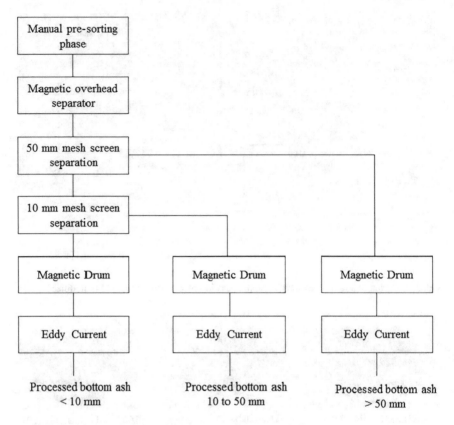

Fig. 3 Schematic representation of the dry separation process [16]

IBA are typically treated by a dry separation process in order to recover ferrous and non-ferrous fractions. The simplified flow-chart of this treatment process is shown in Fig. 3.

Treatment of IBA in Maturation and Processing Plant

The activities in a MPPs aim at completing the process of maturation of IBA before recovery. These installations are equipped with magnetic drums and eddy current separators to further remove ferrous and non-ferrous metals contained in IBA. Storage areas are dedicated to maturation, a process designed to reduce the polluting potential of heavy metals and other undesired chemicals. In most cases, IBA can mature between 2 and 4 months, and up to 12 months in certain cases [17]. The material is left exposed to air and rain so that unstable chemical species can evolve into secondary, more stable compounds, mainly thanks to hydration and carbonation reactions. Figure 4 shows the appearance of IBA after 4 months of maturation.

Fig. 4 IBA appearance after
4 months of maturation in
MPP

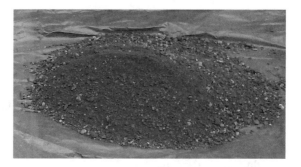

3 Factors Influencing the Composition of IBA

The physical, chemical and mineralogical properties of IBA at the outlet of MWIP
and/or MPP are influenced by several parameters, namely: the selective sorting
of household waste upstream, the choice of incinerator, the cooling mode, the
maturation process and the type of magnetic separators used.

3.1 Selective Collection of Household Waste

Selective collection makes it possible to recover materials for which recycling options
are available and economically viable.

A comparative study carried out on the average composition of household waste
provided insights on the major differences in the composition of waste to be incinerated [18]. Where separate collection was carried out, glass and paper/cardboard
fraction fell by 10 and 8% respectively, while the presence of other materials (e.g.
textiles) increased on average by 30%.

Studies carried out on several MWIPs confirmed the significant correlation
between the composition of IBA at the outlet of MWIP and the composition of
household waste upstream [19]. It has been observed, for example, a decrease in
the proportions of silicon in the composition of IBA where selective collection is
practiced, as a result of the reduction of glass scrap in household waste [19].

3.2 Choice of Incineration Technology

The way household waste is incinerated has a significant influence on the composition
and properties of IBA. Incineration is a complex technique to master because it
involves several essential parameters [18, 20]:

(i)   storage and composition of incoming waste;
(ii)   feeding to the furnace;
(iii) combustion temperature;
(iv)   combustion of waste, with oxygen supply;
(v)   process time in the furnace.

Currently, the most common waste incineration techniques are:

(i)   grate furnace technology;
(ii)   rotating furnace technology;
(iii) fluidised bed technology.

3.2.1   Grate Furnace Technology

This technique is the most used in household waste incineration plants. In the grate furnace plants (Fig. 5), the waste is introduced and combusted for a period of two to three hours at temperatures between 750 and 1,000 °C. A grid is used to allow the passage of air through the ignition layer. These furnaces are suitable in the case of incineration of medium to large capacity (from 3 to 6 t waste/h). The technologies differ according to the grid (fixed or mobile grid), to the mixture of waste and to the injection of air to achieve better combustion. These parameters have important effects in the production of ash and in the presence of organic matter [18].

3.2.2   Rotating Furnace Technology

The rotating furnace system consists of a cylinder slightly inclined to the horizontal. These furnaces are used mainly for industrial waste treatment, because melting of metals would otherwise deteriorate the grid. The waste is introduced into the furnace with a longitudinal aeration, allowing the optimisation of the mix waste/air, at a temperature of 1,200 °C. The extraction of IBA and ash is obtained by gravity, under the effect of the slope (Fig. 6). The introduction of air through the waste bed ensures the desired combustion quality. These furnaces are typically better suited for plants with medium waste volume capacities (i.e., about 3 t/h) and involve a shorter process time compared to grate furnace technology.

3.2.3   Fluidised Bed Technology

A fluidised bed furnace is a closed vertical system, cylindrical in shape, containing a bed of sand at a temperature between 750 and 800 °C. The principle of this technique is to perform the combustion of solid products in a bed of inert materials that is suspended thanks to the injection of hot air. As mentioned, this bed is often a mixture of sand to which a small fraction of waste (5%) is added. The solid mix is made fluid

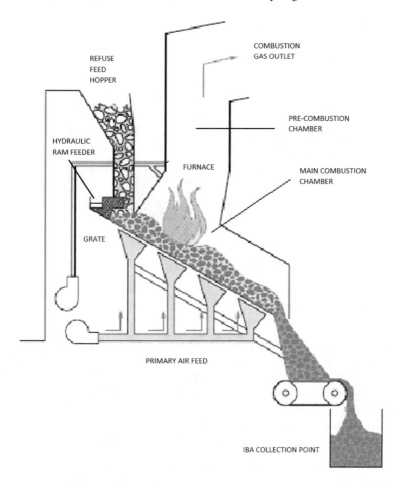

Fig. 5 Grate furnace diagram [21, modified]

by air injection (vertical, horizontal, at the base or in the walls of the oven). The fluidised bed technique was initially developed to burn coal, however, it has been adapted in recent years to the treatment of household waste [20]. There are three types of fluidised bed furnace technologies: dense fluidised bed, rotary fluidised bed, and circulating fluidised bed. These technologies differ in the way, position and velocity at which air is injected into the furnace. Fluidised bed technology provides a very efficient waste combustion and thus a low rate of unburnt material [20].

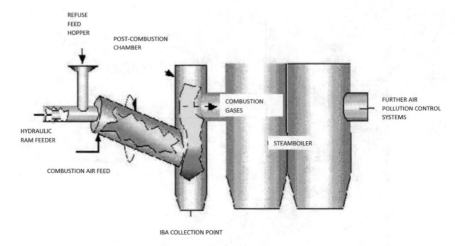

Fig. 6 Rotating furnace [18, modified]

3.3 Cooling Processes

At the exit of the combustion chamber, IBA are cooled either with water or air. The water cooling process is the most used in MWIP. Recent studies have shown that air cooling of IBA can also be advantageous, although this process generates significant quantities of dust that requires special treatment. Air cooling method is widely adopted in Japan, and can be found in some incineration plants in Europe, especially in Switzerland (e.g., KEZO incineration plant) [22].

The main benefit from the adoption of the air cooling method are [22]:

(i)   To increase the maximum recovery rate of ferrous and non-ferrous metals by 45 and 50% respectively compared to the water cooling process.
(ii) To limit the triggering of chemical reactions that are inevitable in the presence of water.
(iii) To significantly reduce the leachate concentrations in landfill centres.
(iv) To reduce the concentrations of Total Organic Carbon (TOC) in IBA.
(v)   To reduce the IBA transport cost (20% less due to lack of water).

Table 1 provides a comparison between the two cooling systems at the outlet of the incinerator.

3.4 Maturation Processes

Maturation is a process in the development of IBA that reduces the environmental pollution potential of the material. Thermodynamically unstable compounds evolve into more stable secondary compounds in the presence of carbon dioxide, but also

Table 1 Comparison of cooling systems [22]

Criteria	Water cooling	Air cooling
Cooling process	Fast	Slow
Chemical reactions with water	Yes	None
Loss of metals due to oxidation	Yes	None
Weight of IBA	15–20% increase	None
Dust emissions	Low	High
Adhesion of mineral matter to metals	High	Very low
Corrosion of metals	Moderate	None

oxygen or water. This operation can be carried out in a conventional way, i.e., IBA exposure to air for 1–12 months, during which the carbonation of lime by atmospheric carbon dioxide (atmospheric conditions) takes place, although total stabilisation is practically impossible to achieve [16].

The reaction can be improved by increasing the level of carbon dioxide in the environment where IBA are exposed (accelerated carbonation). This solution has higher cost implications because it will entail additional expenses (pump, construction of a carbonation chamber). Nevertheless, the exhaust gases coming out of the incinerator plant could be used as a source of carbon dioxide in order to mitigate the extra cost. This system has a double advantage because it will allow to treat the fumes by reducing the carbon dioxide emitted, while improving the quality of IBA in a relatively short time [16]. The work of Arickx et al. [23] shows that accelerated carbonation on IBA fine fraction (0–2 mm) significantly reduced the concentrations of leachate of certain compounds in few weeks, compared to natural maturation (during which the concentrations of undesired chemicals remained relatively high).

4 Characterisation of IBA

4.1 Physical Properties

IBA can be described as heterogeneous, dark grey particles, with the presence of ferrous, non-ferrous and unburnt materials (Fig. 4). The choice of magnetic separators (removal of ferrous and non-ferrous metals) has an impact on the physical characteristics of the resulting IBA. Table 2 summarises the physical properties of IBA from different incineration plants in Europe.

It can be observed that IBA shows properties similar to those of natural aggregates (sand and gravel), although absorption coefficients and fire loss values are typically higher.

Table 2 Physical properties of IBA sourced in several EU countries

Properties	France	Netherlands	Denmark	Italy	Spain
Grain size (mm)	(0/31.5) (0/20) [15]		Well graded (0/50) [3]		Well graded (0/40) [3]
Passing 0.075 mm (%)	4–12 [15]	0.6–3.2 [6]	8–9 [3]		16 [3]
Loss on ignition (LOI) (550 °C)			1.7–2.4 [3]	2.9–8.5 [24]	2–9 [3]
Specific gravity	2.5 [10]	2.5–2.8 [6]	2.7 [3]	2.4–2.5 [24]	2.5 [3] 2.2–2.5 [25]
Bulk density (t/m^3)	1.1–1.2 [10]	2.2–2.4 [6]		1.1–1.4 [24]	1.6–2.3 [25]
Los Angeles (%)	35–50 [15]		45–47 [3]		40–42 [3]
Water absorption fraction >4–5 mm (%)	10 [10]	4.8 [6]	5.6–7.7 [3]		3.5–6.2 [3] 2.7 [25]
Water absorption fraction <4–5 mm (%)	15 [10]	5.6–7 [6]	12.3–16.1 [3]		15.8 [25]
Maximum dry density after compaction (t/m^3)	1.4–1.8 [15]		1.8 [3]		1.6 [3]
Optimum moisture content at compaction (%)	15–22 [15]		15 [3]		16 [3]

The high LOI values can be explained by the presence of organic matter that was not consumed completely during incineration. The links of this parameter with the choice of furnace, the combustion temperature and the duration of incineration are obvious. However, recently marketed furnaces allow to achieve average LOI equal to 2% [3], which is below the threshold recommended by French and Dutch regulations for the IBA to be recoverable (5%) [6, 17].

IBA is a porous material containing fine particles, thus resulting in much higher water absorption coefficients than natural sands and gravel. This property should be considered when designing the formulation of cementitious materials, in order to estimate a correct water dosage.

IBA physical properties make it an alternative resource to natural granular materials in construction work, as long as the proportion of TOC remains below the regulation thresholds (i.e., <5%).

4.2   Chemical Properties

The chemical properties of IBA are influenced mainly by the composition of household waste entering the incineration plant [19].

4.2.1   Elemental Composition

The elemental composition of IBA is an important parameter for understanding the chemical behaviour of the material. Table 3 shows the average elemental composition of IBA, highlighting the main elements (>10,000 mg/kg), the minor elements (1,000 mg/kg to 10,000 mg/kg) and the trace elements (<1,000 mg/kg).

Although the IBA chemical composition varies across different incineration plants, the main elements found are typically silicon, calcium, aluminum, iron, magnesium, sodium and potassium [3, 19]. Table 4 provides the relative proportions of main chemical constituents reported for IBA from several incineration plants in Europe.

Table 3 Average elemental composition of IBA [10, 18]

Main elements	Minor elements	Trace elements	
O: oxygen	Mg: magnesium	Sn: tin	Co: cobalt
Si: Silicon	Ti: titanium	Sb: antimony	Ce: cerium
Fe: iron	Cl: chlorine	V: vanadium	Ag: silver
Ca: calcium	Mn: manganese	Mo: molybdenum	Hg: mercury
Al: aluminium	Ba: barium	As: arsenic	B: boron
Na: sodium	Zn: zinc	Se: Selenium	Br: bromine
K: potassium	Cu: copper	Sr: strontium	F: fluorine
C: carbon	Pb: lead	Ni: nickel	I: iodine
	Cr: chrome		

Table 4 Main chemical constituents in IBA [3]

Constituents	Mass proportions (%)
Silicon	16.8–27.4
Calcium	5.12–10.3
Iron	2.11–11.5
Magnesium	0.19–1.18
Potassium	0.72–1.16
Aluminium	3.44–6.48
Sodium	2.02–4.80

4.2.2 Mineralogical Composition

The mineralogical analysis of the incineration clinker provides further elements for the prediction of its behaviour, both from the point of view of the evolution of the solubility of the minerals that compose it as well as from the point of view of the expected mechanical properties. Mineralogical studies on IBA showed that it generally consists of three types of minerals:

(i)   Minerals present in waste and not modified by incineration.
(ii) Minerals formed during incineration.
(iii) Minerals that form right out of the furnace, either during cooling and/or during the maturation phase.

Table 5 shows the main mineral species identified by literature studies carried out on non-hazardous waste IBA.

In most cases, the predominant mineral phases found in non-hazardous waste IBA are quartz (SiO_2), lime (CaO), corundum (Al_2O_3), hematite (Fe_2O_3), and sodium oxide (Na_2O) [25–28]. The presence of this latter is usually related to the large quantities of glass scrap in IBA [28].

5   Regulatory and Legislative Framework for the Management of IBA

The polluting potential of IBA is a key factor in deciding on its use as well as its storage in appropriate sites. Regulations on the environmental characterisation of IBA have evolved in several countries with the aim of limiting their possible negative impact on the environment. Each country has defined a regulatory and legislative framework on the management of IBA with a view to identify recoverable and non-recoverable materials.

Currently, the recycling of IBA in the construction sector remains a priority for all European countries in order to limit the use of non-renewable natural resources. Countries producing IBA-derived granular materials have established in-country standards and regulations governing the management of IBA. However, the non-uniformity of these standards is an obstacle to the achievement of IBA full valorisation potential. The material can be considered recoverable in one region and non-recoverable in another, because of non-uniform threshold values on the leaching tests. Although some European countries are trying to standardise the legislation on the management of IBA, there are still additional efforts to be made in order to stimulate a real dynamic on the valorisation of IBA as building materials.

A word of caution must be mentioned here: laws and regulations are by their own nature relevant and valid only over specific times. Readers are encouraged to check the current regulations in place. The next sessions are provided for giving a wide background over the regulatory instruments and the methodologies for assessing the safe use of IBA within national environmental laws, the details of which fall outside

Table 5 Key mineral species identified in IBA [10, 18]

Family	Mineral species	Chemical formula	Formation
Silicates	Alite	Ca_3SiO_5	Within the furnace
	Clinopyroxène	$Ca(Fe,Mg,Al)(Si,Al)_2O_6$	
	Géhlénite/akermanite	$(Ca,Na)_2(Al,Mg)(Si,Al)_2O_7$	
	Larnite	Ca_2SiO_4	
	Mullite	Al_2SiO_5	
	Olivine Ca	$(Fe,Mg,Ca) Si_2O_4$	
	Plagioclase	$(Ca,Na)(Si,Al)_4O_8$	
	Pseudowollastonite	$CaSiO_3$	
	Quartz	SiO_2	
	Talc	$Mg_3(OH)_2Si_4O_{10}$	
Oxides	Corundum	Al_2O_3	Within the furnace
	Hematite	Fe_2O_3	
	Hercynite	$FeAl_2O_4$	
	Magnetite	Fe_3O_4	
	Rutile	TiO_2	
	Spinel	$MgAl_2O_4$	
	Wüstite	FeO_2	
	Zincite	ZnO	
Hydroxides	Portlandite	$Ca(OH)_2$	Cooling
	Boehmite-Bayerite-Gibbsite	$Al(OH)_3$	
	Goethite	$Fe(OH)O$	
Carbonates	Calcite	$CaCO_3$	Maturation
	Siderite	$FeCO_3$	
Chlorides	Halite	$NaCl$	Within the furnace
	Nantokite	$CuCl$	
	Sylvite	KCl	
Sulfates	Anhydrite	$CaSO_4$	Within the furnace
	Gypsum	$Ca(SO_4)2H_2O$	
Phosphates	Apatite	$Ca_5(PO_4)_3(OH,F,Cl)$	
Sulphures	Pyrrhotite	FeS	Within the furnace
Metals	Aluminum metal	Al	Residual fragments
	Copper metal	Cu	Residual fragments
	Iron metal	Fe	
	Graphite	C	
Other	Neoformed lenses	(Si, Na, Ca, Al, Fe)	

this contribution. For deeper discussion, readers are referred to detailed reviews available in the literature (see e.g. [1]).

5.1 Legislation on the Management of IBA in the European Union

The EU framework for the incineration of waste can be found in Directive 2010/75/EU on industrial emissions and in its implementation by in-country national regulations [1]. Any waste incineration plant operated in a member state needs to adhere to the minimum requirements set out in the above mentioned directive. It is mandatory to adopt operating conditions that can ensure that produced IBA has either TOC contents lower than 3 wt% or LOI less than 5 wt%. The directive also requires to minimise the amount and potential hazard of combustion residues, and to have recycling strategies in place, after having assessed IBA physical and chemical properties, as well as their polluting potential, including the determination of the total soluble fraction and soluble fraction of heavy metals [1].

IBA are classified as waste according to Directive 2001/118/EC 2001, and thus subjected to the Commission Decision 2014/955/EU. The document includes the List of Waste (LoW) that defines waste types and hazardous classification (hazardous—marked with '*'—or non-hazardous). The codes that are relevant for IBA are:

- 19 01 02—ferrous materials removed from bottom ash
- 19 01 11*—bottom ash and slag containing hazardous substances
- 19 01 12—bottom ash and slag other than those mentioned in 19 01 11.

Interestingly, if it is possible to prove that IBA does not contain hazardous substances, then it can be classified as non-hazardous waste. Therefore, IBA has to be tested against the 15 hazardous properties (HPs) laid down in Commission Regulation (EU) No 1357/2014 (see Table 6), and the presence of persistent organic pollutants (POPs) and their concentration have to be assessed and compared to the limits specified in Regulation (EC) No 850/2004.

Member States may characterise a waste as hazardous by HP 15 based on other applicable criteria, such as an assessment of the leachate. In-country experimental protocols for carrying out leaching tests have been developed by each Member State.

The European Union requires Member States to carry out leaching tests according to the standards EN 12457/1 2002, EN 12457/2 2002, EN 12457/3 2002, and EN 12457/4 2002, with a liquid-to-solid (L/S) ratio ranging from 2 to 10 l/kg. prEN 14405 is to be followed in testing percolation with L/S = 0.1 l/kg [6].

The leaching tests are broadly divided into two groups, i.e., batch and column leaching tests. Table 7 summarises the experimental procedures for carrying out leaching tests in some European countries.

There is still no legislation on the recycling of IBA for the incineration of non-hazardous waste within the Community of Member States in the European zone [3,

Table 6 Hazard properties from Commission Regulation No 1357/2014

Hazard property (HP)	Description
HP1: Explosive	Waste which is capable by chemical reaction of producing gas at such a temperature and pressure and at such a speed as to cause damage to the surroundings. Pyrotechnic waste, explosive organic peroxide waste and explosive self-reactive waste is included
HP2: Oxidising	Waste which may, generally by providing oxygen, cause or contribute to the combustion of other materials
HP3: Flammable	Waste which can ignite under the conditions laid out in the regulation
HP4: Irritant—skin irritation and eye damage	Waste which on application can cause skin irritation or damage to the eye
HP5: Specific Target Organ Toxicity (STOT)/Aspiration Toxicity	Waste which can cause specific target organ toxicity either from a single or repeated exposure, or which cause acute toxic effects following aspiration
HP6: Acute Toxicity	waste which can cause acute toxic effects following oral or dermal administration, or inhalation exposure
HP7: Carcinogenic	Waste which induces cancer or increases its incidence
HP8: Corrosive	Waste which on application can cause skin corrosion
HP9: Infectious	Waste containing viable micro-organisms or their toxins which are known or reliably believed to cause disease in man or other living organisms
HP10: Toxic for reproduction	Waste which has adverse effects on sexual function and fertility in adult males and females, as well as developmental toxicity in the offspring
HP11: Mutagenic	Waste which may cause a mutation, that is a permanent change in the amount or structure of the genetic material in a cell
HP12: Release of an acute toxic gas	Waste which releases acute toxic gases (Acute Tox. 1, 2 or 3) in contact with water or an acid
HP13: Sensitising	Waste which contains one or more substances known to cause sensitising effects to the skin or the respiratory organs
HP14: Ecotoxic	Waste which presents or may present immediate or delayed risks for one or more sectors of the environment

HP15: Waste capable of exhibiting a hazardous property listed above not directly displayed by the original waste

Table 7 Experimental standards for leaching tests in selected countries [3, 6]

Country	Test standards	Protocol	L/S	Other conditions
Austria	EN 12457-4	Rotation	10	–
Republic Czech (2002)	Properties of waste leaching according to The Order of the Ministry of the Environment No. 383/2001	Rotation	10	Rotation around an axis with a speed of 5–10 rpm for 24 h at 15–20 °C
Denmark	CEN prEN 12457-3	Rotation	2	Rotation around an axis, with a speed of 5–10 rpm for 6 h at 22 °C (15–25 °C)
Finland	prCEN/TS14405 (baseline characterization) CEN EN 12457-3	Percolation Rotation	10	Featured in the National Guide for Pre-Landfill Waste Assessment
France	NF EN 12457-2	Rotation	10	Rotation around an axis, with a speed of 5–10 rpm for 24 h
Germany	DIN 38414-S4 DIN EN 12457-4, only for the production of leachates	Rotation	10	L/S 10, 24 h
Italy	UNI10802 after Decree 5 April 2006, n. 186 EN12457-2	Sequential rotation	10	L/S 10, 24 h
Netherlands	NEN 7384 (or NEN 7373)	Columns	10 (0.1–10)	
Spain	DIN 38414-S4 (Catalonia)	Rotation	10	24 h
Sweden	No official test but EN 12457 is usually applied	Rotation	2 and 10	Test series L/S = 2+ L/S = 8
Switzerland	TVA analysis of eluates	Rotation	10	Test series with L/S = 10; pH constant 5–6

6]. Hereafter, some of the regulations for IBA management in selected EU countries are discussed.

5.1.1 France

Legislation Framework and Required Chemical Quality

France is one of the largest producers of non-hazardous waste IBA in Europe (on average 3 million tonnes in 2010, or 20–25% of the tonnage incinerated). A study carried out in 2012 by the national network AMORCE allowed to identify 127 household waste incineration plants and 71 maturation and production platforms in operation, disseminated across the nation [29]. Before 1st July 2012, the use of IBA from incineration of non-hazardous waste was allowed in road construction if compliant with the requirements set out in the government document ("circular") dated 9th May 1994 [17]. This circular defined three categories of IBA:

(i) IBA with a low leachable fraction, known as "V" type (recoverable), which could be used in road technology for the subbase of road structures or car parks and embankments up to three metres high.
(ii) Intermediate IBA, known as "M" type (which must be processed with maturation before being considered for reuse).
(iii) IBA with a high leachable fraction, called category "S" (to be disposed of in a non-hazardous waste storage facility) [17].

Following a wide consultation exercise involving a large number of stakeholders, aimed at providing guarantees of environmental acceptability in response to the objections of environmental associations, a new regulation concerning IBA has been put in place.

This regulation is based on:

(i) The Decree No. 2011-767 of 28 June 2011 and the Order of 25 July 2011 that specify the conditions under which IBA are considered non-recoverable and must therefore be sent to a non-hazardous waste storage facility without paying the general tax on polluting activities.
(ii) The Decree dated 18 November 2011 (replacing the 1994 circular from 1 July 2012) on the recycling of IBA, which set out the conditions under which IBA can be recovered in road construction and provided an application guide [30].

These documents are based on a SETRA guide (Service d'Etudes sur les Transports, les Routes et leurs Aménagements) published in March 2011 on "The environmental acceptability of alternative materials in road construction" [30].

The decree 18/11/2011 introduced additional constraints such as the restriction of possible uses, the addition of new parameters to be analysed and the tightening of the thresholds to be respected. According to this decree, the authorised applications are limited within surfaced road works of types 1 and 2 defined below [4].

(i) Utilisation in works "Type 1"

3IBA can be used for the construction of road subbases that do not exceed three metres high, or on the shoulder of paved road structures.

Table 8 Limit values for intrinsic levels of pollutants [4]

Parameter	Threshold
TOC (total organic carbon)	30 g/kg dry matter
BTEX (benzene, toluene, ethylbenzene and xylenes)	6 mg/kg dry matter
PCBs (polychlorinated biphenyls, congeners: 28, 52, 101, 118, 138, 153 and 180)	1 mg/kg dry matter
Hydrocarbons (C10 to C40)	500 mg/kg dry matter
PAHs (polycyclic aromatic hydrocarbons)	50 mg/kg dry matter
Dioxins and furans	10 ng I-TEQ (OMS2005)/kg dry matter

Calculation of the I-TEQ with a concentration equal to zero for any congener of concentration below the limit of quantification (LQ)

(ii) Utilisation in works "Type 2"

Use of IBA in Type 2 works includes the construction of embankments not exceeding six metres high, or its use in the construction of road shoulder, as long as the road is surfaced. Uses of more than three meters and not more than six meters in height in underlay of pavement or shoulder, also fall under road uses of types 2 [4].

Tables 8 and 9 specify the threshold values to be respected in order to valorise IBA in road construction.

Economic Considerations and Barriers to IBA Reutilisation

There is not a harmonised approach for the trading/management of IBA. At least three different approaches are adopted by MWIP, MPP, or other bodies in charge of IBA management:

(i)   Give IBA to main contractor companies free of charge;
(ii) Sell IBA at an average rate of EUR 2 excluding VAT/tonne;
(iii) Pay contractor companies to dispose of IBA (on average 10.15 euros excluding VAT/tonne).

The percentage of IBA recovered annually is estimated at an average of 60%. However, this figure does not consider potentially recoverable IBA that does not find a suitable value chain for its reuse. Estimates for 2010 suggest that 208,426 tonnes of fit-for-reuse IBA were not recovered (20%), and in 2011 this figure rose to about 27% (or 275,157 tonnes).

As discussed, the main sector for IBA utilisation is road constructions. A very small proportion of IBA are used in the cement sector (0.2%), however, there is no statistics on the use of IBA in the manufacture of mortars/concretes.

IBA that are not reused (either for poor chemical quality or for the lack of a suitable value chain) are landfilled. Before the decree 28 June 2011 and the decree 25 July 2011, all landfilled IBA were subjected to the general tax on polluting

Table 9 Limit values for leaching behaviour [4]

Parameter	Limit value to be respected expressed in mg/kg dry matter	
Road type	V1	V2
Arsenic	0.6	0.6
Barium	56	28
Cadmium	0.05	0.05
Chrome total	2	1
Copper	50	50
Mercury	0.01	0.01
Molybdenum	5.6	2.8
Nickel	0.5	0.5
Lead	1.6	1
Antimony	0.7	0.6
Selenium	0.1	0.1
Zinc	50	50
Fluorides	60	30
Chlorides*	10,000	5,000
Sulfates*	10,000	5,000
Soluble fraction*	20,000	10,000

* For chlorides, sulfates and the soluble fraction, it is appropriate to comply with either the values associated with chlorides and sulphates or the values associated with the soluble fraction

activity (GTPA). Since the publication of these regulations, non-recoverable IBA are currently exempt from this tax. The cost of landfilling (without transport or GTPA) is very variable (between 27 and 150 euros excluding VAT/tonne), on average 66 euros excluding VAT/tonne. The GTPA due for recoverable IBA varies according to the parameters set out in the Customs notice of 28 March 2012, ranging from 11 to 20 euros per tonne.

When summing up the cost of landfill, transport and GTPA, the cost of landfilling IBA exceeds 90 EUR/tonnes. Compared to the average price of 10.15 euros/tonne that MWIP pays to the contractor companies, the cost of landfilling is significant.

MWIP operators and local authorities face many difficulties for the valorisation of IBA. The main barriers identified are:

(i) The problem of the image of IBA as "waste" and not "product" (notion of social acceptability).
(ii) Strong competition from natural gravel or slag.
(iii) Regulations that limit possible applications and impose increasingly stringent quality criteria.
(iv) The opposition from environmental associations to the use of IBA.

(v)   The environmental restrictions on use, in particular those linked with the presence of groundwater.

Awareness campaigns for public and private stakeholders in public works have been carried out in order to mitigate some of the above-mentioned barriers.

5.1.2   The Netherlands

The waste management in the Netherlands is governed by the Environmental Management Act, and the current legislation covering the use of IBA is the document dated 13/12/2007, no. DJZ2007124397 "Regeling van 13 December 2007, no. DJZ2007124397, houdende regels voor de uitvoering van de kwaliteit van de bodem (Regeling bodemkwaliteit)".

Only waste that cannot be recycled or incinerated is directed to storage centres [6].

In order to avoid environmental problems caused by waste (contamination of surface water and groundwater by leaching), several environmental requirements have been developed in the Netherlands. The first regulation on the quality of materials was first introduced by the Building Materials Decree (BMD) of 1995–1999, which was then replaced in 2007 by the Soil Quality Decree (SQD). These two regulations refer to the principles of the Soil Protection Law (WBB 1986), the Waste Disposal Act (1993/31/EU 1993) and the Surface Water Protection Law (WVO 1969) [6]. The SQD imposes certain rules (leachate limit values, physico-chemical composition of materials) for the use building materials, soil-type materials and dredged materials in order to limit their effect on soil and surface water. Table 10 shows the leaching threshold values according to BMD and SQD regulations.

When materials satisfy the limit values of category N1, no precautions are required to prevent contact with water. When the leaching values exceed the threshold of category N1, but comply with those of category N2, the waste is recycled as a material of category N2, provided that measures are taken for avoiding contact with groundwater.

For monoliths (i.e. moulded materials), the SQD recommends the application of the NEN 7375 standard (diffusion test), while for granular materials it recommends the NEN 7373 (NEN 7373 2004) or NEN 7383 standard (column test) [6].

The column test is recommended for applications related to granular materials. The material, whose grains are preliminary crushed to a size less than 4 mm, is poured in a mould with internal diameter equal to 5 cm and length equal to 20 cm. Demineralised water enters the column upwards with an flow rate of 2 cm/h. The eluates are collected in different fractions according to the desired L/S ratio. According to the SQD, the maximum value between L/S must be equal to 10. Seven fractions are collected, corresponding to the ratio L/S 0.1 (K1), L/S 0.2 (K2), L/S 0.5 (K3), L/S 1 (K4), L/S 2 (K5), L/S 5 (K6), L/S 10 (K7). The duration of this test is about 21 days [6].

Table 10 Leachate limit values according to BMD and SQD [6]

BMD (1998)			SQD (2007)		
Substance	N1	N2	Monolithic materials (mg/m^2)	Granular materials (mg/kg)	IBC (mg/kg)
Cu	0.72	3.3	98	0.9	10
Mo	0.28	0.84	144	1	15
Sb	0.045	0.41	8.7	0.16	0.7
Br	4.4	44	670	20	34
As	0.88	7	260	0.9	2
Three	5.50	55	1,500	22	100
Cd	0.032	0.061	3.8	0.04	0.06
Cl	600	8,790	110,000	616	8,800
Cr	1.3	12	120	0.63	7
Co	0.42	2.3	60	0.54	2.4
F	13	96	2,500	55	1,500
Hg	0.018	0.075	1.4	0.02	0.08
Pb	1.9	8.1	400	2.3	8.3
Ni	1.1	3.5	81	0.44	2.1
Se	0.044	0.094	4.8	0.15	3
SO$_4$	750	22,000	165,000	1,730	20,000
Sn	0.27	2.3	50	0.4	2.3
In	1.6	32	320	1.8	20
Zn	3.8	14	800	4.5	14

IBC: Applications requiring impervious layers and protection measures for avoiding groundwater contamination

The diffusion test simulates the leaching of inorganic compounds from monolithic and moulded materials under aerobic conditions as a function of time over a period of 64 days. The samples are immersed in a liquid (demineralised, neutral pH of the water) and the leachate concentrations are measured over the time. Based on the results of the diffusion test, the quantities leached per unit mass are measured for each component analysed [6].

Where the concentrations of the eluates are higher than the limit values for IBC applications (Table 10), the waste must be disposed of in appropriate disposal centres. Hazardous waste must be sent to landfills of category C1 to C3. The allocation in a category depends on the leachate concentrations of the different compounds present in the waste, according to NEN 7343 for column test with an L/S ratio equal to 1 l/kg. The limit values are shown in Table 11. If the concentrations of the eluates are below the limit values, then the waste is classified in category C3. If at least one element has a concentration equal to or greater than the corresponding limit values,

Table 11 Landfill limit values in the Netherlands [6]

Substance	As	Ba	Cd	Cr	Cu	Hg	Mo	Ni	Pb	Sb	Se	Zn	Cl^-	$F-$	SO_4
L/S = 1 mg/kg	9	60	0.2	30	10	0.1	3		10	25	0.8	0.3	40	50,000	280	80,000

the waste must be classified in category C2. Category C1 is a very specific category for hazardous waste with critical properties, e.g., containing mercury [6].

5.1.3 Germany

The principles on waste management in Germany are governed by the law "Kreislaufwirtschaftsund Abfallgesetz" which entered into force in 1996. The role of this law was to ensure the recycling of materials and consequently to preserve natural resources. It also provides guidelines for the rational disposal of waste in the environment. The Act clearly defines the terms "deposit" and "recycling". The principle is to avoid any landfill [3]. In addition, a previous law called "BundesImmissionschutz-Gesetz" which entered into force in 1990, recommended in its various regulations that the use of waste be allowed, if all its impacts on the environment can be assessed and evaluated [3].

In Germany, waste is classified into 7 categories (Z0, Z1.1, Z1.2, Z2, Z3, Z4 and Z5) according to its physico-chemical composition and leachate concentrations [6]. The leaching test is carried out according to the DIN 38,414-S4 standard, involving sample stirring for 24 h and an L/S ratio equal to 10 l/kg. In addition to the various categories mentioned above, a special category for IBA was created, since the leachate concentrations did not comply with the limit values of category Z2 [6, 31]. When recycling is not possible, IBA are landfilled as non-hazardous waste [31]. Table 12 summarises the threshold limit values for the recycling of IBA in road constructions.

Table 12 Threshold values in leaching test (DIN 38414-S4, L/S = 10 l/kg) for the reuse of IBA in Germany [31]

Categories	Limit value (mg/kg)
Cd	0.05
Cr	2
Cu	3
Hg	0.01
Ni	0.4
Pb	0.5
Zn	3
Cl^-	2,500
SO_2^{4-}	6,000
pH	7–13

5.1.4   Belgium

The regulatory and legislative framework for the management of IBA in Belgium is produced by the two regions Flanders and Wallonia.

In the Flanders region, the current legislation for the use of IBA is the VLAREMA-2012. IBA are not considered hazardous waste and therefore can be recycled in various construction applications, provided that the leaching concentrations of inorganic compounds are below the provided limit values defined in the Flemish standard for granular materials [31]. The leaching test refers to the NEN 7343 standard (column test).

When leachate concentrations comply with the limit values, IBA can be recycled as aggregates in monolithic/bound applications. If leachate concentrations are above the threshold limits, IBA may be recycled in monolithic applications provided that the diffusion test, in accordance with NEN 7345 is carried out on monolithic samples and that the obtained concentrations comply with the immersion limit values for soils [31]. IBA complying with the leaching limit values can also be used as granular materials in foundations (fraction 6–50 mm), replacing natural aggregates in brick making (2–6 mm) and as a foundation underlay (fraction 10–40 mm) in certain specific applications. The 0–0.1 mm fraction is directed to landfills [31].

Landfilling remains the only choice when recycling is no longer an option. Fly ash is considered hazardous waste and must be landfilled in Category I landfill centres (waste consisting mainly of heavy metals). Where leaching concentrations, in accordance with DIN 38414-S4, are non-compliant with limit values for the deposit of hazardous waste or non-hazardous waste, the waste must be stabilised with specific additives such as cement before landfilling [31]. Unlike fly ash, IBA are sent to Category II landfills [31].

In Wallonia, the use of IBA is regulated by a governmental law of 14/06/2001 "Arrêté du Gouvernement wallon, 14/06/2001 (Wallonia)". Incineration residues can be classified as hazardous or non-hazardous waste according to the total heavy metal content. IBA is considered non-hazardous waste while fly ash is classified as potentially hazardous. As a result, it is mandatory to recycle IBA, typically as granular materials in foundations and concretes [31]. However, leaching tests must first be carried out on IBA to assess their pollutant potential for leaching, in accordance with NEN 7343 (column test).

Wallonia has established three main landfill categories: Class 1, which receives hazardous waste; Class 2, for non-hazardous industrial waste stocking and municipal waste; and Class 3, for inert waste. However, there are also special categories of landfills, i.e., Class 4 for dredged sludge and Class 5 for exclusive use by waste producers [31].

Table 13 summarises the limit values for leached elements to be respected for the reuse of IBA.

Table 13 Limit values of leached elements for IBA recycling (L/S = 10 l/kg) [31]

Chemical species	Leaching (mg/kg)	
	Flanders	Wallonia
Al	–	2,000
As	0.8	0.8
Cd	0.03	0.03
Co	–	0.25
Cr	0.5	0.5
Cr^{VI}	–	0.05
Cu	0.5	5
Hg	0.02	0.02
K	–	1,700
Mo	–	1.8
Ni	0.75	1.8
Pb	1.3	2.2
Sb	–	0.3
Ti	–	2.4
Zn	2.8	4.0
Cl$^-$	–	6,000
F$^-$	–	20
SO2^{4-}	–	4,000

5.1.5   Denmark

The recovery of IBA has been regulated in Denmark since 1983. The law No. 655 of 27 June 2000 on "The use of residues and soil in construction works and buildings" entered into force on 1 January 2001, defining the procedure for the assessment of upgraded materials impact on the environment. The current legislation refers to the Order N.1672 (2016). The law recommends that the opportunity of recycling residues must be evaluated on the basis of the polluting potential (leaching test) rather than on the pollution class [3]. Residues may be used without permission if the intended user notifies the administrative authorities with a minimum notice of four weeks. The administrative authorities may, in turn, oppose the project and require that a more in-depth environmental study is carried out. In view of this status, inorganic residues and soils are classified into three categories with different applications. IBA are not listed in category 1 (unpolluted soils) because of their potentially high content of inorganic pollutant [3].

The leaching test is carried out in accordance with the CEN prEN 12457-3 standard of the rotary test with an L/S ratio equal to 2 l/kg. The concentrations of the eluates shall not exceed the limit values for the different categories mentioned in Table 14.

Table 15 shows the different applications of IBA in buildings and construction works in Danish regulations.

Table 14 Leaching limit values in Denmark [3]

Chemical	Category 1 + 2		Category 3	
	mg/l	mg/kg	mg/l	mg/kg
Chlorides (Cl)	150	300	3,000	6,000
Sulfates (SO$_4$)	250	500	4,000	8,000
Sodium (Na)	100	200	1,500	3,000
Arsenic (As)	0.008	0.016	0.05	0.1
Barium (Ba)	0.3	0.6	4	8
Lead (Pb)	0.01	0.02	0.1	0.2
Cadmium (Cd)	0.002	0.004	0.04	0.08
Chrome (Cr) total	0.01	0.02	0.5	1
Copper (Cu)	0.045	0.09	2	4
Mercury (Hg)	0.0001	0.0002	0.001	0.002
Manganese (Mn)	0.15	0.30	1	2
Nickel (Ni)	0.01	0.02	0.07	0.14
Zinc (Zn)	0.1	0.2	1.5	3

Table 15 Applications for use of IBA in construction [3]

Buildings and construction works included in the new statute	Applications for Category 2	Applications for Category 3
Applications regulated by the new statute		
Routes	Yes	Yes
Sidewalks	Yes	Yes
Floors and foundations	Yes	Yes
Parking	Yes	No
Embankment	Yes	No
Applications regulated by the Environmental Protection Act		
Dikes Dams Embankments Landscaping Maritime constructions Other land applications	These applications must be approved according to the Danish Environmental Protection Act	

The leaching limit values of category 2 are very strict and the implementation of the statute is accompanied by an obligation for the managers of incineration plants to find ways and means to improve the properties of IBA before any use. On the other hand, since the entry into force of the 2001 Law, IBA have not been able to meet the criteria of category 2 (Table 16); hence their exclusive use for applications of category 3 [3].

Table 16 Leaching limit values in Catalonia [3]

Chemical species	mg/l	mg/kg
As	–	–
Cd	0.005	0.05
Cr total	0.2	2
With	0.3	3
Pb	0.05	0.5
Zn	0.3	3

Approximately 50% of applications involving the use of IBA are regulated by the Danish Environmental Protection Act. For example, marine applications (port construction and dock protection) are not included in the 2001 Act but rather governed by the Environmental Protection Act, simply because of their non-compliance with category 2 requirements [3].

The general conditions for the use of inorganic residues (IBA) in construction work and buildings can be summarised by the following points [3]:

(i)   All applications using these residues must be waterproof (waterproofness is mandatory for category 3).
(ii)   The minimum distance from a drinking water borehole is 30 m.
(iii) Residues shall be placed above the highest level of the water table.
(iv)   The maximum leaching average thickness of the residue layer shall be 1 m.

In general, major construction works must seek environmental approval in accordance with the Environmental Protection Act, prior to any operation. In this case, the applicant must imperatively include the use of the residues while ensuring that the environmental estimate of all its activities will be evaluated in accordance with the said law [3].

From an overall point of view, about 70–90% of the annual production of IBA in Denmark is used in road constructions in base underlays or pavement-shaping materials [32].

5.1.6   Italy

In 2020, there were 37 MWIP in Italy. Incineration plants for urban waste produced 167,310 t of IBA, fly ash and hazardous waste (EER 190111*, 190113* and 190115*) and 1,076,515 t of IBA and non-hazardous slag (EER 190112, 190114 and 190116) consisting of 74% non-hazardous IBA, 14.8% hazardous waste from fume abatement processes and 11.2% fly ash, IBA and hazardous slag (data 2019). In 2019, non-hazardous IBA and slag were mainly sent to recycling/recovery (about 874,000 t) while about 87,000 t were landfilled or sent to other waste treatment. 58,000 t of hazardous IBA and slag have been sent for recycling/recovery. About 20,000 t were shipped in Germany, out of which 90% were recovery and 10% were landfilled.

Waste management is regulated by Decree No. 152 of 3 April 2006 (modified by the Decree No. 46 of 4 March 2014, implementing the Directive 2010/75/EU), which revised and superseded the Decree No. 22 of 5 February 1997, known as the "Ronchi decree". The law implementing the European Union's waste directive (2000/76/CE) is the Decree no. 133 11 May 2005. IBA that do not contain hazardous substances (in accordance with leaching tests duly carried out according to the standards in force) can be used either as raw materials in the production of cement, or as aggregates in the production of concrete, landscaping, and also in embankment construction [3].

The leaching test for recycling IBA were carried out in accordance with the decree of 5 February 1998, now superseded by the Decree 5 April 2006, n. 186 requiring the test to be carried out according to the standard UNI10802 (UNI EN 12457-2). The pollutant concentrations of the eluate are analysed and compared to the permissible limit values given by the criteria for the use and landfill of IBA [3].

According to Italian Ministerial Decree No. 201 of 3 August 2005, IBA classified as hazardous waste must be directed to landfill centres for hazardous waste, or for non-hazardous waste after adequate treatment to make them inert (vitrification, cement immobilisation or similar). Waste must comply with the leaching test requirements of European Standard 12457/4. The test consists of immersing a representative sample in demineralized water for 24 h with an L/S ratio equal to 10 [3].

With the entry into force of Legislative Decree 116/2020, implementing Directive 2018/851/EU, the fraction of metals (ferrous and non-ferrous) separated following the incineration of municipal waste can be taken into account in the calculations for achieving reuse and recycling targets, provided that the recovered metals meet the quality criteria established by the EU Decision 2019/2014.

A crucial issue concerning the correct recycling/reuse of IBA concerns their correct classification in the light of the changes made with EU Regulation 997/2017. The classification of IBA is currently carried out on the basis of EU Regulation 1357/2014 (as well as Decision 2014/955/EU with reference to EC Regulation 850/2014 in regards to waste containing dibenzo-p-dioxins and polychlorinated dibenzofurans) which defines the criteria for the assessment of the hazardous characteristics of waste from HP1 to HP15.

5.1.7   Spain

About half of the 12 waste incineration plants in Spain (in 2020) are located in the Catalan region. The management on the use of IBA is regulated by the standards of the Government of Catalonia since 1996 [1, 3]. However, most IBA are landfilled. Several studies have been undertaken to evaluate the possibility of reusing IBA, mainly in road construction. The leaching test is carried out according to DIN 38414-S4, with an L/S ratio equal to 10. The limit values set by the Catalan standard are shown in Table 16.

5.1.8   Sweden

In most Swedish regions, the abundance of natural aggregates does not favour the use of non-hazardous waste IBA in construction work. However, the government has introduced a tax of 40 euros per tonne of materials delivered in dump centres in a move to limits the landfilling of IBA [3].

In 2020, there were still no national regulations on the use of IBA as materials in construction, only the guidance that was provided in 2010. It is up to the local and regional authorities to decide whether or not the use of IBA can be accepted in construction works. Most often, constructions incorporating IBA are not allowed due to environmental concerns. However, in some cases, IBA can be used under the condition that they do not cause negative impacts on the environment. Thus:

(i)   Construction sites using IBA must not be in the proximity of drinking water points.
(ii)   Soils must have low hydraulic conductivity (clay).
(iii) The works must be watertight (covered).

In addition, water monitoring stations need to be setup to monitor groundwater quality. These installations involve placing piezometers on sites that record water quality data for at least five years, and allow for an assessment of the impact of IBA on their immediate environment [3].

The development of the guidelines by the Environmental Protection Agency is meant to allow a rational and ecological use of waste in civil works. The Swedish Geotechnical Institute in collaboration with the National Directorate of Roads worked on the development of a manual for the use of alternative materials in the construction sector [3].

5.2   Legislation on the Management of IBA in the World

5.2.1   China

China is among the largest producers of solid waste in the world [33]. The management of this waste is governed by the "Law on the Prevention of Environmental Pollution" [33]. The incineration of 30% of the waste is one of the 2030 targets in China. In 2006, out of 148 million tonnes of waste collected, 91.4% was landfilled, 6.4% was incinerated and 2.2% was composted [34]. Given that in 2007 there were 366 technical landfills versus only 66 incineration plants, household waste landfill was the most common method of waste management. In the last decades, a strong shift in waste management has been planned, witnessed by the increase in the number of waste incineration plants to about 300 and counting, driven by the country energy policy [35].

The residues from the waste incineration are of two types: IBA and fly ash. IBA (which represent 25% of the tonnage incinerated) are recycled as materials for making bricks after special prior treatment and also in road constructions [35]. Although IBA are reused in various construction applications, some incineration plants do not implement operations that can recover metal particles from these materials [34]. As for fly ash, which represents on average 1–5% of the tonnage incinerated, it is treated with cement before being landfilled in designated sites [34, 35].

5.2.2 United States

Currently, the recycling of residues from the incineration of household waste is not a widespread practice in the USA [36]. These residues (IBA and fly ash) are mainly directed to the appropriate landfills and arranged in layers which are then coated with specific protective materials. Preliminary studies have been carried out with the aim of recovering these residues in the field of construction (geotechnical applications in subbase pavements, embankments, cementitious applications). Although the conclusions of these studies are positive, the absence of federal regulations and the non-homogeneous approach in IBA reuse in different states prevent the development of recovery strategies for incineration residues [36].

In addition, the Environmental Protection Agency has developed characterisation tests named "Toxicity Characteristic Leaching Procedure (TCLP)" in order to be able to quantify the rate of release of contaminants into the environment. These leaching tests allow to classify waste into two categories, namely "hazardous" and "non-hazardous" waste. Incineration residues must meet the environmental requirements of TCLP to be considered non-hazardous. Alternative leaching tests (Waste Extraction Test—WET) were proposed as it was observed that TCLP test overestimated the potential for leaching of residues compared to real conditions. Therefore, each state has the possibility to refer to environmental tests other than TCLP, sometimes with much more restrictive threshold values [36]. Table 17 presents the U.S. Environmental Protection Agency's environmental limit values for waste characterisation.

Table 17 Permissible leaching values according to US EPA [36]

Element	US EPA cut-off values (mg/l)
As	5
Ba	100
Pb	5
Cd	1
Cr	5
Hg	0.2
Se	1
Hg	5

5.3   Closing Remarks

From a global point of view, the analysis of regulatory and legislative framework on the management of IBA in the countries where these granular materials are produced shows significant divergences and non-homogeneity. In addition, the experimental protocols for evaluating the environmental quality required for possible use of IBA in construction, while avoiding environmental pollution, are not consistent. The lack of harmonisation is thus one of the main barriers in exploiting the potential of reuse of IBA.

6   IBA Recycling Opportunities

6.1   Benefits from Valorisation

The valorisation of IBA as secondary granular materials in construction dates back many years. It has undeniable advantages, namely:

(i)   The preservation of natural resources (use of ferrous and non-ferrous metals present in IBA and the mineral fraction as a substitute for either sand or natural gravel).
(ii)   A possible positive economic return.
(iii)   To solve the storage problem and reduce the risk of contamination on the environment (reduce the materials to be stored and avoid the creation of new landfills).

Table 18 provides an overview of the value chains for IBA as secondary granular materials in different European countries [37], suggesting road construction as the most common application. In addition, IBA are also used in the manufacture of concrete in partial or total substitution of aggregate natural material.

Due to their physico-chemical, mineralogical and geotechnical properties, IBA can be recycled in several other applications.

6.2   Reuse in Road Construction

Several types of materials can constitute the different sub-layers of pavements (foundation layer, shape layer, base layer), provided that they meet the desired properties. The use of IBA to substitute virgin materials can be considered when these have physical and geotechnical properties comparable to natural aggregates [38]. In France, IBA for the incineration of non-hazardous waste is used as alternative materials mainly as road embankments and for the layers constituting the body of the roadway [15]. 21,000 tonnes of IBA were used for the Stade de France (12,000 tonnes in

Table 18 Recycling of IBA as secondary materials in Europe [37]

Country	Recycling applications
Austria	No recycling, except in landfills as layering materials
Belgium	Aggregates in road constructions and concrete products
Denmark	Subbase layer of pavements and dikes, marine structures (dams, ports), construction material for car parks
France	80% in road constructions
Germany	Pavement subbase layer, surfacing material in landfills
Italy	Road construction, construction of landfills
Netherlands	Underlayment of pavements and dikes, noise barriers, concrete structures
Portugal	Road constructions, pavement subbase layer, surfacing material in landfills
Spain	Road constructions, pavement subbase layer, surfacing material in landfills
Sweden	Filling materials in landfills
United Kingdom	55% recycled in road construction in 2011

embankment of the railway SNCF station platform and 9,000 tonnes in subbase layer for the station/RN link) and 117,000 tonnes were used for the Euro-Disney construction site [39]. Several other examples of recycling IBA in road construction in different French regions are listed in an AMORCE technical report dated 2014 [40]. An interesting case on the valorisation of IBA in road construction is the "IBA" road. It is an experimental road built in 2013 in Frétin in the Hauts-de-France region, and of which a fraction of IBA (0–6.3 mm) was used partly directly under the asphalt with a traffic structure under a T3 traffic (100 PL/D/direction) [41].

The main concern rising for the use of IBA in road construction is the contamination of surface water and groundwater by leaching of heavy metals. Pre-treatments by washing (water or alkaline solutions) or stabilisations by vitrification [42] are possible solutions before any use. However, measures are being taken in some countries to minimize the impact of the release of pollutants (heavy metals in particular) into the environment. For example, in the Netherlands, in order to avoid groundwater contamination from IBA, embankments are generally raised to at least one metre above high water and IBA are packed in bentonite clay soils and protected by high-density polyethylene films [43].

Studies have shown that IBA can serve as secondary granular materials mainly such as:

(i)   Pavement cover screed and embankments in landfills [27, 44].
(ii) Post-exploitation mine site covering materials [27, 44].
(iii) Substitute sub-layer aggregates [16, 27, 38, 45, 46].

6.3   Application in Ceramics

IBA can be used as raw materials in the production of ceramics [14]. The mineralogical composition of IBA (SiO_2, CaO, Al_2O_3) shows that they could partially replace clay for the production of ceramics without preliminary processing. Studies have shown the feasibility of recycling IBA in the manufacture of ceramic tiles. These findings suggested that the incorporation of up to 20% by weight of IBA in ceramics does not substantially change the mineralogical and thermal properties of ceramic tiles [47].

6.4   Application in Cement Production

The minerals present in anhydrous cement are very similar to those found in IBA, although in different mass proportions. Thus, IBA can be used as raw material in the manufacture of raw cement [14] or as a partial or total substitute for anhydrous cement in cementitious applications.

During the decomposition of calcium carbonates ($CaCO_3$) into lime (CaO), a significant amount of carbon dioxide is released, with obvious environmental hazards. Since IBA already contain lime rather than calcium carbonates, the level of carbon dioxide released during the manufacturing process can be reduced [14]. However, the high concentration of chloride and sulfate ions can affect the quality of the final product [14].

Several studies have shown the possible use of ground IBA as a partial or total substitute for cement in the production of cementitious materials [5, 24, 26, 48–53]. Juric et al. [48] recommended replacing cement with IBA by up to 15% by weight and using this type of binder for non-structural concrete elements. Li et al. [50] observed that the mechanical properties of cementitious matrices containing IBA decrease proportionally according to the substitution rate, and a substitution of 30% allows to have compressive strengths at 28 days of 38.9 MPa, higher than the minimum characteristic value of the cement class (32.5). Al-Rawas et al. [26] showed that the substitution of 20% cement by IBA gives higher compressive strengths compared to the reference value at 14 and 28 days of cure. Nevertheless, the substitution of cement by IBA affected the fresh properties of concrete. Low workability was observed and was proportional to cement substitution rates [26, 50]. In addition, some chemicals present in IBA (zinc, lead, copper) can delay the start of cement setting [49, 54] and consequently slow down the hydration reaction of cement compared to conventional (i.e. no substitution) Portland cement [14, 50].

6.5   Application in the Manufacture of Mortars and Concretes to Replace Natural Aggregates

The use of IBA in concrete is not yet regulated in most European countries. The Netherland is the only country to regulate IBA recycling in cementitious applications through a directive "CUR-Aanbeveling 116" which entered into force in October 2012 [55].

This Dutch directive recommends, for example, environmental tests (mainly leaching tests) on the monolith containing the IBA before they are recycled into concrete, as it best illustrates the situation in which the material is used [56, 57]. For the recycling of IBA or recycled aggregates as a partial replacement for natural aggregates in cementitious applications, the "CUR-Aanbeveling 116" Directive recommends the following [55]:

(i)   The use of IBA-derived aggregates is excluded in the case of prestressed concrete.

(ii) The use of recycled aggregates is recommended for strength classes 12/15; 16/20.

(iii) The replacement of natural aggregates by recycled aggregates may be up to 20% (volume) in reinforced concrete and up to 50% in unreinforced concrete and concrete products.

(iv) The use of recycled aggregates in plain concrete is recommended in all exposure classes.

(v)   The use of recycled aggregates in reinforced concrete is recommended for all exposure classes with the exception of environmental classes XA2 and XA3 (subject to medium and heavy chemical attacks).

(vi) The replacement of natural aggregates by recycled aggregates is limited to 20% (volume) for exposure to class XF (subject to attacks due to freeze–thaw cycles).

(vii) The use of CEM III/B and CEM II/BV cement is mandatory for environmental exposure classes XD and XS (corrosion induced by chlorides of non-marine origins and chloride-induced corrosion of seawater).

Previous studies in the scientific literature demonstrated that non-hazardous waste IBA can be used as a partial or total substitute for the granular skeleton in the manufacture of mortars and concretes [6, 26, 28, 48, 51, 54, 58–70]. The mechanical strength of concrete containing IBA has been found to be lower than that of standard concrete [28] and therefore its use is restricted to non-structural concrete [61, 63, 65]. Conversely, some results show that the mechanical strength of concretes containing IBA may be equal to or even greater than those of standard concretes [6, 67]. These values obtained are influenced by the different treatments carried out on IBA.

Saikia et al. [58] observed that the partial substitution of 25% by weight of the fraction 0.1–2 mm treated chemically (0.25 mol/l Na_2CO_3) and thermally (vitrification at 675 °C) improves the compressive strength of mortars compared to those of the same fraction that have not undergone any chemical and heat treatment. Pera et al. [28] reported that the 4–20 mm fraction of IBA treated with sodium hydroxide could replace natural gravel by up to 50% by volume in the manufacture of concrete, without affecting the durability of the materials and giving compressive strengths in the order of 25 MPa at 28 days of cure. Similarly, Zhang and Zhao [49] concluded that the rate of substitution of natural gravel by pre-washed IBA in the manufacture of concrete must not exceed 50% by weight in order to have mechanical properties comparable to those of standard concretes. Keppert et al. [60] recommended that the substitution rate of natural sand by untreated IBA should be less than 10% so that the mechanical strengths of concrete specimens are not significantly affected. Ferraris et al. [51] showed that vitrification-treated IBA (1,450 °C) are not suitable as a replacement for natural sand in concrete production. However, cementitious materials retain their mechanical properties with a replacement of natural gravel by up to 75% by volume compared to reference concretes. Al-Rawas et al. [26] showed that substituting natural sand with IBA up to 40% by weight gives compressive strengths at 28 days higher than those of concrete made with natural sand. Nielsen et al. [67] demonstrated that by substituting 50% of natural gravel with treated IBA, compressive strength at 28 days increases by 20% for hollow concrete blocks, and by 34% for solid concrete blocks compared to those of reference concretes. Muchova [6] concluded that the substitution of 30% of natural gravel by IBA treated according to a wet process developed in the Netherlands made it possible to obtain compressive strengths equal to 41.5 MPa at 28 days, against 36.2 MPa when only natural materials are used in the manufacture of concrete [6].

The major observation made throughout the literature is that the optimal substitution rate of aggregate and the strengths obtained at different days of curing are influenced by the treatments carried out on the aggregates of IBA (washing by alkaline solutions, solidification/vitrification, magnetic and mechanical separation, to cite few). Table 19 provides a non-exhaustive list of the mechanical properties of mortars based on treated IBA-derived sands replacing natural sand (fraction less than or equal to 5 mm).

The low mechanical properties observed in concrete when natural aggregate (sand or gravel) is partially or totally substituted by IBA are presumably due to the presence of certain chemical elements [6, 28, 61, 67]. The use of IBA in mortar or concrete mixes can lead in some cases (and under certain curing conditions) in the development of undesired chemical reactions inside the matrix and, consequently, can cause swelling pathologies on concrete structures [25, 71, 72].

Typical reactions likely to occur are:

(i)   The alkali silica reaction.
(ii) The formation of aluminium hydroxides and hydrated aluminates.
(iii) The formation of secondary ettringite.
(iv) The hydration of lime and magnesium oxides.

Table 19 Mechanical properties of cementitious materials containing IBA-derived sand

Fraction (mm)	Treatment	Substitution rate	w/c	Compressive strength @ 28 days (MPa)	Compressive strength @ 60 days (MPa)	Refs.
0.1–2	Baseline (no substitution)	0%	0.5	54	57	[58]
	Untreated	25% (mass)		30.2	33.1	
	Washed with water					
	Chemical (0.1–0.25 Na$_2$CO$_3$)					
	Thermal (675 °C)			35.1	37.1	
	Thermal (675 °C) Chemical (0.25 Na$_2$CO$_3$)			40.5	43.9	
	Chemical (0.25 Na$_2$CO$_3$) Thermal (675 °C)			29.7	31.9	
	Superplasticizer Thermal (675 °C) Chemical (0.25 Na$_2$CO$_3$)			49.7	53.6	
0–5	Baseline (no substitution)	0%	0.6	44.7	45.8	[51]
	Vitrification (1450°C) Grinding Sieving	25% (volume)		40.0	42.7	
		50% (volume)		35.7	38.3	
		75% (volume)		33.6	34.8	
		100% (volume)		30.7	32.2	
0.075–4.75	Baseline (no substitution)	0%	0.7	30.5	n.d	[26]
	Not specified	10% (mass)		34.8		
		20% (mass)		36.4		
		30% (mass)		31.3		
		40% (mass)		35.5		

6.5.1 Alkali-Silica Reaction

The alkali-silica reaction (ASR) is the result of the reaction of amorphous silica from aggregate (e.g., glass debris in IBA) with the hydroxide and alkaline ions in the pore solution of concrete [72–74]. This reaction gives rise to the formation of a gel that develops inside the glass inclusions and along the pores of the cementitious matrix, thus causing the cracking of the concrete and in some cases its destruction (Fig. 7).

The gel formed is mainly composed of silicon, calcium, potassium and sodium [72, 73]. The high concentration of glass debris in the cementitious matrix can accelerate the process of its formation. Three simultaneous conditions are necessary for the triggering of ASR [75]:

(i)   The presence of potentially reactive aggregate.
(ii) A high concentration of alkaline elements in the pore solution.
(iii) Environmental conditions of high relative humidity (RH $\geq$ 80%).

The presence of glass particles in concrete can also lead to a pozzolanic reaction, which will improve the mechanical properties and durability of concrete [71, 72]. For obtaining this positive effect, the glass needs to be in a very fine, powdery form such as a binder. The powdered glass will allow the silica to react with the lime produced by the reaction of the cement, thus creating compounds with binding properties. This pozzolanicity will increase with the fineness of the particles [73]. Studies have shown that equivalent or higher compressive strengths can be achieved using up to 40% glass with a fineness of 540 m^2/kg, compared to a reference mix without glass [73]. Similarly, the use of up to 30% glass substitution in concrete does not cause ASR when the alkaline content of the mortar and/or concrete is less than 3 (Na_2O_{eq} $\leq$ 3) [73].

However, several preventive actions need to be applied in order to curb the ASR in concrete, such as the use of silica fume, fly ash, metakaolin and even finely ground glass particles (180–540 m^2/kg) [73]. Idir et al. [73] demonstrated that the simultaneous use of coarse and fine glass fractions in concrete reduces the occurrence of

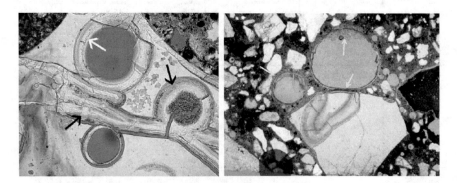

Fig. 7 Silica gel formation [72]. Left: gel produced from the glass particle outwards. Right: gel produced also from the surrounding of the glass particle inwards

ASR. The Dutch directive "CURAanbeveling 89" provides a set of measures and different precautions to be taken when manufacturing concrete in order to avoid the appearance of the alkali-silica reaction [76]. The absence of swelling in the test specimens might not mean that ASR did not take place. This reaction can appear in the form of efflorescence on the sample faces and only further microscopy analyses with a scanning electron microscope (SEM) would confirm or disprove the presence of ASR in the specimen [73].

6.5.2 Formation of Aluminium Hydroxides and Aluminates

Aluminum dissolves due to the presence of OH^- ions in water-based alkaline solutions (when the pH is greater than 10) to form aluminates. This reaction is accompanied by a steady development of gas (H_2) and occurs near or around aluminum particles [71, 72].

$$Al + 2OH^- + H_2O^- \rightarrow (AlO(OH)_2)^- + H^2 \uparrow \quad (pH > 7)$$

The release of hydrogen gas that occurs during the fresh phase of cement paste is the main source of the very high porosity of concretes containing IBA compared to standard concretes [72]. Figure 8 shows some forms of porosity identified in concrete containing IBA.

When the pH drops in the range 9–10, there is the formation of an aluminum hydroxide gel around the aluminum particles, responsible for the swelling and appearance of cracks on the cementitious materials [71, 72].

$$(AlO(OH)_2)^- + H_2O^- \rightarrow Al(OH)_3 + OH^- \quad (pH = 9 - 10)$$

$$Al(OH)_3 + OH^- -\rightarrow (Al(OH)_4)^- \quad (pH > 10)$$

Fig. 8 Different forms of porosity identified in IBA-based concrete [72]. Left: pores in a linear arrangement. Right: rounded shape of porosity surrounding Al metal particles

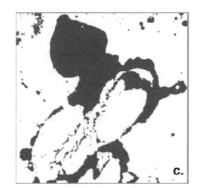

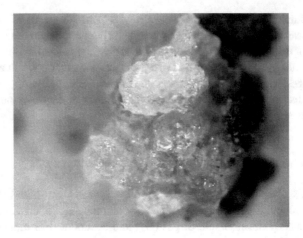

Fig. 9 Aluminum hydroxide gel formed [71]

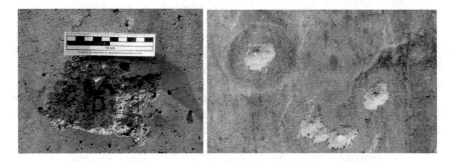

Fig. 10 Crumbling and spalling of concrete surface. Left: [72]. Right: [67] reproduced with permission from ISCOWA

Aluminum hydroxide appears in three distinct forms: an amorphous form $Al(OH)_3$ and two crystalline forms namely bayerite (α-$Al(OH)_3$) and boehmite (γ-$AlOOH$) [72]. While the amorphous form appears as a gel (Fig. 9), bayerite sometimes appears as a well-formed plate as an aggregate in the form of crystals inside the pores.

In addition, if these compounds form near the surface of a concrete element, the damage can appear in the form of crumbling or spalling, see Fig. 10.

6.5.3 Formation of Secondary Ettringite

Ettringite is a mineral species composed by calcium sulfate and hydrated aluminum. At a very high pH level, the compound ($(Al(OH)_4)^-$) formed from aluminum metal can combine with sulfate and calcium ions to form expansive ettringite products, which significantly reduce the compressive strengths of cementitious materials [58]

Fig. 11 Formation of secondary ettringite [67] reproduced with permission from ISCOWA

if these are produced when the concrete has already hardened (secondary ettringite formation). The formation of a secondary ettringite layer in cementitious materials is conditioned by the concentration content of sulfate ions. Secondary ettringite appears in the cracks inside the matrix as an irregular mass and in the pores in the form of circular crystals (Fig. 11).

Similarly, sulfates present in IBA can form expansive sulfate minerals in hardened concrete (delayed ettringite), which can deteriorate the durability of concretes [58]. The formation of delayed ettringite occurs especially in the case where the concrete at a young age is subjected to temperature greater than 65–70 °C [77].

6.5.4 Hydration of Lime and Magnesium Oxides

Free calcium and magnesium oxides cause instability during their transformation into hydroxide [71].

$$CaO + H_2O- \rightarrow Ca(OH)_2$$

$$MgO + H_2O- \rightarrow Mg(OH)_2$$

This swelling is observed after cement setting and is due to the development of massive hydrates such as portlandite ($Ca(OH)_2$), whose growth exerts pressure on the entire structure and causes its expansion. It is related to the amount of non-hydrated free lime available after the start of setting. If this amount is negligible, the swelling may be negligible [75]. In addition, the pathology of swelling caused by this reaction appears less important than the one caused by ettringite and aluminum hydroxide gel [71].

7 Improved IBA Treatment Processes

Among all the chemical reactions that can occur in the cementitious matrix, the oxidation reaction of aluminum metal is the main cause of swelling in concretes containing IBA [71, 72].

Therefore, in order for significant volumes of IBA to be recycled in cementitious applications, advanced treatment processes that can allow the efficient removal of metal particles (ferrous and non-ferrous, mainly aluminum metal and all other undesirable materials) are required. The recovery of ferrous and non-ferrous fractions can also provide an economic value that supports the viability of IBA recycling in a wider perspective. In this section, the main current methods for removal of ferrous and non-ferrous fractions are described.

The techniques commonly used for the treatment of IBA can be grouped into two categories: dry separation processes and wet separation processes [78].

7.1 Dry Separation Processes

Dry separation processes are those where water does not interact in the separation process between the mineral fraction and materials to be removed.

7.1.1 Heros (Granova) Process

This separation technique was developed in the Netherlands. Heros, one of the subsidiaries of REMEX (a company specialising in the treatment of residues from waste incineration), produces sustainable secondary aggregates branded "Granova" that are used in building and civil construction [57, 79].

The process of developing Granova aggregates can be described as follows. The coarse input is directed to different screens where it is optimally sieved and separated. Mixed materials and metal particles that could not be removed by the screens are removed manually. Ferrous metals are extracted using different magnets while non-ferrous metals are extracted using eddy current separators. Subsequently, the industrial by-product is directed to an air separator to extract all the organic matter present in the materials. At the end of this process, the IBA is stored for an appropriate period (maturation) to reduce heavy metal concentrations, and eventually it is ready as a substitute for natural aggregates in construction.

No statistical data could be obtained on the recovery rate of ferrous and non-ferrous metals through this process. 500,000 tonnes of "Granova" aggregates are produced annually in the Netherlands for their use as secondary materials in roads, earthworks, acoustic protection walls, landfill constructions and concrete structures [56, 80]. Heros has developed a Granova technical data sheet showing compliance to CE marking and KOMO certification [56]. KOMO certification is a label established

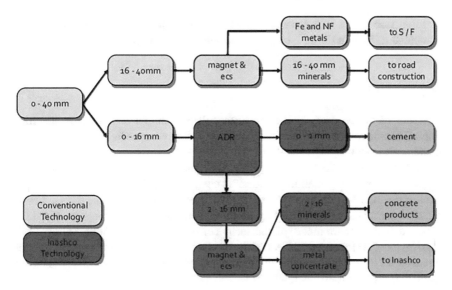

Fig. 12 Synoptic of the ADR-Inashco separation process [83]

by the Independent Dutch Certification Institute that certifies the conformity of the characteristics contained in the product data sheet.

7.1.2   Inashco ADR Procedure

The Inashco Advanced Dry Recovery (ADR) process is a technology created and developed in the Netherlands by a joint effort between Delft University of Technology and the Fondel Group, aiming at creating a novel extraction technique for non-ferrous metals contained in the fine fractions of IBA. This process is mainly composed of two units: an ADR separator and a recovery unit [81, 82].

ADR Separator

The ADR separator is a treatment unit that uses kinetic energy to break the existing bonds in particles. Through the ADR, there is a physical separation between fine and large particles. Thus, the separation process (magnetic drums and conventional eddy current separators) can be carried out more efficiently [8]. The unit can be either associated with an existing conventional processing plant or stand-alone (Fig. 12) [7].

The ADR unit allows physical separation of wet materials (on average 20% [7]) up to 2 mm without any form of drying or addition of water and chemical adjuvants [8]. Previous studies have shown that the recovery rate of aluminum metal in the

2–8 mm fraction of IBA increased to 89% after treatment through the ADR unit, compared to 37% in the case of a conventional treatment process [8].

Recovery Unit

The role of the recovery unit is to further the separation and improve the fraction of non-ferrous metals, aiming at making it refined enough to be reused in the metallurgical sector [82].

In its separation process, the Inashco ADR manages directly dust and gas emissions, and limits the formation of undesired chemical reactions and the concentration rates of potentially polluting metals. This will significantly reduce the storage period required for maturing IBA [81, 82]. In compliance with the European standard EN 12620 and the requirements of the technical directive "CUR Aanbeveling 116", Inashco has developed a datasheet of the product on the basis of the mineral fraction obtained for its use in the manufacture of concrete structures in the Netherlands and labelled by KOMO certification [82].

7.2 Wet Separation Processes

In wet separation processes, the water plays an essential role in the different stages of the treatments for separating the mineral fraction, metal particles and soluble salts.

7.2.1 Indaver Process

The treatment of IBA by the Indaver process is a technique that relies mainly on the screening, washing and separation of granular and metallic materials (Fig. 13) [37, 84]. It optimises the extraction of organic matter and soluble salts contained in IBA through a washing process.

The largest particles/elements are removed preliminarily from the raw materials. The IBA is then screened and washed on a sieve with opening of 50 mm, in order to separate particles with a grain size greater than 50 mm. Ferrous metals are extracted from this fraction with the use of magnets, and the remaining materials are directed once again to the incinerator. In a second step, particles smaller than 50 mm pass through a basin filled with water, in order to separate the aggregates from the organic matter. A second screening and washing stage is then carried out for sorting particles in the fractions 6–50 mm, 2–6 mm and less than 2 mm. Ferrous and non-ferrous metals are extracted from granular fractions 6–50 mm and 2–6 mm respectively by magnetic bars and eddy current separators. Eventually, particles smaller than 2 mm are divided into two granular fractions: the sandy fraction 0.1–2 mm and the fraction less than 0.1 mm (sludge), which is landfilled without any form of stabilization [84].

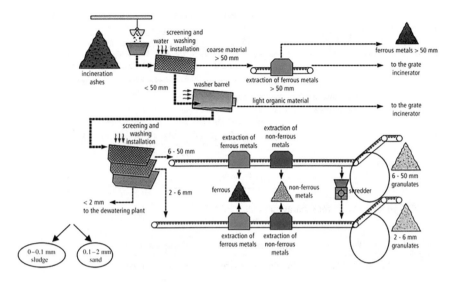

Fig. 13 Scheme of treatment of IBA by the Indaver process [84]

Materials with granular fractions 6–50 and 2–6 mm are stored three months in suitable spaces where weathering will complete the maturation [84]. After this period, leachate concentration rates are compared to limit values for ensuring the suitability of processed IBA as secondary granular materials in the construction sector.

7.2.2 AEB Wet Process

This wet separation process is a technique that has been tested in Amsterdam at the Afval Energie Bedrijf (AEB) incineration plant. It is mainly based on a physical separation according to the size, density, magnetic and conductive properties of the particles [6, 78]. The synoptic of this wet separation process is shown in Fig. 14.

The process subdivides the IBA into several fractions (0–2, 2–6, 6–20 and 20–40 mm) in order to optimise the recovery of metals [6]. Each granular fraction is treated separately by appropriate equipment to effectively recover metal particles up to 0.1 mm [7]. The fraction 0–45 μm is landfilled because it contains high proportions of organic matter and heavy metals [6]. Specific separation equipment was used for the 0–2 and 2–6 mm granular fractions while standard magnetic drums and eddy current separators were used for the larger fractions (6–20 and 20–40 mm). Table 20 shows the various specific equipment used for the treatment of fractions of IBA in fractions 0–2 and 2–6 mm.

The treatment synoptic of the 0–2 mm fraction is shown in Fig. 15.

The separation of ferrous metals is carried out over the fraction 2–6 mm using a kinetic gravity separator. The prototype used consists of a metal shell divided into two cylinder-shaped parts, the first having an internal diameter of 100 cm and a height

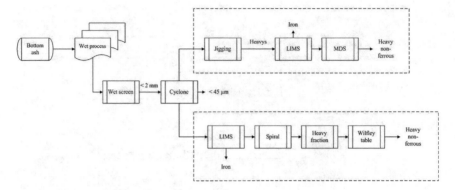

Fig. 14 Synoptic of the wet separation process of AEB [7]

Grain size	Equipment
Fraction 0–2 mm	Hydrocyclone Gravity separator Low intensity magnetic separator Vibrating table
Fraction 2–6 mm	Kinetic gravity separator Wet eddy current separator

Table 20 Different equipment of the AEB process used in the fractions 0–2 and 2–6 mm [6]

Fig. 15 Principle of separation mode of the fraction 0–2 mm [6]

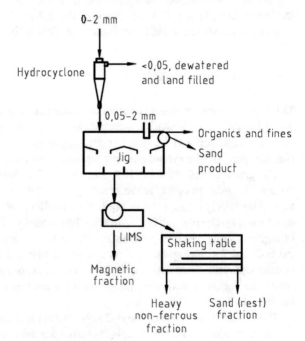

Fig. 16 Kinetic gravity separator [6]

of 100 cm and the second part having a cylindrical diameter of 60 cm. A blade wheel with 6 compartments where particles are collected is mounted coaxially inside the cylinders (Fig. 16). The physical principle of this application relies on the maximum velocity of particles falling into the water, which is function of their density, size and shape. The water is injected at the level of blades and move at a constant angular speed. When the IBA particles are introduced by a feed hopper located above the separator, the speed of rotation of the blades and the height of fall of the particles will determine the compartment in which particles with the same properties (density, size, dimension) will end. Heavy metal particles will be collected in the first compartment, light metals, mineral matter and glasses will be deposited in the second and third compartments, while light particles and organic materials will be collected in the last three compartments [6].

Separation of non-ferrous metals is carried out using a wet eddy current separator. The principle of operation of the eddy current separator by the wet way differs significantly from the standard "dry" method, although the same separator is used. The difference is that the materials are slightly moistened (10–15%) at the inlet, and the rotor in the separator spins in the opposite direction of the conveyor, unlike that of the standard "dry" version. The moment caused by the magnetic field of the rotor "breaks" the contact between the non-ferrous metal particles and the conveyor belt. Non-ferrous metal particles detached from the conveyor belt are collected in front of the separator while the remaining materials (mineral fraction, etc.) remain on the conveyor belt, and are removed downstream [6].

7.2.3 Process HVC-Dolman

The HVC-Dolman process is a wet extraction method based on three phases: granular separation, reduction of polluting potential, and recovery of non-ferrous metals [85].

IBA are screened in two fractions (4–15/20 mm and a fraction 63 μm–4 mm), and then washed to remove dissolved soluble salts and organic matter. Each clean fraction is directed for its use as a substitute for natural aggregate. The obtained sludge (very fine fractions) is dried before being disposed of in appropriate sites [85].

7.2.4 Other Separation Techniques

The scientific literature describes other complementary techniques that can increase the recovery rate of non-ferrous particles in the fine fractions of IBA.

- Separation by the Magnus effect.

This separation process consists of passing materials close to a fast rotor in order to create a selective rotation of the non-ferrous particles, which divert them from the material flow by the Magnus effect. It is suitable for particles with diameter between 0.2 and 2 mm [86].

- Reverse Foucault current separation.

This separator, unlike the standard eddy current separator, is composed of a magnetic pole that rotates counter clockwise [87]. Previous studies demonstrated that the recovery rate of metal particles could be significantly improved when the magnetic pole rotates in the opposite direction [88].

A comparative study of the different treatment processes for IBA along with their metal particle recovery rates is shown in Table 21.

Concluding, it appears that improved treatment processes allow to optimise the removal of ferrous and non-ferrous metals, delivering a better quality mineral fraction recyclable in construction. In addition, wet separation processes with washing stages significantly improve the quality of the mineral fraction, removing soluble salts (chlorides and sulfates) and organic matter. Nevertheless, this process inevitably entails additional costs for the treatment of wastewater polluted with soluble salts, compared to dry treatment processes.

Currently, there is no specific treatment for IBA with a fraction of less than 2 mm in dry separation processes. The AEB wet separation process, through specific equipment, makes it possible to extract metal particles and organic fraction efficiently.

8 Coal Bottom Ash (Furnace Bottom Ash—FBA)

According to the World Coal Association [89], coal is still the world's largest single source of electricity, and its contribution to worldwide energy production is projected to be 22% in 2040 globally, with some countries possibly having a higher dependence on coal electricity production (up to 39% in South East Asia).

Coal Combustion Residues (CCRs), or Coal Combustion Products, is the waste from coal-fired power stations. Main CCRs are fly ash (FA), furnace bottom ash (FBA), boiler slag (BS), fluidised bed combustion ash (FBCA), and flue gas desul-furization (FGD). An estimate about production and reutilisation of CCRs by major

Table 21 Summary of different treatment processes for IBA [7]

Technologies for the treatment of IBA	Equipment	Treatment for fraction <2 mm	Recovery rate of ferrous and non-ferrous metals (%)	References
Standard process	– Dry screening – CF separator – Drums Magnetic	None	Al/IBA (1%)	[2, 45]
ADR-Inashco process	– Dry screening – Separator ADR – Air separator – Magnets	None	N.F./N.F. (89%)	[7, 8]
Wet AEB process (2007)	– Water screening – Cyclones – CF separator – Kinetic separator – Drums Magnetic	Gravity separators Magnets	Al/IBA (1.62%)	[7, 45, 78]
Indaver process	– Water screening – Magnets – CF separator	None	Fe/IBA (9%) N.F./IBA (1.6%)	[37, 84]
HVC process (pre-treatment)	– Water screening – Cleaning – Magnets – CF separator	None	Fe/IBA (0.7%) N.F./IBA (0.5%)	[37, 85]
Magnus process			Al/IBA (1.77%)	[45]
Astrup et al. (2007)			Fe/IBA (3.6–6.9%)	[45]
Lamers (2008)			Fe/IBA (8%)	[45]
Barcellesi (2008)			Fe/IBA (8.01%)	[45]

producer Countries reported in the literature suggests a global CCRs production exceeding 1.2Gt, with an average utilisation rate in the range of 64% [90]. However, the utilisation rate for CCRs ranges from about 20 to nearly 100%, with marked differences among different CCR products. In Europe, about 50% of the total CCRs are utilised in the construction industry, while about 41.5% are used for the restoration of mining/quarrying works. Only 8.5% of the total CCRs are temporarily stockpiled (1.8%) or landfilled (6.7%) [91].

Table 22 Physical parameters of FBA

Test	References	
Grain size distribution	Gravel to sand	[97]
Grain density	2.2/2.8 (t/m^3)	[93]
Apparent density	0.7/0.9 (t/m^3)	[97]
Porosity	30–55%	[98]

8.1 Physical and Chemical Characterisation

8.1.1 Physical Characterisation

FBA is usually described as a coarse, partly vitrified product consisting of incombustible minerals occurring in the fuel [92–96]. The grain size of FBA lays in the class of sandy gravel, with a low solid density ranging between 1.8 and 2.2 t/m^3, a bulk density in the range of 0.7–0.9 t/m^3 and a compacted dry density about 1.2 t/m^3, see Table 22 FBA permeability can be estimated in the range 10^{-2}–10^{-5} m/s.

The low porosity values have been justified in the literature as the result of a combination between a macro-porosity (i.e. inter-grain porosity) and a micro-porosity (i.e. grain structure) due to the vesicular texture ("popcorn" particles), which degrade quickly under loading or compaction [95, 99].

8.1.2 Chemical and Mineralogical Characterisation

The chemical composition of FBA is influenced by the quality and mineralogical nature of the coal utilised in the power station. FBA is typically characterised by a high fraction of Si and Al (about 90% of the oxide composition obtained with XRF analysis) [99]. Other elements that are typically encountered are Fe, Mn, Mg, Ca, Na, K, Ti, P and S. Trace elements of Zn, Cd, Cu, Pb, Mo, Ni, As, Se and B have been reported in the literature and are of interest when landfilling is considered, due to the potential pollution hazard represented by these chemicals [92–96, 99–107].

The mineralogical composition reported in [99], obtained through XRD analysis coupled with Rietveld refining method and using corundum as internal standard, suggested a large fraction of amorphous phases (about 61%), whereas Mullite $Al_{4,56}Si_{1,44}O_{9,72}$ (22%) and Silica SiO_2 (11%) were the main crystal structures detected, and Hematite Fe_2O_3 and Rutile TiO_2 were found as minor crystal components (<1%). By comparing the chemical and the mineralogical compositions, it has been suggested that SiO_2 might be largely present in its amorphous phase, as well as K and Mg.

The amorphous (glassy) nature of FBA was confirmed by SEM images, see Fig. 17.

Fig. 17 SEM image of IBA [99]

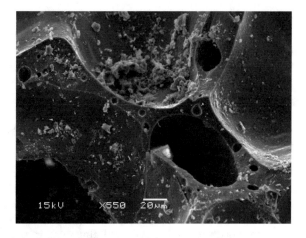

8.2 Environmental and Health Considerations

The main threats to human health from heavy metals are associated with exposure to Pb, Cd, Hg and As. However, environmental regulations for materials, waste disposal, and wastewater typically include in the list of hazardous elements also Cr, Cb, Cu, Mo, Ni, Sb, Se, Sn, V, and Zn [108–110]. For these reasons, detailed chemical analysis must be carried out, and contamination mechanisms must be verified when considering reuse strategies. Environmental risks from CCRs are mainly related to land stocking issues and to water pollution by heavy metals and sulfate/nitrate residues [92, 104].

Meij et al. [108] investigated the leaching properties of bricks and concrete manufactured incorporating FA, and reported an increase of As, Mo and SO_x in the leachate when compared to baseline values (i.e. without FA), whereas results for V and Cr results were less definitive. Interestingly, the concentrations of investigated elements were lower than the limits dictated by the Dutch Building Material Decree, and thus it was concluded that FA could be utilised in building material production. Similar results were reported by Liu et al. [111], who assessed the leaching behaviour of compressed bricks produced with FA. The concentration of pollutants from bricks caused by rain was found to be negligible, and bricks passed the TCLP test recommended by EPA.

The combustion of coal in power stations can also result in an accumulation of naturally occurring radioactive materials (NORM) in the ash, and thus radioactive emissions of FBA need to be checked according to current guidelines and regulations [112].

The position of the environmental protection agency in United States (EPA) on the possibility of reusing CCRs has been changing over the time, influenced by available technical studies, general economic trend, political conditions, and pressure from civil society [96, 106, 113–115]. Following the large spills occurred near Kingston (Tennessee) in 2008, and Eden (North Carolina) in 2014, which caused widespread environmental and economic damage to nearby waterways and properties, the Environmental Protection Agency started a revision of existing regulation to help ensure protective coal ash disposal. In 2010 the Environmental Protection Agency (EPA) has proposed to reconsider these residuals as special wastes when they are destined for disposal in landfills or surface impoundments, or to regulate the disposal of such materials by issuing national minimum criteria. However, in order to support the beneficial reuse of CCRs, in 2015 EPA provided a definition to distinguish between beneficial use and disposal. The beneficial use of CCRs needs to be evaluated according to the following criteria:

(i)   The CCRs must provide a functional benefit.
(ii) The CCRs must substitute for the use of a virgin material.
(iii) CCRs must meets product specifications and/or design standards; and
(iv) When CCRs use involves placement on the land of 12,400 tons or more in non-roadway applications, the emissions in water, soil and air needs to be comparable to or lower than those from analogous products made without CCR, or that releases will be below relevant regulatory and health-based benchmarks for human and ecological receptors.

Recent revisions limited the evaluation described at point (iv) to locations where a geological risk exists. However, due to the intrinsic strategic nature of the use of coal in the energy mix, and the possibility of managing the combustion residues with strategies other than landfilling, the debate is still open under the influence of civil society activists, industrial lobbies, and environmental groups.

8.3   Reuse Pathways

The reutilisation and/or recycle of CCRs has been extensively investigated worldwide [92], but with significant differences as far as the end-product value is concerned (see Tables 23 and 24).

Whilst FA is widely recycled by the cement industry, and in many Countries it has become a traded commodity rather than a by-product, the reuse of FBA, which represents about 10–30% of total CCR production, is less successful. About 40% of FBA is reused as concrete aggregate [93], embankment/filling material, snow and ice control, waste stabilisation, or blasting grit/roofing granules [116]. In developing countries, valorisation of coal combustion waste is still vastly underexploited, although some experience has been gained assessing the environmental impact reduction for a 125 MW power station in Senegal [117], and investigating the use of FBA as constituent for the production of stabilised compressed blocks [99, 118].

The chemical nature of FBA (rich in aluminosilicate) and its predominantly amorphous nature might imply that fine grinding of FBA could, in principle, deliver a material whose reactivity and utilisation potential are not dissimilar to those of fly ash.

Table 23 Commons reuse and recycle of CCRs [99]

Sector	Applications
In the construction sector	As a pozzolanic additive for the substitution of clinker and clay in Portland cement production, as partial substitution for cement and sand in concrete, as fine aggregate in blocks and as a raw material in the production of lightweight and insulating aggregates
In infrastructure sectors	As fill material to amend low-quality materials for paving roads, marine works, landscape rehabilitation, foothills stabilisation. In foundations and backfilling in mining works. As an impervious layer in the restoration of landfills and the containment of contaminated soils
In agriculture	As an ingredient in growth bed mixtures for plants and as a soil improvement. As improvement for fertiliser made from municipal sewage sludge
In industry	As filler in plastic, paint, asphalt and sealing materials and as a carrier for controlled–released fertilisers. As a raw material in ceramic glass materials and refractory materials. As filtration mean for water filtering
In the metal market	As a secondary raw material for the production of metals

Table 24 Special reuses and recycling pathways for CCRs

Country	CCR type	Application	References
India	CCRs	Various brick productions	[92]
Japan	Ashes	Development of artificial Z-sand for Civil Engineering applications	[119]
USA	FA	Non-fired bricks	[111]
Australia	FA	1,000 °C fired bricks	[120]
USA	FA	Fired bricks	[94]
Taiwan	FA	Fired and non-fired bricks using ashes mixed with pond sludge	[100]
Niger	FBA	Compressed block using FBA and local soil (laterite), stabilised with mineral binders	[99, 118]
Turkey	FA	Fired bricks	[101]
Russia	Ashes and mining wastes	Construction ceramic materials	[121]
Israel	FA	Cement and fine sand replacement in concrete	[122]
Turkey	FA, FBA, furnace slag	Cement and fine sand replacement in concrete	[123]
USA	FBA and FA	Improvement of rammed earth constructions	[124]
Sweden	FBA	Light fill material	[97]
Spain	FA and slag	Light weight aggregates	[102]
Netherlands	FA	Bricks and concrete	[108]

9   Conclusions

The objective of this chapter was to review the production, characterisation and possible reutilisation of bottom ash. A distinction has been introduced between incineration bottom ash (IBA), which is produced in municipal waste incineration plants, and furnace bottom ash (FBA), which is produced during the combustion of coal in power plants. These are the two main categories of bottom ash of practical interest. Regulations and legislative framework for their use have also been briefly recalled. The take-home messages are:

(i) IBA is a well-known alternative granular material in the world and the main producers in Europe are Germany, France, the Netherlands and Denmark. The waste sorting during collection, the choice of incinerator technology, the cooling mode and the maturation time are the main parameters that influence the composition of IBA. The main chemical components identified are silicon, calcium, aluminum, iron, magnesium and potassium. The main mineral phases detected are quartz, calcite, corundum and iron oxides. The physical and geotechnical parameters of IBA are similar to those of natural gravel and sand, with the exception of water absorption coefficients and LOI values, which are higher for IBA.

(ii) Each country producing IBA has defined a regulatory and legislative framework, both for its use as alternative secondary materials and for its deposit in landfills. The non-standardisation of regulations in Europe and sometimes within regions in the same country (e.g., USA or Belgium) is a barrier in the valorisation of IBA in the construction sector.

(iii) Nowadays, established recycling pathways for IBA in the construction sector are available. The main option for IBA recycling in Europe relates to road construction, which is regulated by several legislative instruments, aimed at avoiding hazards from leaching of heavy metals into the environment. The Netherlands was the first country in 2012 to develop guidelines for the use of IBA in cementitious materials. In addition, several studies showed that, depending on their chemical and mineralogical composition, IBA can be used in the manufacture of ceramics, cement, and as aggregates for partial to total substitution of natural aggregates.

(iv) Certain material pathologies have been identified when IBA are used into the cementitious matrix. Alkali-silica reaction, secondary ettringite, and the formation of aluminum gel are considered the most critical. All these pathologies can lead to poor mechanical properties and low durability of concretes containing IBA. The formation of aluminum gel is the most severe pathology among all those identified in the scientific literature.

(v) Conventional IBA treatment techniques, commonly used in most incineration plants and maturation plants, are not suitable for the recovery of ferrous and non-ferrous metals in fine fractions. As a result, novel technologies have emerged to improve the quality of IBA ahead of its use in cementitious materials. Dry and wet separation processes can provide satisfactory recover of metallic aluminium from IBA.

(vi) Although involving production volumes significantly lower than IBA, bottom ash from the coal combustion (FBA) is a by-product of technological interest. Several reutilisation options have been already trialled in research works as well as in the engineering practice. The quality and consistency of FBA are generally higher than those of IBA due to the more homogeneous quality of the feedstock fuel.

References

1. Blasenbauer D, Huber F, Lederer J, Quina MJ, Blanc-Biscarat D, Bogush A, Bontempi E, Blondeau J, Chimenos JM, Dahlbo H, Fagerqvist J (2020) Legal situation and current practice of waste incineration bottom ash utilisation in Europe. Waste Manag 102:868–883
2. European Coal Combustion Products Association (2022) Production and utilisation of CCPs in 2016 in Europe (EU 15). https://www.ecoba.com/evjm,media/ccps/ECO_stat_2016_EU15_tab.pdf. Accessed 04 Apr 2022
3. Crillesen K, Skaarup J, Bojsen K (2006) Management of bottom and ash from wte and plants: an overview of management options and treatment methods. Technical report, Working Group on Thermal Treatment of Waste, ISWA
4. A.M. Ministre de l'écologie du développement durable des transports et du logement. Arrêté ministériel du 18 novembre 2011 relatif au recyclage en technique routière des mâchefers d'incinération de déchets non dangereux, 18 novembre 2011. Version consolidée au 24 août 2015, 8 p
5. Bertolini L, Carsana M, Cassago D, Curzio AQ, Collepardi M (2004) MSWI ashes as mineral additions in concrete. Cem Concr Res 34(10):1899–1906
6. Muchova L (2010) Wet physical separation of MSWI bottom ash. PhD thesis, TU Delft, the Netherlands
7. Hu B, Rem P, van de Winckel T (2009) Fine heavy non-ferrous and precious metals recovery in bottom ash treatment. In: Vieira JMP, Ramisio PJ, Silveira ANI (eds) Turning waste into ideas, pp 1–8 (Lisbon, Portugal: ISWA/APESB2009)
8. Vries WD, Rem PC, Berkhout P (2009) ADR: a new method for dry classification. In: Vieira JMP, Ramisio PJ, Silveira ANI (eds) Turning waste into ideas, pp 1–10 (Lisbon, Portugal: ISWA/APESB2009)
9. Agence de l'Environnement et de la Maitrise de l'Energie (ADEME). Chiffres-clés: Déchets. ADEME Editions 2014 80 p, ADEME
10. Bröns-Laot G (2002) Evaluation environnementale de la valorisation de mâchefers d'incinération d'ordures ménagères en remplissage de carrière. Thèse de doctorat, Institut National des Sciences Appliquées de Lyon
11. Agence de l'Environnement et de la Maitrise de l'Energie (ADEME) (2015) Chiffres-clés: Déchets. ADEME Editions 2015 96 p, ADEME
12. Agence de l'Environnement et de la Maitrise de l'Energie (ADEME) (2012) L'incinération des déchets ménagers et assimilés. Technical Report 6 p
13. (CEWEP) Confederartion of European Waste-to Energy Plants. Accessed 22 Sep 2016
14. Lam CHK, Ip AWM, Barford JP, McKay G (2010) Use of incineration MSW ash: a review. sSustainability 2(Iss 7):1943–1968 (ISSN 2071-1050)
15. Minane JR (2017) Contribution à l'élaboration de sables de mâchefers améliorés en vue d'une valorisation dans la formulation de matériaux cimentaires. Thèse de doctorat, Université des

Sciences et Technologies de Lille, Ecole Mines-Télécom Lille-Douai - Département Génie Civil et Environnemental
16. RECORD (2015) Qualité et devenir des mâchefers d'incinération de déchets non dangereux etat des lieux et perspectives. Rapport technique 134 p, n 13–0241 1A
17. Ministère de l'Environnement Français. Circulaire dppr/sei/bpsied n°94-iv-1 du 09/05/94 relative à l'élimination des mâchefers d'incinération des résidus urbains, 9 Mai 1994. 16 p
18. Kaibouchi S (2004) Mâchefers d'Incinération d'ordures Ménagères: Contribution a l'étude des mecanismes de stabilisation par carbonatation et influence de la collecte sélective. Thèse de doctorat, Institut National des Sciences Appliquees de Lyon
19. Rendek E, Ducom G, Germain P (2007) Influence of waste input and combustion technology on MSWI bottom ash quality. Waste Manag 27(10):1403–1407. In: Wascon 2006 6th international conference: developments in the re-use of mineral waste
20. SENAT (2015) L'incinération. http://www.senat.fr/rap/o98-415/o98-41513.html. Accessed 24 Feb 2015. Rapports d'information
21. ENSEEIHT (2017) Etude d'implantation de filières de traitement des déchets en guadeloupe, Consulté 10 Février 2017. Rapport d'études
22. Bourtsalas A (2012) Review of WTE ash utilization processes under development in northwest Europe. Technical report
23. Arickx S, Van Gerven T, Vandecasteele C (2006) Accelerated carbonation for treatment of MSWI bottom ash. J Hazard Mater 137(1):235–243
24. Filipponi P, Polettini A, Pomi R, Sirini P (2003) Physical and mechanical properties of cement-based products containing incineration bottom ash. Waste Manag 145156(23):12
25. Ginés O, Chimenos JM, Vizcarro A, Formosa J, Rosell JR (2009) Combined use of mswi bottom ash and fly ash as aggregate in concrete formulation: environmental and mechanical considerations. J Hazard Mater 169:643–650
26. Al-Rawas AA, Hago AW, Taha R, Al-Kharousi K (2005) Use of incinerator ash as a replacement for cement and sand in cement mortars. Build Environ 40(9):1261–1266
27. del Valle-Zermeño R, Formosa J, Chimenos JM, Martínez M, Fernández AI (2013) Aggregate material formulated with {MSWI} bottom ash and {APC} fly ash for use as secondary building material. Waste Manag 33(3):621–627. Special Thematic Issue: Urban Mining Urban Mining
28. Pera J, Coutaz L, Ambroise J, Chababbet M (1997) Use of incinerator bottom ash in concrete. Cem Concr Res 27(1):1–5
29. CEREMA (2014) Gestion des mâchefers d'incinération de déchets non dangereux. Application de l'arrêté ministériel du 18 novembre 2011: bilan des pratiques, Direction territoriale Centre-Est Rapport d'enquête
30. SETRA (Service d'études sur les transports les routes et leurs aménagements) (2012) *Acceptabilité environnementale de matériaux alternatifs en technique routière. Les mâchefers d'incinération de déchets non dangereux (MIDND)*. Direction générale de la prévention des risques (DGPR), du Ministère de l'écologie, du développement durable et de l'énergie (MEDDE) and Centre d'études techniques de l'équipement de Lyon (CETE de Lyon), 110, rue de Paris - 77171 SOURDUN - France
31. Van Gerven T, Geysen D, Stoffels L, Jaspers M, Wauters G, Vandecasteele C (2005) Management of incinerator residues in Flanders (Belgium) and in neighbouring countries. A comparison. Waste Manag 25(1):75–87
32. Astrup T (2007) Pretreatment and utilization of waste incineration bottom ashes: Danish experiences. Waste Manag 27(10):1452–1457
33. Chen X, Geng Y, Fujita T (2010) An overview of municipal solid waste management in china. Waste Manag 30(4):716–724
34. Zhang DQ, Tan SK, Gersberg RM (2010) Municipal solid waste management in China: status, problems and challenges. J Environ Manag 91(8):1623–1633
35. Efstratios Kalogirou. Waste-to-energy research and technology council
36. An J, Kim J, Golestani B, Tasneem KM, Al Muhit BA, Nam BH, Behzadan AH (2014) Evaluating the use of waste-to-energy bottom ash as road construction materials. Technical report, Department of Civil, Environmental, and Construction Engineering - University of Central Florida, 4000 Central Florida Blvd. Orlando, FL

37. Van Brecht A, Konings A (2011) Innovative and BREF proven material recycling of MSWI bottom ashes. In: 2nd international slag valorization symposium, p 14, 18–20 April 2011
38. Forteza R, Far M, Seguì C, Cerda V (2004) Characterization of bottom ash in municipal solid waste incinerators for its use in road base. Waste Manag 24(9):899–909
39. Miquel G, Poignant S (1999) Les techniques d erecyclage et de valorisation des déchets ménagers et assimilés. Rapport technique, Rapport de l'Assemblée Nationale n °1693 et du Sénat n 415
40. AMORCE (2014) Recueil d'exemples de chantiers ayant valorisé des mâchefers. Rapport technique Réf. AMORCE: DT65, Réf ADEME: 8481, 30 p, ADEME, AMORCE – 18, rue Gabriel Péri – CS 20102 – 69623 Villeurbanne Cedex
41. Mines Douai Département Génie Civil & Environnemental (2014) Etude de granulats de mâchefers améliorés & suivis mécanique et environnemental d'une route expérimentale à base de mâchefers amliorés. Rapport d'étude 30 p, Neo Eco Recycling
42. Toraldo E, Saponaro S, Careghini A, Mariani E (2013) Use of stabilized bottom ash for bound layers of road pavements. J Environ Manag 121:117–123
43. Noulin F (2008) Mâchefers: prendre la mesure des retards et des enjeux. déterminer le coût environnemental de la « valorisation » non réglementée d'un déchet. In: La Lettre Indusrie - déchets
44. Schreurs JPGM, van der Sloot HA, Hendriks Ch (2000) Verification of laboratory–field leaching behavior of coal fly ash and MSWI bottom ash as a road base material. Waste Manag 20(2–3):193–201
45. Grosso M, Biganzoli L, Rigamonti L (2011) A quantitative estimate of potential aluminium recovery from incineration bottom ashes. Resour Conserv Recycl 55(12):1178–1184
46. Becquart F, Bernard F, Abriak NE, Zentar R (2009) Monotonic aspects of the mechanical behaviour of bottom ash from municipal solid waste incineration and its potential use for road construction. Waste Manag 29(4):1320–1329
47. Andreola F, Barbieri L, Corradi A, Lancellotti I, Manfredini T (2001) The possibility to recycle solid residues of the municipal waste incineration into a ceramic tile body. J Mater Sci (4869–4873):36p
48. Juric B, Hanzic L, Ilic R, Samec N (2006) Utilization of municipal solid waste bottom ash and recycled aggregate in concrete. Waste Manag 26(12):1436–1442
49. Zhang T, Zhao Z (2014) Optimal use of MSWI bottom ash in concrete. Int J Concr Struct Mater 8(2):173–182
50. Li X-G, Lv Y, Ma B-G, Chen Q-B, Yin X-B, Jian S-W (2012) Utilization of municipal solid waste incineration bottom ash in blended cement. J Clean Prod 32:96–100
51. Ferraris M, Salvo M, Ventrella A, Buzzi L, Veglia M (2009) Use of vitrified mswi bottom ashes for concrete production. Waste Manage 29(3):1041–1047
52. Tang P, Florea MVA, Spiesz P, Brouwers HJH (2016) Application of thermally activated municipal solid waste incineration (mswi) bottom ash fines as binder substitute. Cement Concr Compos 70:194–205
53. Minane JR, Becquart F, Abriak NE, Deboffe C (2017) Upgraded mineral sand fraction from MSWI bottom ash: an alternative solution for the substitution of natural aggregates in concrete applications. Procedia Eng 180:1213–1220
54. Sorlini S, Abbà A, Collivignarelli C (2011) Recovery of MSWI and soil washing residues as concrete aggregates. Waste Manag 31(2):289–297
55. CUR-Aanbeveling (2012) Aec-granulaat als toeslag-material voor beton. CUR BOUN & INFRA
56. REMEX Mineralstoff GmbH (2014) Agrégats pour béton. http://www.granova.eu/fr.html. Accessed 15 Dec 2014
57. REMEX Mineralstoff GmbH (2014) Opportunités pour les agrégats de mâchefers d'usine installation ordures ménagères. http://www.granova.eu/
58. Saikia N, Mertens G, Van Balen K, Elsen J, Van Gerven T, Vandecasteele C (2015) Pre-treatment of municipal solid waste incineration (mswi) bottom ash for utilisation in cement mortar. Constr Build Mater 96:76–85

59. Siddique R (2010) Use of municipal solid waste ash in concrete. Resour Conserv Recycl 55:83–91
60. Keppert M, Pavlík Z, Černý R, Reiterman P (2012) Properties and of concrete and with municipal and solid waste and incinerator and bottom ash. In: 2012 IACSIT Coimbatore conferences, vol 28, p 5
61. Scories BM (1986) d'ordures incinérées comme granulat pour béton. Bulletin du Ciment 5557(7)
62. Tay J-H, Tam C-T, Chin K-K (1982) Utilization of incinerator residue in concrete. Comervotion Recycl 5(2/3):107–112
63. Qiao XC, Ng BR, Tyrer M, Poon CS, Cheeseman CR (2008) Production of lightweight concrete using incinerator bottom ash. Constr Build Mater 22(4):473–480
64. Cioffi R, Colangelo F, Montagnaro F, Santoro L (2011) Manufacture of artificial aggregate using mswi bottom ash. Waste Manag 31(2):281–288
65. Chang N-B, Wang HP, Huang WL, Lin KS (1999) The assessment of reuse potential for municipal solid waste and refuse-derived fuel incineration ashes. Resour Conser Recycl 25(3–4):255–270
66. Kuo W-T, Liu C-C, Su D-S (2013) Use of washed municipal solid waste incinerator bottom ash in pervious concrete. Cement Concr Compos 37:328–335
67. Nielsen P, Quaghebeur M, Laenen B, Kumps R, Van Bommel P (2009). The use of mswi-bottom ash as aggregate in concrete. Limitations and possible solutions. In: Vision on technology (WASCON), p 26. WASCON
68. Van Wegen G, Hofstra U, Spreerstra J Upgraded MSWI bottom ash as aggregate in concrete. SGS INTRON
69. Courard L, Degeimbre R, Darimont A, Laval A-L, Dupont L, Bertarnd L (2002) Utilisation des mâchefers d'incinération d'ordures ménagères dans la fabrication des pavés en béton. Mater Struct/Matériauxe et Constructions 35:365–372
70. Saikia N, Cornelis G, Mertens G, Elsen J, Van Balen K, Van Gerven T, Vandecasteele C (2008) Assessment of pb-slag, MSWI bottom ash and boiler and fly ash for using as a fine aggregate in cement mortar. J Hazard Mater 154(1–3):766–777
71. Pecqueur G, Crignon C, Quénée B (2001) Behaviour of cement-treated MSWI bottom ash. Waste Manag 21(3):229–233
72. Müller U, Rübner K (2006) The microstructure of concrete made with municipal waste incinerator bottom ash as an aggregate component. Cem Concr Res 36(8):1434–1443
73. Idir R, Cyr m, Tagnit-Hamou A (2010) Peut-on valoriser massivement le verre dans les bétons? etude des propriétés des bétons de verre. Technologie Valorisation 16(5/6):8 p
74. Shayan A, Aimin Xu (2004) Value-added utilisation of waste glass in concrete. Cem Concr Res 34(1):81–89
75. Dupain R, Saint-Arroman JC (2009) *Granulats, sols, ciments et bétons: Caractérisation des matériaux de génie civil par des essais au laboratoire*, volume 4 e édition actualisée. Ecole Française du Béton
76. CUR-Recommendation 89 (2003) Measures to prevent damage to concrete by alkali-silica reaction (asr). CUR building & infrastructure, 2nd, revised edition
77. Pathologie: L'ettringite (2017) Mars – Avril. Accessed 11 June 2017
78. Muchova L, Rem PC (2007) Wet and dry separation—management of bottom ash in Europe. Waste Manag World 8(6):46–49
79. REMEX Mineralstoff GmbH (2014) Concevoir des environnements. http://www.granova.eu/fr.html. Accessed 15 Dec 2014
80. REMEX Mineralstoff GmbH (2014) Incinerator bottom ash aggregate (IBAA): opportunities and developments, p 28. REMEX, September 2014
81. INASHCO (2015) Ash recycling. http://www.inashco.com/en/ash-recycling. Accessed 27 Jan 2015
82. Waste Management World (2015) Rising from the ashes. http://www.waste-managementworld.com/articles/print/volume-10/issue-6/features/rising-from-the-ashes.html. Accessed 20 Jan 2015

83. INASHCO (2011) From ashes to metals. In: Confederation of European waste-to-energy plants—EAA seminar, 9p, 5th–6th September 2011
84. Vandecasteele C, Wauters G, Arickx S, Jaspers M, Van Gerven T (2007) Integrated municipal solid waste treatment using a grate furnace incinerator: the Indaver case. Waste Manag 27(10):1366–1375
85. HVC and Boskalis Dolman (2016) Wet extraction of wte plant bottom ash. Accessed on April 2016
86. Fraunholcz N, Rem PC, Haeser PACM (2002) Dry magnus separation. Miner Eng (15): 45–51
87. Biganzoli L (2012) Aluminium recovery from MSWI bottom ash. PhD thesis, Politecnico di Milano, Department of Civil and Environmental Engineering
88. Zhang S, Rem PC, Forssberg E (1999) The investigation of separability of particle smaller than 5 mm by eddy current separation technology. Part I: rotating type eddy current separators. Magn Electr Sep 9: 233–251
89. World Coal Association (2022) World coal statistics. https://www.worldcoal.org/coal-facts/coals-contribution/. Accessed 8 April 2022
90. Harris D, Heidrich C, Feuerborn J (2019) Global aspects on coal combustion products. In: Proceedings of the world of coal ash (WOCA), St. Louis, MO, USA, pp 13–16
91. ECOBA (2017) Production and utilisation of CCPs in 2016 in Europe (EU 15). https://www.ecoba.com/evjm,media/ccps/ECO_stat_2016_EU15_tab.pdf. Accessed 10 Apr 2022
92. Asokan P, Saxena M, Asolekar SR (2005) Coal combustion residues—environmental implications and recycling potentials. Resour Conserv Recycl 43:239–262, Elsevier
93. Siddique R (2010) Utilization of coal combustion by-products in sustainable construction materials. Resour Conserv Recycl 54:1060–1066, Elsevier
94. Chou M, Botha F (2003) Manufacturing commercial bricks with Illinois coal fly ash. In: International ash utilisation symposium, Center for Applied Energy research, University of Kentucky, Paper #25
95. Kassim TA, Simoneit BRT, Williamson KJ (2005) Recycling solid wastes as road construction materials: an environmentally sustainable approach. Handb Environ Chem, vol 5, Part F. 1:59–181. https://doi.org/10.1007/b98264, Springer
96. U.S. Department of Transportation (1997) User guidelines for waste and byproduct materials in pavement construction. U.S. Department of Transportation, Federal Highway Administration, Publication Number FHWA-RD-97-148
97. Rogbeck J, Knutz A (1996) Coal bottom ash as light fill material in construction. Waste Manag 16(1–3):125–128, Elsevier
98. Saxena M, Asokan P (2000) Rehabilitation of backfill area using fly ash with suitable plant life at abandoned Gorbi mine, Northern Coal Field, Singrauli, India. Report, Regional Research Laboratory, Bhopal and CLI Coal Handling Co. Pvt. Ltd., New Delhi, India
99. Vinai R, Lawane A, Minane JR, Amadou A (2013) Coal combustion residues valorisation: research and development on compressed brick production. Constr Build Mater 40:1088–1096
100. Hsu Y-S, Lee B-J, Liu H (2003) Mixing reservoir sediment with fly ash to make bricks and other products. In: International ash utilisation symposium, Center for Applied Energy research, University of Kentucky, Paper #89
101. Fatih T, Umit A (2001) Utilisation of fly ash in manufacturing of building bricks. In: International ash utilisation symposium, October 22–24, 2001, Center for Applied Energy research, University of Kentucky, Paper #13
102. Anieto M, Acosta A, Rincon JM, Romero M (2005) Production of lightweight aggregates from coal gasification fly ash and slang. In: International ash utilisation symposium, Center for Applied Energy research, University of Kentucky
103. Acosta A, Aineto M, Iglesias I, Romero M, Rincón JMa (2001) Physico-chemical characterization of slag waste coming from IGCC thermal power plant. Mater Lett 50:246–250. https://doi.org/10.1016/S0167-577X(01)00233-6
104. Omana J, Dejanovic B, Tuma M (2002) Solutions to the problem of waste deposition at a coal-fired power plant. Waste Manag 22:617–623

105. Vassilev SV, Vassileva CG (1997) Geochemistry of coals, coal ashes and combustion wastes from coal-fired power stations. Fuel Process Technol 5I:19–45
106. EPA (1999) Report to congress on remaining fossil fuel combustion wastes: waste characterization. https://www.epa.gov/sites/default/files/2015-08/documents/march_1999_report_to_congress_volumes1and2.pdf
107. Alam MGM, Allinson G, Laurenson LJB, Stagnitti F, Snow ET (2002) A comparison of trace element concentrations in cultured and wild carp (Cyprinus carpio) of Lake Kasumigaura, Japan. Ecotoxicol Environ Saf 53:348–354
108. Meij R, Kokmeijer E, Tamboer L, te Winkel H (2001) Field leaching of bricks and concrete containing coal fly ash. In: International ash utilisation symposium, October 22–24, 2001, Center for Applied Energy Research, University of Kentucky, Paper #97
109. Stosnach H, Gross A (2007) Analysis of heavy metals in sewage samples. Lab report XRF 423, Bruker AXS Microanalysis GmbH, Berlin, Germany
110. EPA (1992) EPA test method 1311—TCLP, toxicity characteristic leaching procedure. Provided by Environmental Health & Safety Online www.ehso.com
111. Liu H, Banerji SK, Burkett WJ, VanEgelenhoven J (2009) Environmental properties of fly ash bricks. World of Coal Ash (WOCA), May 4–7, 2009, Lexington, Kentucky, USA
112. Sas Z, Vandevenne N, Doherty R, Vinai R, Kwasny J, Russell M, Sha W, Soutsos M, Schroeyers W (2019) Radiological evaluation of industrial residues for construction purposes correlated with their chemical properties. Sci Total Environ 658:141–151
113. EPA (1993) Final regulatory determination on four large-volume wastes from the combustion of coal by electric utility power plants. Federal register, vol 58, No 151, Rules and Regulations, Environmental Protection Agency (EPA), Part V, 58 FR 42466
114. EPA (2020) Hazardous and solid waste management system: disposal of coal combustion residuals from electric utilities; A holistic approach to closure Part A: deadline to initiate closure. A rule by the environmental protection agency on 08/28/2020
115. EPA (2020) Hazardous and solid waste management system: disposal of coal combustion residuals from electric utilities; A holistic approach to closure Part B: alternate demonstration for unlined surface impounds. A rule by the environmental protection agency on 11/12/2020
116. American Coal Ash Association (2020) Coal combustion product (CCP) production & use survey report. https://acaa-usa.org/wp-content/uploads/2021/12/2020-Production-and-Use-Survey-Results-FINAL.pdf. Accessed 8 Apr 2022
117. BAD (2009) Sendou 125 MW Centrale à Charbon, Sénégal. Banque Africaine de Développement, Référence No. P-SN-F00-004
118. Lawane A, Minane JR, Vinai R, Pantet A (2019) Mechanical and physical properties of stabilised compressed coal bottom ash blocks with inclusion of lateritic soils in Niger. Sci Afr 6:e00198
119. Ohnaka A, Hongo T, Ohta M, Izumo Y (2005) Research and development of coal ash granulated material for civil engineering applications. In: World of coal ash (WOCA), April 11–15, 2005, Lexington, Kentucky, USA
120. Kayali O (2005) High performance bricks from fly ash. In: World of coal ash (WOCA), April 11–15, 2005, Lexington, Kentucky, USA
121. Lemeshev VG, Gubin IK, Savel'ev YuA, Tumanov DV, Lemeshev DO (2004) Utilization of coal-mining waste in the production of building ceramic materials. Glass Ceram 61(9–10) (translated from Steklo i Keramika, No. 9, pp 30–32)
122. Ravina D (1997) Properties of fresh concrete incorporating a high volume of fly ash as partial fine sand replacement. Mater Struct/Matériaux et Constructions 30:473–479, RILEM
123. Haibin L, Zhenling L (2010) Recycling utilization patterns of coal mining waste in China. Resour Conserv Recycl 54:1331–1340, Elsevier
124. Dockter BA, Eylands KE, Hamre LL (1999) Use of bottom ash and fly ash in rammed-earth construction. In: International ash utilisation symposium, Center for Applied Energy Research, University of Kentucky, Paper #56

Spatialising Urban Metabolism: The Supermarket as a Hub for Food Circularity

Emma Campbell, Greg Keeffe, and Seán Cullen

1 Introduction

As human populations rise and rapidly urbanise, resource consumption and waste continue to intensify in urban contexts. The challenge of climate change and planetary biocapacity will require cities to urgently reconnect with the global hinterland servicing them. How resources are sourced, manipulated, moved, consumed, and disposed of remains a major focus in geo-politics, spurred by the UN Sustainable Development Goal Twelve 'Responsible Consumption and Production' [24]. Here, 'responsible' can be understood in a couple of ways. First, as more efficient consumption and production of resources. Second, towards better management of resource flows at the scale of building, neighbourhood, city, region, country, and beyond. Research reveals that human populations use more resources and produce more waste than can be replenished or processed, overshooting planetary biocapacity by around 54% [10]. Despite the mindset of 'otherness' described by Thomson and Newman [27], cities have a metabolism that contributes to this overshooting condition. Cities also significantly influence local, regional, and global environments. This is recognised within UK climate change legislation as transboundary emissions or impacts. For example, The Environment Act for Wales 2016 requires an estimate of the greenhouse gas emissions relating to Welsh consumption of goods and services [15], recognising that, when a Welsh consumer chooses dairy milk over oat milk at a supermarket, this has worldwide implications on resource availability, water, and air quality both pre-and post-transaction. The concept of urban metabolism, which

E. Campbell (✉) · G. Keeffe · S. Cullen
Queen's University, Belfast, UK
e-mail: e.j.campbell@qub.ac.uk

G. Keeffe
e-mail: g.keeffe@qub.ac.uk

S. Cullen
e-mail: sean.cullen@qub.ac.uk

© The Author(s), under exclusive license to Springer Nature Switzerland AG 2023
M. Tribaudino et al. (eds.), *Minerals and Waste*, Earth and Environmental Sciences Library, https://doi.org/10.1007/978-3-031-16135-3_8

sits within the field of Landscape Urbanism, observes this state of interconnection between cities and wider planetary ecosystems. Grounded in systems thinking, the concept frames cities as living organisms that are constantly in flux [20–22, 28, 29]. Just like a human digestive system, what a city eats changes its performance, impacts how much waste it produces, and influences future resource demand. As a result, cities are constantly creating ripples, and sometimes waves of change beyond their boundaries.

This chapter explores a spatial approach to visualise and reprogram urban metabolism using systems thinking and research-by-design methodologies, explored through a range of drawing techniques such as diagrams, collages, and maps. It questions how spatial interventions might leverage profound system shifts in both urban and wider ecosystem metabolisms. Focusing on food, food waste and food packaging flows, it evaluates the role of UK supermarkets as a key node, actor, and spatial interface in the production of waste and imagines how supermarkets might harness their integral position in food systems to catalyse circular economies, thereby increasing resource efficiency and eliminating waste.

2   Urban Metabolism Approaches

Seeing how resources flow through a spatial condition, such as a city, increases understanding of how and why it operates as it does and offers new perspectives on how the performance of inward and outward flows might be optimised. Current approaches to the visualisation of urban metabolism include Life Cycle Analysis (LCA), Material Flow Analysis (MFA), and Input–Output Analysis (I-OA) [19]. All rely on data accounting methodologies to reveal the scale and impact of resource flows. Each of these approaches work in different ways to generate visualisations defined by three primary parameters of operation:

1. Flow type: Studies may choose to focus on just one flow, while others may also look at a collection of flows in order to assess their independencies. Studies may look at different scales, including: system flows (e.g. energy); product flows (e.g. a carton of milk); nutrient flows (e.g. phosphorus). Studies may also examine particular flow characteristics such as flows that are biotic and biodegradable (e.g. a cucumber), or abiotic and non-biodegradable (e.g. a television).
2. Time frame: As the city is in constant flux, current urban metabolism practices measure flows either at a point in time or across a timeframe. While it is not explored within the scope of this chapter, improvements in data management through digital-twin technologies are likely to lead to the standardisation of real-time urban flow visualisation [8, 12] to provide instant feedback on flow optimisation.
3. System boundary: As ecosystems are intrinsically connected, it can be difficult to visualise urban flows without setting out a territory of investigation. These

boundaries may be spatial, for example, they may look at flows within a particular neighbourhood. They may follow a flow-type, such as the journey of an avocado across a supply chain, or they might look at flows more generally within a timeframe, such as across a weekend or a decade.

Each of the current approaches to visualising urban metabolism are designed to capture different types of system boundaries and flows. For example, LCA tends to focus on specific products or services, following resources from production to consumption and disposal. In this case, the system boundaries span beyond the city to the countries in which resources are sourced or disposed of. In MFA studies, system boundaries are normally more geographically fixed, for example, a neighbourhood, city, or region [14].

While these methodologies offer a way to understand the city by what flows through it, they are largely data-driven and often represent findings in abstract, non-spatial, and diagrammatic visualisations, such as Sankey diagrams and flow charts. Athanassiadis [1] states that studies are hindered by the availability, collection, and management of data and, thus, often focus on the description of flows and not their relationship to social, economic, or spatial factors. Keeffe and Cullen [13] argue that the abstract nature of these approaches makes it difficult for architects and urban designers to engage with invisible resource flows as they reshape the built environment. They posit that developing better ways of communicating the scale and impact of urban flows on the built environment enables opportunities for acute system acupuncture [7]. Meadows [16] describes these as leverage points, which if adjusted, can rewire the whole system's programming and help designers to become aware of the invisible actors and networks at work across the territories in which they design in.

3 Feeding Hungry Cities

The urban metabolism of food flows in UK cities is largely influenced by supermarkets which, since the mid-1990s, have remained the dominant food-shopping model. Today, supermarkets are the primary interface between local food consumers and global food producers, with nine retailers controlling over 95% of the UK's grocery market [25]. Thomson and Newman [27] describe contemporary cities as 'extractive engines', while Girardet [11] defines them as a 'Petropolis', on a feeding frenzy of cheap energy. UK cities fit this description, relying on a global hinterland to supply cheap resource inputs, such as food, while absorbing their outputs, such as plastic waste [26]. Sitting within linear economies, they pull resources into urban environments for consumption, with the remaining waste moved to the city's boundaries for management. Resource extraction, production, processing, conveyance, and management spaces sit almost exclusively outside the urban zones. At the global scale, this has led to regional centralisation of production for global markets, such as the production of soy in Brazil. While at the urban scale, cities have become reserved

Fig. 1 Supermarket 'throughput' moments

primarily for the role of consumption, and what was once the hinterland for local supply, urban edges have become service spaces for global inputs and outputs [4, 17]. The supermarket shopping experience reflects this current state of play. This shopping type with big-box store, dedicated carpark, trolley, and checkout, emerged as a logistics-focused model to facilitate the efficient 'throughput' of global foods to local consumers (Fig. 1) [30]. Supported by the development and refinement of refrigeration, container shipping, packaging, and barcodes, supermarkets offer low-prices, choice, and all-year seasonality through a scaled-up, centralised, spatial model [6].

Almost all urban food and food packaging flow through this space on its journey from farm to fork to the domestic bin. For example, around twenty supermarkets serve a population of just over 336,800 residents in the low-density city of Belfast, Northern Ireland (Fig. 2) [3]. As a result, supermarkets have significant influence in leveraging food consumption, production, and waste at global and local scales. Decisions made by consumers inside supermarkets actively shape the actions of food and waste management systems serving them. Despite this, the generic supermarket shop floor focuses solely on the display and transaction of foods, with storage, delivery, and disposal spaces strategically hidden behind the back-of-house wall. Operating on tight-profit margins and large economies of scale, supermarkets employ a range of spatial and psychological tactics to increase consumption, from buy-one-get-one-free (BOGOF) offers to the installation of long, tall aisles that promote disorientation, disconnection, and impulse [5]. Aided by the development of a consumer culture post-WWII, and rising car and fridge ownership since the mid-1970s, supermarket dominance has also solidified the transformation of homes into a store for goods. While saving consumers time and effort, this has also contributed to the UN's estimation of around a third of all consumable food going to landfill [9].

At a system scale, supermarkets also influence raw material and nutrient flows globally (Fig. 3). On a warm weekend, consumers might decide to have barbeques and buy more beef burgers and packets of lettuce than they normally would. Further back in the supply chain, this increases the consumption of soy feed, water, fertilisers, and pesticides, causing in turn, increased production of nitrogen, phosphorus, carbon dioxide, and ammonia. Similarly, after the barbeque, consumers have used more plastic packaging than they normally might, causing increased consumption of oil to make that plastic. Within a few days of consumption, global nutrient flows are lost as they meet local sewage systems, and once treated, they merge with wider water, ground, and air systems without boundaries. Discarded packaging meets a similar fate, first sorted by local waste management systems, then becoming invisible as it moves to sorting and processing locations internationally.

Fig. 2 Mapping Belfast's supermarket infrastructure

4 Framing Supermarket Food Flows: Input–Output–Interface

As most of the UK's food flows through supermarkets, urban food inputs are carefully monitored and controlled. Food moves quickly through agile, just-in-time supply chains, with large orders constantly amended to account for fluctuations in demand. Across the year, whole, fresh foods are sourced wherever it is summertime so that consumers can buy asparagus or a watermelon in any season. Though complex, flows up to this point are well-managed to ensure food arrives on the shelves before it reaches critical sell-by dates. Where food is sourced from, how and when it arrives, what it is packaged in, how it is stored and displayed; supermarkets have made the route food takes from farm to shelf quick, agile, and efficient. However, once food is purchased, how and where it flows to becomes less clear (Fig. 4). From homes, consumed food waste, and nutrient profiles within that waste, move through sewage networks and into local, then global, water courses. While most unconsumed food either goes to landfill or to local councils for treatment such as composting and incineration. Equally, packaging moves to either landfill or to local recycling centres, and in many cases, is shipped for processing elsewhere. How food, food waste, and

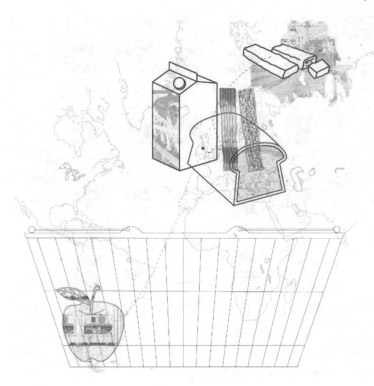

Fig. 3 Mapping staple food flows and intangible sub-flows

packaging flows is dictated by a complex landscape of conflicting policies, balancing efficiency, speed, environment, and economics.

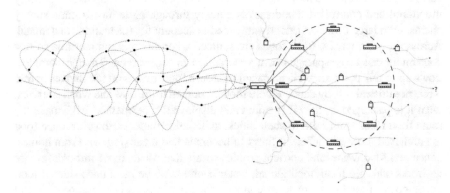

Fig. 4 Supermarket as orchestrator of urban food inputs, home as a frontier for urban food outputs

There is a clear priority imbalance in the flow of supermarket inputs and outputs and this is largely due to boundaries of responsibility. Designed for monetary transactions, the supermarket checkout also transfers the responsibility for waste. Once food leaves the store, consumers take on the responsibility for disposal in their homes, supported by council waste collection services. Beyond the transaction, supermarkets do not need to consider the waste produced as a result of their business decisions. It is for this reason that supermarkets prioritise the pre-consumption role of packaging. This is best explained through the example of yoghurt packaging where easy-peel (but difficult to recycle) lids are used to enhance the consumer experience. This example reveals that little consideration is given to the waste incurred in a system that focuses on consumer convenience. Operating a for-profit model, increased sales equate to higher profits and the impact of resource use and waste production takes lower priority. The lack of responsibility for waste works well for supermarkets because their model relies on built-in waste and single-use packaging is a good example of this. The vast number of products available in a typical supermarket is supported by single-use packaging, which make food easy to move, keep it fresh for longer, build consumer trust and enable convenience all at once.

How and where food flows on its way to supermarkets and from homes remains unclear for individual consumers. While acting as an interface for the responsibility for waste, the convenience and speed of the supermarket shopping experience creates an interface between consumers and supply chains too (Fig. 5). Packaging serves as the main form of communication between the farm or factory and the consumer. Often information about provenance remains unclear or lists the countries in which a product is sourced across a year making it difficult for consumers to trace where the food they buy actually comes from. Information is even less clear for ready-meal products, which rely on multiple, invisible supply chains for raw ingredients. Consumers are largely uninformed about the impact of their choices as they shop. *How much water was embodied in this tomato? How many tomatoes went to landfill during production? How has the tomato's fertiliser affected land and water systems?* These questions are difficult to answer.

How waste food flows from homes remains unclear too because there are so many actors and networks involved, such as local councils, water, and sewage infrastructure and recycling businesses. Without knowledge of pre- and post-consumption impacts, consumers find it difficult to change behaviours and make more sustainable choices. Without a shift in responsibility between producers and retailers, elimination of inherent waste remains without incentive. That said, there is some indication that consumers want retailers to help food systems waste less. In the UK, the BBC'S Our Plastic Feedback campaign encouraged shoppers to bring single-use plastic back to the shop as a form of protest [18]. Soon after, major supermarkets responded by placing packaging bins at the exit point. There has been a gradual removal of unnecessary plastic packaging and supermarkets are beginning to trial small refill zones in stores too, but in general, actions to reduce waste are small and slowly implemented.

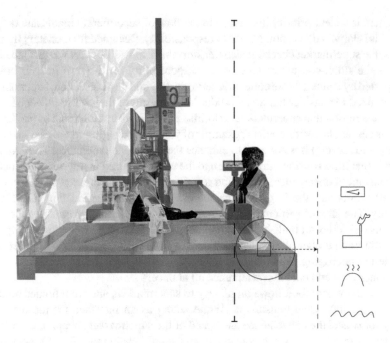

Fig. 5 Supermarket checkout as primary food flow interface

5   Reprogramming the Metabolism of Urban Food Flows: Linear to Circular

Redesigning supermarkets both spatially and operationally through Meadow's concept of system leverage may open opportunities to influence the urban metabolism of food flows in UK cities. As the primary controller of food inputs and key interface with urban consumers, a two-fold approach may be taken to localise the flow of food inputs and waste outputs and make consumers more aware of this as they shop.

Supermarkets sit as key spatial nodes within a linear economy, where resources are taken from the earth, processed, bought, used, and disposed back to the environment. Circular economy principles offer a viable route to eliminate waste through cities [2]. Where linear economies focus on 'throughput', the efficient flow of resources from A to B, circular economies focus on 'roundput', the flow of resources from A to B to A again [30]. McDonough and Braungart's seminal text on circularity, Cradle to Cradle (2009), defines an approach based on the separation of biotic and technological materials and a shift from owning to sharing things. They posit that biotic waste should become food for new productive systems as is the case in nature. Likewise, they suggest that technological materials and products should shift to service models so that, rather than being owned by individuals, they should be rented out by businesses. Designed to be shared and reused, businesses would then be incentivised to create products that are easy to disassemble and repurpose, eliminating overproduction and planned obsolescence as a result. McDonough and Braungart also note that the separation of material flows and shift in ownership work better when resources flow in smaller loops because resources are easier to monitor and manage when they move less.

As with current urban metabolism visualisation approaches, circular economy principles are often imagined in abstract and non-spatial ways, primarily through diagrams, structured design processes and strategies. As is also the case with urban metabolism studies, circular principles require systems thinking so that, for example, a new circular business can fit within wider systems of circularity beyond the boundaries of operation. As a primary interface connecting global food producers to local consumers, imagining how supermarkets might become circular could help to catalyse global and local scale circular economies (Fig. 6).

Acting as a centralised urban node for food and food packaging inputs, what if supermarkets went beyond providing packaging bins to become centralised nodes to manage resource outputs as well as inputs? Supermarkets could reprogram food chains so that they become circular, but the space itself must be reimagined to enable this to happen (Fig. 7). So, how could supermarkets shift from becoming interfaces in a linear food system to nodes in a circular one? Reflecting on the circular economy principles, they would need to ensure biotic and technological flows remain distinctly separate and take ownership of flows, both pre-and post-consumption. To enable this, supermarkets would need to do more than display and sell food. They would need to host all the activities typically found along a supply chain, to become part-farm, factory, distribution centre, shop, kitchen, and recycling hub. Food sold in supermarkets would then flow within smaller system boundaries so that urban food waste could be collected to produce more food or energy. Packaging could be biotic, i.e. biodegradable, or technical, therefore completely separate from food flows and easily reused. As Cradle to Cradle suggests, technological packaging could be a supermarket-owned asset, loaned to the consumer but ultimately the supermarket's responsibility. This concept is already being tested by start-up companies such as Deliver Zero and Dispatch Goods, which provide a reusable container service for fast food delivery businesses [23].

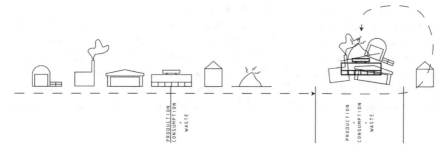

Fig. 6 Reimagining the supermarket food chain to enable urban food circularity

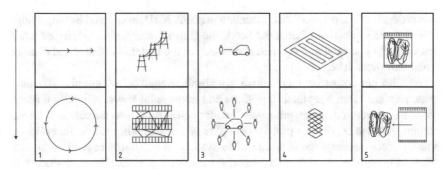

Fig. 7 Linear to circular food urban metabolism matrix. 1. linear to circular economy, 2. centralised to distributed energy, 3. private to shared mobility, 4. traditional to technical agriculture, 5. disposable to reusable packaging

At the city scale, each supermarket node could connect with its neighbourhood to deliver and collect resources to enable the transformation of food waste back to food again. Food could be produced, processed, sold, and recollected within this future urban landscape, supported by a fleet of supermarket-owned mobility vehicles delivering and collecting packaging, food, and food waste, as well as people and locally produced energy. Thus, the seemingly frictionless flow of foods from around the world to urban supermarkets could be mirrored as they flow between supermarkets around cities and back to supermarkets again.

The separation and closed-loop nature of food and packaging flows would lead to a broadening of activities at the shop scale too. The delivery yard could expand to receive flows from nearby producers and consumers. While, beside it, a combined heat and power plant could produce energy for the neighbourhood from organic waste flows. The carpark could also be filled with urban farming units and docked fleet vehicles, remaking the supermarket a node for mobility as well as resource flows. Inside, rentable kitchen units could allow small-scale businesses to process food and prepare meals for local residents.

If replicated across the UK, this new type of food-retail would shift responsibility for food-related waste management from consumers and councils back to supermarkets resulting in profound impacts on the visibility of food systems within cities to effectively reinstate connections between demand and supply. This new form of food-retail would also reframe the role of the consumer, resulting in radically changed operations within supermarkets. With the shift in responsibility for waste directed towards the retailer, food packaging would prioritise resource efficiency and reuse over consumer convenience. Despite this, consumers would still need to play a part in transferring foods to a different end-user for this new form of food-retail to be realised.

6 Conclusion

While existing urban metabolism methodologies focus on abstract visualisations to understand how resources flow through cities, analysis and reprogramming of space can help identify ways to spur physical action towards sustainable consumption and production of resources. This chapter applies systems thinking and research-by-design through drawing to explore how supermarkets manage and influence food flows along complex food chains as well as inside generic shopping spaces. Reviewing how urban metabolisms are revealed within physical spaces opens opportunities to design 'better' flows. For supermarkets, this could mean taking advantage of urban density to become more resource-efficient by mapping existing adjacencies for a 'passing trade' effect and knitting new nodes to host circular economies. Reprogramming the role of supermarkets for urban resource management could profoundly shape how food, food waste, and packaging flows across neighbourhoods, cities, and regions. Though not explored extensively in the scope of this chapter, expanding the functionality of urban food-retail to host food production, processing, and waste management could also influence smaller flow scales, such as how nutrients move and transform within and beyond the city boundaries to impact air, water and soil quality at a planetary scale.

Looking specifically at urban food flows, supermarkets are a recognisable spatial interface in the urban metabolism of UK cities; and, as a generic spatial typology, they operate in the same way despite their location. The propositional waste-free supermarket explored in this chapter uses this generic, repeatable characteristic and capitalises on existing grocery market dominance to imagine an approach to rapidly unlock circularity in any city's urban metabolism. Spatial visualisation provides a way to see how resources flow today and could flow in the future as well as how metabolisms manifest in physical environments. Though radical, the proposition seeks to promote understanding of how urban metabolisms might be altered through the design of physical space and how this might, in turn, change social behaviours and require new infrastructures of responsibility. More broadly, the proposition opens a discourse on how cities should address the challenge of rapid urbanisation, growing populations, planetary boundaries, and climate change; to consider how hungry they can afford to be, and to question the need for an urban resource diet.

References

1. Athanassiadis A (2019) Urban metabolism and open data: opportunities and challenges for urban resource efficiency. Open Cities|Open Data, pp 177–196
2. Baker-Brown D (2019) The re-use atlas. RIBA Publishing, London, p XIII
3. Belfast City Council (2020) Local development plan: 2020–2035 [online]. Belfast. https://www.belfastcity.gov.uk/getmedia/1d2a917d-3d3f-4f74-a29e-c8d360a0c83d/POP008_TP-Pop.pdf. Accessed 2 Dec 2021
4. Berger A (2007) Drosscape. Princeton Architectural Press, New York. Capra F, Luisi P (2018) The systems view of life. Cambridge University Press, New York, p 336

5. Bowlby R (2000) Carried away: the invention of modern shopping. Faber and Faber, London
6. Boyd G (2012) Designing bare essentials: ALDI and the architectures of cheapness. In: Morrow R, Abdelmonem MG (eds) Edge conditions in architecture. Taylor and Francis, London
7. Cullen S, Keeffe G, Logan K, Campbell E (2020) End-of-the-line urbanism: reprogramming the FEW-nexus of the city region for a post-carbon society. In: Proceedings of the 18th international conference on sustainable energy technologies (SET 2019), 20–22 August 2019, Kuala Lumpur, Malaysia, vol 2. 18th International Conference on Sustainable Energy Technologies, Kuala Lumpur, Malaysia, 20/08/2019, pp 386–394. https://nottingham-reposi tory.worktribe.com/output/3936800/proceedings-of-the-18th-international-conference-on-sus tainable-energy-technologies-set-2019-20-22-august-2019-kuala-lumpur-malaysia
8. D'Amico G, Arbolino R, Shi L, Yigitcanlar T, Ioppolo G (2021) Digitalisation driven urban metabolism circularity: a review and analysis of circular city initiatives. Land Use Policy 112:105819
9. FAO (2013) Food wastage footprint [online], p 6. Accessed 25 Oct 2021
10. Footprintnetwork.org (2021) Global footprint network [online]. https://www.footprintnetwork. org/. Accessed 15 Dec 2021
11. Girardet H (2010) Regenerative cities [online]. Worldfuturecouncil.org. www.worldfuturecoun cil.org
12. Hämäläinen M (2020) Smart city development with digital twin technology. In: 33rd Bled eCon-ference—enabling technology for a sustainable society: June 28–29, 2020, online conference proceedings
13. Keeffe G, Cullen S (2021) The flexible scaffold: design praxis in the FEW nexus. In: Roggema R (ed) TransFEWmation: towards design-led food-energy-water systems for future urbanisation. Contemporary Urban Design Thinking, Springer Nature Switzerland, Switzerland, pp 95–106
14. Maranghi S, Parisi M, Facchini A, Rubino A, Kordas O, Basosi R (2020) Integrating urban metabolism and life cycle assessment to analyse urban sustainability. Ecol Ind 112:106074
15. Mayne B (2021) Climate change and the circular economy in Wales [online]. Circular Online. https://www.circularonline.co.uk/research-reports/blog-climate-change-and-the-circular-economy-in-wales/. Accessed 15 Dec 2021
16. Meadows D (ed) (2009) Thinking in systems: a primer. Earthscan, London
17. Pawley M (1998) Terminal architecture. Reaktion Books, London
18. Petter O (2021) This campaign wants people to return their plastic to supermarkets with a special note [online]. The Independent. https://www.independent.co.uk/life-style/plastic-sup ermarkets-campaign-hugh-fearnley-whittingstall-shoppers-a8973836.html. Accessed 17 Dec 2021
19. Pincetl S, Bunje P, Holmes T (2012) An expanded urban metabolism method: toward a systems approach for assessing urban energy processes and causes. Landsc Urban Plan 107(3):193–202
20. Roggema R (2019) City of flows: the need for design-led research to urban metabolism. Urban Plan 4(1):106–112
21. Roggema R (2020) Designing the sustainable city. In: Roggema R (ed) Designing sustainable cities. Contemporary urban design thinking. Springer, Cham
22. Roggema R (2014) Towards enhanced resilience in city design: a proposition. Land 3(2):460–481
23. Schupak A (2021) 'The cusp of a reuse revolution': startups take the waste out of takeout [online]. The Guardian. https://www.theguardian.com/us-news/2021/oct/22/takeout-delivery-food-waste-free-reusable-containers-startups. Accessed 17 Dec 2021
24. Sdgs.un.org (2021) UN [online]. https://sdgs.un.org/goals. Accessed 15 Dec 2021
25. Statista (2021) Great Britain: Grocery market share 2017–2021 [online]. https://www. statista.com/statistics/280208/grocery-market-share-in-the-united-kingdom-uk/. Accessed 17 Dec 2021
26. Steel, C. (2013). Hungry city: how food shapes our lives. Vintage, London, pp 111, 112, 141–143
27. Thomson G, Newman P (2018) Urban fabrics and urban metabolism—from sustainable to regenerative cities. Resour Conserv Recycl 132:218–229

28. Thün G, Velikov K, Ripley C, McTavish D, Fishman R, McMorrough J (2015) Infra eco logi urbanism, 1st edn. Park Books, Zurich
29. Van den Dobbelsteen A, Keeffe G, Tillie N, Roggema R (2012) Cities as organisms. In: Roggema R (ed) Swarming landscapes: the art of designing for climate adaptation. Springer, Dordrecht, pp 195–206
30. Webster K (2017) The circular economy: a wealth of flows. Ellen MacArthur Foundation Publishing, Isle of Wight

A Brief Glance on Global Waste Management

Astrid Allesch and Marion Huber-Humer

1   Introduction

Waste management (WM) is a very trans- and interdisciplinary field of research. How wastes are managed, collected, treated and disposed have a significant impact on the public health and environment. WM and its strategies have been clearly broadened and changed during the past 25 years. Further, burden on limited resources will cause environmental degradation and fragility, while resources in dumps are lost posing a threat to our environment. Hence, the UN and European Union currently seek an effort to create a "**sustainable and circular society**" via the **Sustainable Development Goals (SDG)** and the **Circular Economy (CE)** Package.

This paper provides an overview of main **goals, trends, challenges and data availability** in global waste management with the focus to show the importance of a sound database needed to enhance a well-founded and goal-oriented decisions making towards a circular economy (see Fig. 1).

Using wastes as **secondary (raw) materials** will undoubtedly be an essential part of future WM systems. During the twentieth century, it became apparent that the uncontrolled economic growth could not be sustainable without permanent damage of the future environment [60]. To provide a global overview supporting efficient resource management, resource classification frameworks were developed to quantify natural resources and compare potential mining projects [103]. Hence, numerous classification systems for resources (minerals and energy) have been established all over the world. McKelvey [65] used the terms "reserves" and "resources". *Reserves* are defined as identified deposits, recoverable under existing economic and technological conditions. *Resources* include undiscovered deposits (same quality as reserves), as well as stocks currently unrecoverable for either technological or

A. Allesch (✉) · M. Huber-Humer
Department of Water-Atmosphere-Environment, Institute of Waste Management and Circularity, University of Natural Resources and Life Sciences, Vienna, Muthgasse 107, 1190 Vienna, Austria
e-mail: astrid.allesch@boku.ac.at

© The Author(s), under exclusive license to Springer Nature Switzerland AG 2023   227
M. Tribaudino et al. (eds.), *Minerals and Waste*, Earth and Environmental Sciences Library, https://doi.org/10.1007/978-3-031-16135-3_9

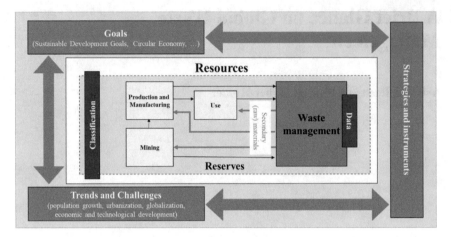

Fig. 1 Overview and contextualization of goals, trends, challenges and strategies on global waste and resource management

economic reasons. In 1997 the UN published the United Nation Framework Classification (UNFC) for Reserves and Resources of Solid Fuels and Mineral Commodities (see Fig. 2), which has been continuously revised and updated. The UNFC [93] is a principles-based system in which products of a resource project are categorized by three criteria of technical feasibility (F), environmental-socio-economic viability (E), and degree of confidence in the estimate (G).

Today, the UNFC [93] provides companies, countries, economic institutions and further stakeholders with an instrument for the sustainable development of mineral and energy resources. UNFC applies to energy resources including renewable energy, gas, oil, minerals, nuclear energy, projects for storage of CO_2, groundwater, and anthropogenic resources, such as secondary resources. For the transition to a CE, a standardized classification methodology for anthropogenic resources appears to be necessary and resource classification can play a key role in assessing the recovery and utilization potential of anthropogenic resources by analogy to natural resources [102].

Driven by industrialization huge amounts of resources (raw material) have been mined over the last centuries to become a part of the anthroposphere and aggregated to urban stocks [16]. Hence a successful transition to a CE, as promoted by the European Commission, requires solid data on the future availability of anthropogenic resources [102] as a basis for "urban mining"—urban mining projects for example include (i) heavy metals such as Pb, Zn and Cd from waste incineration residues (e.g. filter ashes), (ii) increasing the re-use and recycling rate of building and deconstruction materials, and (iii) making concrete available for the industry [85]. Especially industrialised nations have accumulated a wealth of resources in the form of infrastructure, buildings and other goods, which are a valuable stock of secondary resources. This anthropogenic stock should be seen as a potential capital which has to be exploited and managed in a systematic way [81]. Allesch et al. [4]

summarized different studies on the anthropogenic stock and shows that minerals are the major share followed by metals, wood and plastic (see Fig. 3).

For some materials, especially in cities, the anthropogenic resources are already in the range of economically mineable amounts of natural resources [4]. Urban mining is only successful if the economic incentives are high (e.g., gold in jewellery, rhenium in gas turbine blades, lead in batteries). However, without economic instruments, our society seems likely to continue in mining and processing primary metals, using them seldom, and letting them dissipate into the environment [43]. Today, many elements in the periodic table are used to take their advantage of its unique physical

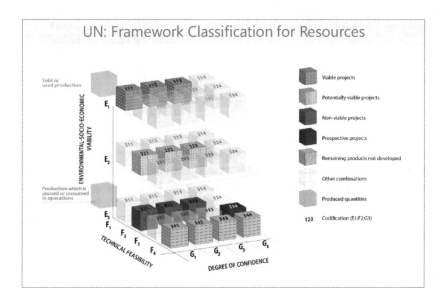

Fig. 2 UN: framework classification for resources [93]

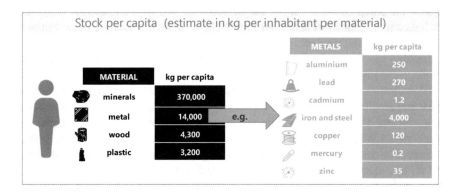

Fig. 3 Stock per capita (global estimate in kg per inhabitant per material) based on Allesch et al. [4]

Table 1 Comparison of specific material values of buildings reported in different studies [kg/m³ gross volume] based on [55]

	Kleeman et al. [55]	Gorg [41]	Rentz et al. [77]	Gruhler et al. [44]	Baccini and Pedraza [7]	Blengini [8]
Minerals (bricks, concrete, gravel, sand)	260–450		100–419			387
Cement asbestos	0–1.5					
Steel	0.1–8.6	2.08–23.22	2–16	0–37		14.6
Aluminium	0.03–0.55					0.013
Copper	0.002–0.5				0.05–0.24	0.023
Wood	0.6–20		2–28			0.44
Various plastics	0.15–5.1					0.96

and specific chemical properties resulting in numerous products which are more reliable and practical. As a consequence recycling has become more challenging and complex, often non-existent or at least inefficient because of limits imposed by thermodynamics, product design, social behaviour and recycling technologies [76]. Urban mining is a practical option for transforming the social behaviour into a CE— especially as natural resources are exhausting, the linear economy (traditionally: take-make-dispose) is no longer wanted and converting to a CE is an answer towards sustainability [16]. Continuous accumulation of materials in cities has led to the exploration of urban mining for secondary resources. Kleemann et al. [55] compared the gross volume for different material fractions (see Table 1).

To achieve effective urban mining, different measures are needed, such as design-for-recycling, mandatory use of secondary raw materials, development of digital material inventories like component recognition using robotics and artificial intelligence including databases about the composition [4]. Non-energy raw materials are linked to all industries across all supply chain stages but they are mainly driven by changes through new energy technologies like wind turbines, solar panels and sustainable mobility (e.g. electric vehicles), in which these materials are currently irreplaceable [64].

Globalisation has led to a uncontrollable and untraceable exchange of resources and waste around the world; hence currently a large amount of waste even hazardous ones, can end up in uncontrolled dumps or treatment plants posing a great risk for future global WM and a threat to the environment. Especially, the increased waste generation is causing greater environmental degradation when polluting water, land and air due to unsustainable waste disposal and waste management [107]. And as the world hurtles toward its urban future, MSW is one of the important by-products of urban lifestyle, which is growing faster due to rapid urbanization [67]. However, there is a very diverse understanding of the term and definition of waste. A reflection and review of the literature demonstrates that there is a great uncertainty concerning the term "waste" with different authors using the term contrarily [89]. A very general classifications of waste is presented by Amasuomo and Baird [5]:

- Physical state: solid, liquid and gaseous
- Source: household, industrial, agricultural, commercial, demolition-construction, mining
- Environmental impact: hazardous and non-hazardous

The European Union defines waste within the Waste Framework Directive as any substance or object which the holder discards or intends or is required to discard [30]. The EU Waste List, which administers waste designations for all EU member states, mainly classifies waste according to the source of waste generation and differentiates between hazardous and non-hazardous wastes [92]. For example, construction and demolition waste (C&DW) is defined as waste generated by construction and demolition activities. Solid waste from the C&D industry is one of the main waste streams in many countries. Further, industrial wastes are generated due to processing and manufacturing raw materials and new products including commercial, agricultural and mining waste. In Austria [104], MSW is defined as mixed and separately collected waste from households, including paper and cardboard, glass, metals, plastics, bio-waste, wood, textiles, packaging, waste electrical and electronic equipment, waste batteries and accumulators, and bulky waste, including mattresses and furniture or as mixed waste and separately collected waste from other sources, where such waste is similar in nature and composition to waste from households. Further, an additional characteristic of waste is addressed [104] as the need for treatment; meaning that collection and treatment of an object is in the interest of the public (so-called objective waste term).

Managing waste is one of society's greatest challenges [60] and different studies show that especially developing countries and countries in transition may struggle to provide effective and efficient systems to the inhabitants. They often face problems beyond the ability of the municipal authority to tackle, mainly due to the lack of organization, financial resources, complexity and the system´s multi-dimensionality [46]. Hence, managing MSW effects the health and the environment [110]. Globally, WM has been a local concern over the past few decades, with local authorities implementing various management practices to control the increasing quantities of solid waste being generated and disposed [60]. On the local level the focus on MSW was and is also a hygienic and sanitation perspective, particularly in low-income countries.

Waste is managed in very different ways at every scale, and because waste is a societal concern, governments play a key role in waste prevention and management, from collection and treatment to indirect controls such as enforcement of waste regulations and laws by local and national authorities [60]. There is a significant presence of the informal sector in waste management, especially in low-middle income cities as formal collection systems for waste and recyclable materials are not developed yet—these activities are intensified in times of economic crises when raw materials are more expensive [36].

In developed countries the focus switched over time from managing MSW for sanitary reasons to manage solid waste to conserve resources—for all types of wastes, from mining waste to MSW. Especially C&DW is one of the major waste fractions in our society [11]. A considerable and increasing quantity of C&DW is produced in the EU and as a result, particular interest must be paid to C&DW management at the EU level having consequences for national policies [18]. Further, mining activity has considerably increased, leading to growing environmental concerns and challenges for all of stakeholders within the sector. The mining industry produces huge quantities of waste that must be managed (strategically) and treated combining economic efficiency and environmental sustainability. Energy requirements, environmental and human health risks, demands for water resources, and the required technology must all be taken into account [6].

2   Waste Management Goals

The European Union defines WM in general as the collection, transport, recovery (including sorting), and disposal of waste, including the supervision of such operations and the after-care of disposal sites, and including actions taken as a dealer or broker [30]. Solid WM is an important topic for sustainable growth involving technical, economic, ecological, social, political, and cultural aspects [14]. Sustainable WM as part of resource management requires clear goals and priorities and the general aim is to improve WM in terms of environmental protection reducing impacts per capita and resource conservation.

Allesch and Brunner [3] pointed out that developed countries like Austria, Germany and Denmark have achieved high principles in WM. However, particularly on a global level there is still considerable need for improvement. WM has to be optimized by further expanding separate and recycling-oriented waste collection, reducing emissions from waste treatment, and creating aftercare-free landfills for (hazardous) residues that have to be discharged from the anthropogenic cycle. Especially for developing countries—essential initial steps are rejecting open dumps or open burning and implementing an extensive waste collection service to achieve sustainable waste management in lower-income countries [78]. WM and environmental sustainability is one important target for every country since it could be linked to the wellbeing and is considered as an essential right and privilege [2].

Global: United Nations (UN) Sustainable Development Goals (SDG)

The UN introduced 17 SDGs for worldwide sustainability achievements and committed to achieve sustainable development in its three dimensions—economic, social and environmental—in a balanced and integrated manner [94]. Proper waste management and treatment plays a central role towards the reduction of uncontrolled dumping, reducing its negative environmental and health impacts [62]. WM is a trans-sectoral subject affecting many areas of sustainability in each of the three domains [78]. In relation to resources and WM, the UN agenda attempts to reduce the negative impacts of urban activities and the use of chemicals that are hazardous for human health and the environment, especially through environmentally sound management and the safe use of chemicals, as well as the reduction and recycling of waste, and a more efficient use of water and energy. The principles of sustainability were introduced within the SDGs to improve social equality, reduce poverty, decrease environmental pollution and secure better living [36]. In particular, the global WM goals for improving sustainability at global level, are mainly mentioned within SDG 11 and SDG 12:

- 11.6 By 2030, reduce the adverse per capita environmental impact of cities, including by paying special attention to air quality and municipal and other WM
- 12.3 By 2030, halve per capita global food waste at the retail and consumer levels and reduce food losses along production and supply chains, including post-harvest losses
- 12.4 By 2020, achieve the environmentally sound management of chemicals and all wastes throughout their life cycle, in accordance with agreed international frameworks, and significantly reduce their release to air, water and soil in order to minimize their adverse impacts on human health and the environment
- 12.5 By 2030, substantially reduce waste generation through prevention, reduction, recycling and reuse

Improvements in WM will substantially contribute to better living conditions and better health of more than 2–3 billion people who currently lack services, prevent plastics entering the oceans, significantly contribute to climate change mitigation, and help restore terrestrial ecosystems. This will create new jobs that support many people on their way out of poverty [78].

Europe: EU Goals

The European Green Deal aims to move Europe towards a circular and climate-neutral economy and hence the European Commission adopted the new Circular Economy Action Plan in March 2020. This Europe's new agenda for sustainable growth is one of the main building blocks of the European Green Deal. The EU's transition to a CE should reduce pressure on natural resources and will create sustainable growth and jobs [29].

Measures that are introduced under the new action plan aim to make sustainable products the norm in the EU, empower consumers and public buyers, generate less waste, make circularity work for people, regions and cities, lead global efforts on circular economy and focus on the sectors that use most resources and where the potential for circularity is high (electronics and ICT, batteries and vehicles, packaging, plastics, textiles, food, water and nutrients, construction and buildings). The new Circular Economy Action Plan is focused on the built environment which has a significant impact on many sectors of the economy, on local jobs and quality of life [29]:

- addressing the sustainability routine of construction products (including minimum requirements for the recycled content—considering safety and functionality aspects)
- promoting measures to improve the adaptability and longevity of goods
- use life-cycle-assessment methods in public procurements
- investigating the appropriateness of setting CO_2 reduction targets
- consider revising the material recovery targets set for C&DW and its specific factions
- promote initiatives to increase the sustainable and circular use of excavated soils, reduce soil sealing, remediate contaminated or abandoned brown sites and
- recycling waste products

The Waste Framework Directive [30] sets out some basic WM principles and requires that waste be managed without:

- endangering human health and harming the environment
- risk to air, water, soil, plants or animals
- noise or odour nuisance
- disturbance to the countryside

The waste hierarchy defines an order of priority among five WM options (see Fig. 4) [31]. Current WM practices are influenced by the waste hierarchy, which recommends an order of priority from most preferred option to the least preferred option of [39].

Further the "polluter pays principle", set in the Framework Directive, requires polluters to pay for the costs of waste prevention and waste management, ensuring that these costs are shown in the price. Implementing these principles, Extended Producer Responsibility (EPR) is used as an instrument for several waste streams in

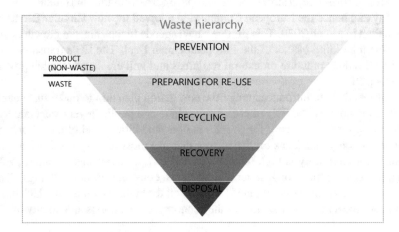

Fig. 4 Waste hierarchy [30]

the EU. It suggests that producers take over their economic operational responsibility for collecting, sorting or even recycling their goods [31].

The CE is at this time one of the most discussed terms among environmental and economic scientists [38]. One accepted definitions of the CE is offered by the Ellen MacArthur Foundation [28]: "A circular economy is one that is restorative and regenerative by design and aims to keep products, components, and materials at their highest utility and value at all times, distinguishing between technical and biological cycles." A review by Kirchherr et al. [54] shows that the CE concept has gained momentum both among scholars and practitioners and that over 100 definitions for CE are available. The review shows that the CE is often used as a combination of reduce, reuse and recycle activities, but often the needed systemic shift is not highlighted. By promoting the adoption of closing-the-loop production patterns within an economic system, CE aims to increase the efficiency of resource use, with special focus on urban and industrial waste, to achieve a better balance and harmony between economy, environment and society [40]. The CE aims at extending the life time of materials and promoting to maximize the recycling quota while the environmental impact and resource use is decreased [90]. CE can be defined as a closed-loop economy including three characteristics [42]:

• reducing the use of non-renewable and toxic materials
• reusing products (and services) by design offering system and business models
• recycling the waste into new resources for further consumption

3 Waste Management Data

Giving an overview of total waste generation would be essential, but in reality it is almost impossible to do this with sufficient certainty [99]. A solid WM and optimization of resource recovery need consistent data on waste generation and its composition. However, worldwide standardised and commonly accepted waste characterisation methodologies, are missing. This limits both comparability and applicability of the global waste data [25]. The consistency of WM data is influenced by many factors, including vague definitions (incomplete or inconsistent), lack of data and methodology description, inconsistent units and estimations based on simple and not described assumptions. Quantifying global pollution from waste, partly due to this lack of fine-scale data, is difficult [15] and hence data on solid waste and resources are only available for developed countries, and on the global level only for MSW. Further, recycling quotes have been defined in many different ways (at different life stages of a product or material) and sometimes the term recycling quotes is just left undefined [43].

People are disposing growing quantities of waste, and the waste composition is becoming more complex (this applies especially for plastic and electronic consumer products) and the world is urbanizing extraordinary, posing a challenge to urban areas and cities, to manage huge amounts of waste in an environmentally and socially acceptable manner [95]. In developing countries, most of the generated waste is landfilled or dumped effecting the public health and environment in a negative way [21]. Especially MSW is known to be an environmental and health challenge but simultaneously an economic source on which many individuals make a living through waste picking, re-using and recycling [110]. In addition, in highly populated countries such as China, India, Turkey, Mexico, and Brazil, almost 90% of the solid waste is landfilled or dumped—including organic waste as the major share [21].

Global MSW generation (amount and compositions)

Waste is a growing issue linked directly to the way society produces and consumes [99]. The generation of waste has been studied at different regional levels and the quantity as well as the quality of waste differs greatly between and within regions [19]. Tisserant et al. [90] calculated that the total globally amount of waste generated was in the year 2007 about 3.2 billion tonnes. Wilson and Velis [100] estimated that the total solid waste in 2010 from households, commerce, industry and construction, is 7–10 billion tonnes. Of this, around 2 billion tonnes are MSW, for which municipal governments have taken responsibility. At least 33% of the generated MSW—extremely conservatively estimated—is not managed in an environmentally safe manner. Based on Kaza et al. [51] the global MSW (residential, commercial, and institutional waste) generation in 2016 was estimated to have reached 2 billion tonnes. Countries in the Pacific, East and Central Asia and Europe are generating 43% of the global waste. The Middle East, North Africa and Sub-Saharan Africa regions produce the least amount of waste—together accounting for 15% of the world's waste. Pacific and East Asia are generating the most waste (absolute)—about 470 million tonnes of waste are generated in 2016 in Pacific and East Asia (generating the most) compared to about at 129 million tonnes in the Middle East and North Africa region (generating the least). Waste generation has an overall positive correlation with economic development as shown in Fig. 5.

Average waste generation per capita across countries varies substantially, from 0.1 kg per capita per day to 4.5 kg per capita per day [51]. The gross domestic product per capita income in developed countries is correlating with the waste generation [19]. Further, Sharma and Jain [84] pointed out that the considerable difference between

Fig. 5 Global waste generation by income level based on [51]

countries in the MSW generation rate highly depends on the country's urbanization, industrialization, population growth and economic development. Das et al. [19] provide an overview of MSW generation in different countries. For example:

- USA (0.64 million tonnes per day)
- India (0.50 million tonnes per day)
- China (0.48 million tonnes per day)
- Germany (0.14 million tonnes per day)
- Mexico (0.13 million tonnes per day)
- Japan (0.10 million tonnes per day)

Most developed countries are trying to implement the WM hierarchy and focus on the reduction of generated MSW quantities. In most developing countries MSW problems have become very serious due to dependence on open waste dumping, inadequate waste infrastructure and the informal sector [84].

When looking forward, global waste is expected to grow to 3.4 billion tonnes by 2050 and there is generally a positive correlation between waste generation and income level [51]. However, it must be taken into account that in high-income, industrialized countries, where formal WM has a high coverage, these quantities are also documented in the official statistics, which is not the case in developing countries, where mainly the informal WM sector is active. This is also a major factor why per capita waste shows such low generation (in the statistical data) in developing countries. Further, reliable data on the composition of waste are crucial for planning and assessing of WM systems as well as for improving resource recovery [25]. Waste composition is generally determined by a standardized audit, in which samples are taken (directly from the waste generators or from treatment and disposal sites), sorted and weighed [51]. At global level, the major waste category of MSW is green and food waste (up to 44%). Dry recyclables (paper and cardboard, plastic, metal and glass) amount to about 38%. In general, the waste composition varies considerably by income level [51] (see Fig. 6).

Global MSW treatment and disposal

Tisserant et al. [90] calculated that the total amount of waste generated worldwide in 2007 was approximately 3.2 billion tonnes, of which 1 billion tonne was recycled or reused, 0.7 billion tonnes were incinerated, gasified, composted, or used as aggregates, and 1.5 billion tonnes were landfilled. Kaza et al. [51] stated that around the world, 37% of waste is disposed of in landfills, 33% is openly dumped, 19% undergoes materials recovery and 11% is treated through state-of-the-art incineration. Although about 33% of waste is openly dumped, many governments are starting to recognize the costs and risks of open dumps and focussing more on sustainable waste disposal method. A sanitary landfill is a state-of-the-art landfill which ensures disposing of wastes by minimizing pollution hazard [19] compared to open dumps. But open dumping is still very common in developing country (posing risk to human health and environment) [87]. Improper waste management through open dumping and burning contributes to many environmental problems, such as human health risks, ozone depletion, global warming, ecological damages, etc. [52].

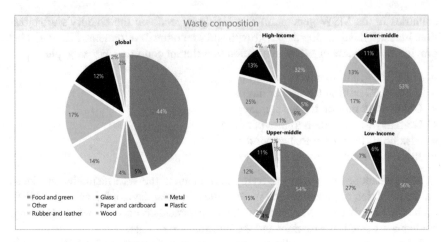

Fig. 6 Global SW composition based on [51]

Although sanitary landfills are more in the focus, Maalouf et al. [62] concluded that the share of uncontrolled disposal (dumping and burring) will continue to rise at least until 2028. Evidently, the global community continues to face serious challenges regarding the implementation of the UN SDG 12, target 12.4, by 2020. The analysis demonstrates that infrastructure capacity must increase by four folds to eliminate uncontrolled disposal practices. Natural disasters and C&D activities generate massive amounts of waste and although efforts to recycle and reuse are increasingly made, it is estimated that about 35% of these quantities are directly landfills without any pre-treatment [66].

EU waste generation (amount and compositions)

MSW accounts for only about 10% of total waste in the EU but has a very high political profile due to its complex composition, its distribution and its link to consumption habits. In 2019, MSW generation vary considerably from 280 (Romania) to 844 (Denmark) kg per capita (see Fig. 7). The variations reflect differences in consumption patterns and economic wealth, but also depend on how municipal waste is defined and how waste data is reported [34]. Differences especially occur in the past, as countries report either the collected or the treated amount of waste as the total waste generated.

C&DW is generated when infrastructures and buildings are renovated or when they reach their end-of-lifetime. C&DW—one of the largest waste streams—have a significant potential for waste reduction in general but especially for decreasing primary raw material consumption [59]. In 2018, the total waste generated in the EU (all economic activities and households) amounted to 2,340 million tonnes (5.2 kg per capita) [35]. In the year 2018, the construction sector in the EU is responsible for more than 35%, followed by mining and quarrying (27%), manufacturing (11%), waste and water services (10%) and households (8%); the remaining was waste generated

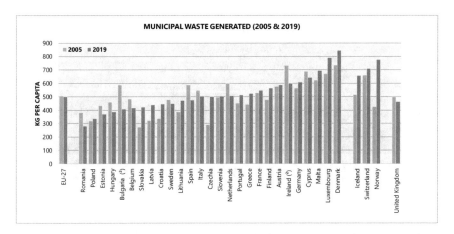

Fig. 7 European Union—municipal solid waste generation 2005 and 2019 in kg per capita [34]

from other economic activities [29]. In the year 2018 in the EU 28 approximately 370 million tonnes of mineral waste from C&D are generated.

Estonia and Finland generate the highest levels of total waste per inhabitant capita (more than 25 tonnes per capita in 2018) which is almost four times higher than the EU average (7 tonnes per capita) (see Fig. 8). Waste from mining and from construction and demolition is most often categorised as mineral waste which results in 2018 to about 5.2 tonnes per capita mineral waste (about three quarter the total waste). EU Member States which had higher shares of mineral waste had relatively large mining activities, such as Bulgaria, Finland or Romania and/or large construction and demolition activities, such as Luxembourg. In these states mineral waste accounted for about 85–90% of all waste generated. A major share of the EU raw materials requirements is covered by import and hence the European Commission launched the Raw Materials Initiative as a response to ensure a sustainable supply of raw materials and to develop a local industry [91]. In total, the construction sector is responsible for more than 35% of the generated waste and the greenhouse gas (GHG) emissions from extraction, manufacturing, construction and renovation of buildings are estimated to be about 10% of the total GHG emissions [33].

EU solid waste treatment and disposal

In the year 2018, 2.200 million tonnes of waste were treated in the EU (import is included, export is not included). As shown in Fig. 9 in the year 2018, about 55% of the waste was recovered (of the total treated waste)—recycled (38%), backfilled (11%) or energy recovered (6%). The remaining 45% was either landfilled (38%), incinerated without energy recovery (1%) or otherwise disposed (6%).

More than half of the amount of waste was not reused, recovered or recycled and hence the EU loses significant quantities of resources (potential secondary raw materials). Though many critical raw materials (CRM) have high recycling potential

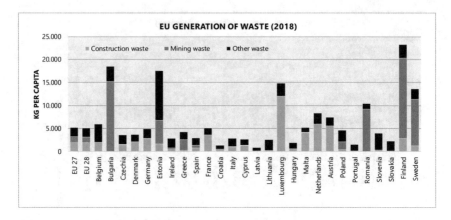

Fig. 8 European Union—waste generation in 2018 in kg per capita [35]

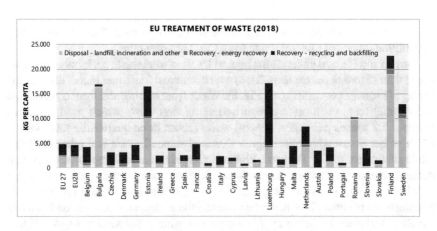

Fig. 9 European Union—waste treatment in 2018 in kg per capita [35]

the end-of-life recycling input rate (EOL-RIR) of CRMs is generally low—despite encouraging governments to move towards a CE [64] (see Fig. 10).

4   Waste Management: Grand Trends and Challenges

Solid WM is a universal issue affecting every single person in the world. Individuals and governments making decisions about WM are affecting the daily health, productivity, and cleanliness of communities [51]. While hygienic considerations have been high on the list of priorities in recent years, the increasing complexity and

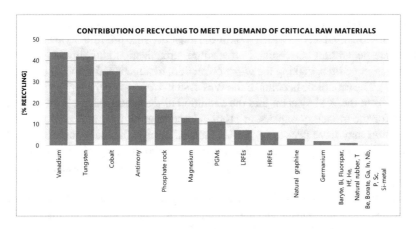

Fig. 10 Contribution of recycling to meet EU demand of CRM based on [64]. Data from: Deloitte [22] and Deloitte [23]

volume of waste have become major issues of WM in our societies [10]. **Urbanization, economic growth and increasing population** have caused intensive resource consumption consequently releasing large amounts of waste to the environment. Globally many waste and resource management systems lack a holistic approach to cover the whole product chain of design, material extraction, production and manufacturing, consumption and WM [86]. Rapid growth (urbanization and industrialization) is ongoing, particularly in developing countries like China and India with the goal to attain developed status. This would result in globally more urban growth and more waste leading to more difficulties for developing countries to tackle these enormous quantities of future waste including a proper disposal of solid waste [87].

On a global perspective, globalisation and urbanisation are key-challenges for the future WM, causing changes from regional to world-wide strategies. Today, **more than 50 % of the world's population lives in cities** and urbanization is increasing rapidly. By 2050, as many people will live in cities as there was in the year 2000. Moreover, the per capita **amount of MSW is growing even faster than the rate of urbanization**. Ten years ago, about 0.64 kg of MSW per person and day were generated (as a mean value on a global perspective), but today these amounts have increased to 1.2 kg per person and day. Although global rates are highly uncertain and variable, they point out future challenges. Especially in developing countries with barely any financial and economic resources for a sound and professional WM, the disposal in uncontrolled landfills/dumps as well as on sanitary landfills will continue.

An integrated municipal solid WM plan and implementation plays a critical role in reducing mortality and morbidity rates as it includes sustainable management of hazardous and non-hazardous waste generated from residential and other commercial sources within the community [2]. The complexity and amount waste associated with modern economy poses serious risk to ecosystems and human health [61].

As countries develop from **low-income to middle or high-income levels**, their WM situations evolve. Growth in wealth and movements to urban areas are increasing the waste generation per capita. Furthermore, urbanization and population growth create larger centres, making the waste collection, treatment and disposal more difficult [51]. The economic benefits of the WM systems depend on the types of resources gained from the WM systems and the condition of the waste market is an important factor for the monetary expansion [106]. Tisserant et al. [90] indicated that waste generation may have peaked in the UK and other high-income countries and that waste generation rates could decline from 2.4 in 2008 to 2.3 kg/day per capita by 2025. Nevertheless it shows that a high GDP correlates with high waste generation rates [19].

Particularly the construction sector plays a crucial role in the future, since a lot of resources are consumed, and pollutants are generated during the construction, operation, and disposal of buildings. Massive building materials are required during the construction phase, often consisting of a large amount of minerals such as clay, limestone, gravel, coal, silica, iron or bauxite. This leads to depletion of natural resources, causing serious environmental problems [16]. The construction industry related questions have been subject to many analysis with respect to the environmental performance in the past decade [17].

Material stocks in cities are growing, which implies that even 100% recycling of construction waste could theoretically replace only a small part of the material needs of the construction sector (e.g. 35% for a case study in Vienna) [56]. To close material cycles in the building sector, knowledge about the use of materials on the one hand and the material input of the WM system on the other hand is essential—while detailed knowledge of waste streams and composition is important to e.g. achieve EU recycling targets [56].

Tomić and Schneider [91] pointed out that socio-economic and legislative developments lead to changes in MSW systems, which results in increased costs of WM services for the citizens. This socio-economic impact can be alleviated through smart planning and parallel integration of energy and material recovery technologies which leads to synergies and decreases WM costs, and thus, increases socioeconomic acceptability of such changes. The mining and processing of ores is associated with the generation of waste and its disposal at the surface or in mines. Mining activities result in many negative environmental and socio-economic impacts. Many changes are taking place on the territory and within the society, such as modifications of soil use, forced resettlement, emissions and pollutions, water shortages and disturbance of groundwater flows, modifications of infrastructure networks, changes in the economic structure [6]. The footprints of these developments are visible in form of heaps, dumps or ponds. Repeatedly the wastes represent an artificial landscape component, often resulting in objects of different shapes and heights or artificial water reservoirs. The waste takes up agricultural and forest land and poses a landslide hazard leading to negative impacts for the environment [57].

Table 2 shows main trends, their effects and challenges for the WM system on the global level, mainly based on trends outlined in EEA [26]. Trends underline the need for action to reconfigure the global systems of production and consumption so that operations within the planetary limits are possible and that the well-being of current and future generations is ensured. These can only be achieved if all stakeholders collaborate along the value chain. Solid WM is a major problem for many urban local bodies, where urbanization, industrialization and economic growth have resulted in increased solid waste generation per person, but where organisational, institutional, legal and financial frameworks are improper or even missing. Achieving sustainability within a regions or countries experiencing rapid population growth and improvements in living standards is challenging [58].

5 Waste Management: Strategies and Instruments

Various instruments and strategies are already in use to reduce post-consumer waste as well as industrial residues and waste in the upstream stages of the value chain, like mining, production, manufacturing and distribution, both in industrial and developing countries. This chapter presents both global strategies in developing /developed countries as well as countries in transition, and further shows strategies on the European level.

Global strategies

Since the 1970s, many directives and laws have been implemented to regulate WM to limit its production and environmental impact. New services, such as garbage collection and kerbside collection points, recycling centres, waste incinerators have been developed and realised [53]. While WM is normally a local service, both national and international (e.g. EU-regulations) administrations play a key role in defining the regulatory framework [51]. Kirakozian [53] shows that, while necessary, regulatory and government options alone do not help to reverse the trend towards increasing waste volumes or to change consumer behaviour. However, economic incentives acting due to a price promote individual behavioural changes. Public awareness, socio-demographic issues and proper WM infrastructure are vitally important [74]. Regions and cities need to be aware of threats and opportunities to improve their capacity in addressing WM challenges [75]. In general, following instruments and steering measures are applicable in WM to control and affect waste prevention and enhance separate waste collection:

- regulatory measures (e.g., bans or commandments)
- economic/financial instruments (like taxes, fees, incentives etc.)
- voluntary activities/measures (e.g., voluntary commitments, contracts between businesses and the state, certifications)
- awareness raising measures (on diverse levels, e.g., from household level to industry)
- infrastructure measures (e.g. enhancement of secondary markets)

Specific recommendations for developing countries are provided by Wilson et al. [99]:

Table 2 Trends and challenges for waste management

		Trends	Implication/opportunities/challenges for waste management
Population and health	World population	• The global population has more than doubled since the 1960s and will reach 9.6 billion in 2050 and virtually all growth is projected to occur in developing countries, and particularly in their cities [26] • A growing population typically leads to increased use of natural resources and pollution, and causes land use changes such as urbanization [26]	• Due to steady economic growth, waste generation is expected to increase, especially in developing countries (to about 3.4 billion tons by 2050) [51] • Still missing or insufficient WM systems in economically developing countries [24] • Sufficient WM is increasingly threatened by population growth and increasing consumption
	Urban and rural population	• In less developed regions, more than 5 billion people will live in urban regions in 2050, while the rural population is declining. In more developed regions about 1 billion will live in urban areas [26]	• Effective WM is a major task and challenge in cities with high population density [58] • The amounts of solid waste are increasing [24], particularly in urban regions, due to strongly increasing number of people—mainly living in cities in developing countries • Lack of adequate waste collection and treatment infrastructure in developing countries including a high spatial inefficiency due to the fact that most future urbanisation is likely to be informal settlements/slums often with inadequate housing, drinking water supply, sanitation, or waste collection and treatment including informal waste pickers
	Age distribution	• In 2050 the world's median age will increase from 28 to 36 and raise the percentage of over 65s to 16% of the global population [26]	• Providing employment and educational chances for younger people will be challenging but critical if social and political stability is to be ensured [26] • Especially training young people in the area of WM is essential to provide a sustainable development
	COVID 19 pandemic	• Globally, on 24 August 2021, there have been 212 357 898 confirmed cases of COVID-19, including 4,439,843 deaths. As of 23 August 2021, a total of 4 619 976 274 vaccine doses have been administered [97]	• Protecting lives has become the main tasks of government decisions and policies at all levels. Waste from isolation and quarantine centres must be treated separately, while the normal waste can be treated regularly. Further, plastic consumption will increase due to the increasing use of personal protective healthcare objects [83] • WM plays an important role in a pandemic in terms of hygienic measures and protection of human health

(continued)

Table 2 (continued)

		Trends	Implication/opportunities/challenges for waste management
	World premature deaths	• The OECD predicts that the number of premature deaths from particulate matter pollution in urban areas could more than double to 3.6 billion by 2050 (mainly in China and India) [26] • Urban air pollution, particularly ozone and particulate matter (PM), is projected to become the world's biggest environmental cause of mortality by 2050 [26]	• Open burning of waste, whether in businesses, homes, individual residences or at dumps, is a major source of air pollutants [98] • The open waste combustion PM2.5 emissions are equivalent to 29% of the total global anthropogenic PM2.5 emissions [98] • Implementation of a sound WM can help decreasing urban pollution from PM
Economic	GDP	• The balance of economic forces will shift over the next decades and after recovering from current crisis, the global GDP could grow at an average rate of approximately 3% per year (over next 50 years). Growth in the 34-nation of the OECD area is projected to be around 2% annually through 2060, with declining rates in many high-income countries [71]	• MSW generation has a positive correlation ($r = 0.539$) with the GDP [106] challenging that about 84% of the waste is collected world-wide and only 15% of the collected waste is recycled • Huge different in coverage rate of formal waste collection schemes between low- and high-income countries [51]
	Middle class consumption	• The global middle class—defined as households with daily available funds of USD 10–100 per person—is projected to increase from 27% of the world population of 6.8 billion in 2009 to 58% of more than 8.4 billion in 2030 [26]	• A growing global middle class is increasingly adopting the resource-intensive consumption patterns of advanced economies, and the shift in overall pollution is rapidly exceeding globally tolerable limits [26] and increased consumption leads to higher waste generation.
Energy and materials	Energy consumption	• Global energy consumption will increase by 31% in the period 2012–2035 [26] • If the world population increases average energy consumption to EU levels, this would mean a 75% increase in global energy consumption, while an increase to US levels would mean a 270% increase [26]	• Waste-to-energy incineration in upper-middle-income countries already increased from 0.1–10%, driven by China's shift to incineration—some countries recovering energy from waste by landfill gas collection and anaerobic digestion [51] • Turning waste into energy could be an additional aspect towards a CE; especially due to the strong correlation of energy-use to economic output this could be regarded as a competitor to material recycling. Hence, waste data, e.g. on composition is crucial for the planning of an integrated waste management systems • Where sanitary landfills are used, the CH4-gas can be captured, converted to power, used for heating, or applied to serve as fuel [51] • However, still a high amount of CH4 generated in landfills is emitted uncontrolled and unused into the atmosphere impacting global climate change.

(continued)

Table 2 (continued)

	Trends	Implication/opportunities/challenges for waste management
Global material use	• Increase of minerals extensively used since 1700 (C, Ca, Fe) to 2000 (over 40 different materials) • Developing regions account for an increasing proportion of global resource use and resource use tends to rise as countries develop economically [26] • Increasing high-tech products lead to increased consumption of various mineral resources and production processes have turned into a complex system using composite and hazardous materials [26]	• The EU is extremely dependent on imports of large quantities of materials used in a modern society, and resource issues (criticality) are high on the political agenda today • For this, proper knowledge of the primary material flows and stocks as well as the human industrial and societal metabolism is required as important background information. A combination of natural and anthropogenic cycles of the elements can help to create more complete 'maps' of Europe's resource base by showing potential future material stocks (in-use stocks) and natural stocks (e.g., in soils) [69] • The waste we produce today (post-consumer) is mainly from diverse sources, expensive to manage sustainably and environmentally damaging [73] • Huge opportunity and challenge for WM to provide clean recycling loops and high-quality secondary resources
Critical raw materials	• Global production of critical raw materials is quite concentrated in a limited number of countries (main countries often in a political unstable situation) [26] • Few countries influence the global prices and supplies	• The use of several materials in individual applications to increase functionality and to promote low-carbon technologies and resource efficiency have increased the demand for many critical elements [69]—big challenge for WM to recover such critical materials from complex products, but an opportunity to become more independent from imports.
Ore grade	• Mining wastes emerge as a result of mine production and increases due to a decreasing ore grade	• Mining wastes are one of the most critical environmental problems resulting from mining operations causing adverse environmental impacts, like water pollution. Other significant environmental impacts of mining waste include visual pollution, ecosystem and land degradation, pulverization and erosion [105]

(continued)

Table 2 (continued)

		Trends	Implication/opportunities/challenges for waste management
Environment	Climate change	• It is unequivocal that human influence has warmed the atmosphere, ocean and land. Widespread and rapid changes in the atmosphere, ocean, cryosphere and biosphere have occurred [48] • Since 2011 GHG concentrations have continued to increase in the atmosphere, reaching annual averages of 410 ppm for carbon dioxide, 1866 ppb for methane, and 332 ppb for nitrous oxide in 2019 [48] • Compared to 1850–1900, global surface temperature averaged over 2081–2100 is very likely to be higher by 1.0–1.8 °C under the very low GHG emissions scenario considered, by 2.1–3.5 °C in the intermediate scenario and by 3.3–5.7 °C under the very high GHG emissions scenario [48]	• GHG emissions resulting from inadequate waste collection, uncontrolled landfilling and incineration. Waste emits methane gas when disposed of in a low-oxygen environment such as dumps or landfills, and emits pollutants and particulate matter when transported and incinerated inefficiently [51] • 1.6 billion tonnes of CO_2-equivalent missions from the MSW management sector are estimated for 2016 and are anticipated to increase to 2.6 billion tonnes by 2050 [51] • GHG emissions from the WM sector are primarily driven by disposal of biodegradable organic waste in uncontrolled dumps and landfills (mainly without landfill gas collection systems) which accounts for at least 630 million t CO_2-equivalents [37]
	Landfills (land use)	• The total global area of the continents is around 14,900 million ha. Depending on the definition and measurement method, built-up areas with settlements and infrastructure occupied a relatively small area around 2005—1–3% of the total area. Without interventions, infrastructures and settlements are expected to expand by around 260 to 420 million ha by 2050, covering about 5% of the global area [9] • Winkler et al. [101] estimates that in just six decades, land-use change has affected nearly a third (32%) of the global land area, about four times the size previously estimated from long-term land-change estimates	• World-wide, most waste is currently dumped or disposed of in some form of a landfill and especially open dumping is prevalent in lower-income countries, where controlled and engineered landfills are not yet available [51] • Closing dumpsites is challenging due to many potential problems associated to the closure of dumps. These issues are usually related to the closure method, how and who pays for it, and what new waste disposal method would be the new best option to replace the open dump situation [49] • It is also a real challenge to close an open dumpsite while it is still in use and without building a new waste disposal facility. Therefore, a new waste disposal facility must be implemented allowing the old site to be completely off-limits to users, thus forcing its full closure once a new, upgraded site becomes available [49] • Selecting suitable sites for the construction of (hazardous) waste facilities is a complex process, because it needs a multidisciplinary approach that incorporates natural, physical-social sciences, politics, and ethics [20] • After removal, the former use of the land as a dump must be noted in the land register and the areas can be treated as fallow land rehabilitation or as passive recreational parks [49]—however the subsequent use of such a site is usually restricted and limited • Particularly the spatial expansion and shape of mining residues and waste has a huge impact on the environment and landscape

(continued)

Table 2 (continued)

	Trends	Implication/opportunities/challenges for waste management
Marine plastic pollution	• More than 40% of the world's population lives in areas within 200 km of the sea, and 12 out of 15 megacities are on the coast—growing human prosperity and rapid industrial development are putting increasing pressure on the oceans [96] • Our more technologically advanced and fast-growing societies are increasingly impacting their environments (local and global), resulting in physical and chemical pollution [96] [1] • Jambeck et al. [50] calculated that 275 million metric tons (MT) of plastic waste was generated in 192 coastal countries in 2010, with 4.8 to 12.7 million MT entering the ocean and creating plastic pollution	• The world's annual plastic production reached nearly 350 million tons in 2018 and 10% of that eventually ended up in the world's oceans, resulting in large concentrations of plastic entering the oceans [109] • Waste (litter) and marine debris have been shown to have negative impacts on public health, wildlife, economy and tourism. About 700 species of wildlife are identified to have interacted with plastic debris, frequently with lethal consequences [82] • Littering, missing or inadequate waste collection and management systems along big river systems and at the coasts—particularly in developing countries—as well as in the shipping industry are still big drivers for marine (plastic) pollution, and thus a huge challenge for WM

- mobilize environmental and climate funds
- 100% waste collection coverage (especially in all mega-cities)
- abolish open incineration of wastes and close dumps by replacing them with controlled treatment facilities
- develop a holistic approach to manage all wastes and residues
- build on existing recycling systems
- promote producer responsibility programmes

Challenges arise in relation to interfaces between optimal planning and dimensioning of solid waste treatment plants and optimal scheduling of waste flows considering ecological and social costs, such as taxes, user fees, investment costs and subsidies [72]. These socio-economic strategies, implemented by only a handful of developed countries in the world, could be expanded to reduce waste generation while decoupling waste generation from economic growth [72]. CE which implies waste prevention, reuse, and recycling, or upper tiers of waste hierarchy, directly contribute to cleaner production because of less overall waste produced and discarded, from both manufacturing and raw material processing [61]. For all countries Wilson et al. [99] recommend to

- improve the access to finance sound waste treatment plants
- reduce waste generation by moving towards a CE and
- improve the reliability and availability of waste management statistics and data (if you don't measure it, you can't manage it).

Globally, many regulatory measures are being proposed to mitigate environmental problems, and the most prominent are extended producer responsibility, environmental taxes, and carbon markets responsibility [45]. Kaza et al. [51] shows that 86% of countries and economies reported that there is official national law or policy governing WM. Many studies emphasize that evolving consumption patterns and the increasing power of an ecological principles are likely to change consumer choices and behaviour [53].

Globally waste segregation is not taking place widely leading to several environmental problems [74]. Schuyler et al. [82] divided possible effective measures into only two variants—market-based economic instruments and command and control measures.

- Economic instruments set monetary incentives or disincentives to influence human behaviour. Both economic incentives such as container deposit or disincentives such as disposal taxes are effective ways to control waste generation and management.
- Command and control instruments are direct regulations related to activities or unwanted articles, such as bans on single-use plastic bags or bans on plastic microbeads.

Both instruments have a strong impact, since waste is not only a complex anthropogenic problem; it is also an expensive problem. Correct consideration of the environmental and economic benefits of WM systems is important as such systems often involve high investment and operation costs in providing an effective waste collection and management system to the citizens [106]. Kirakozian [53] summarized several studies showing that incentive pricing increases the quantity of recycled waste and is positively effecting waste reduction, consequently, acting as a Pigouvian tax (internalization of external effects). Incentive further offers individuals information on the quantity of waste they generate encouraging responsible behaviour and funding of a WM service [53].

European Strategies

Different examples for currently applied regulatory and financial instruments on the European level are available. Many countries have taxes with differing scope. Presently, the majority of non-recyclable MSW is disposed of in landfill resulting in several environmental consequences including soil, air, and water pollution [74].

Hence, many regions and countries are striving to increase the percentage of waste sorted and recycled, thereby reducing the amount of waste going to landfill or incineration [12]. In the year 1995, too much bio-waste was still landfilled in many countries regardless of the fact that such waste could be incinerated or recovered as compost. Landfilling of biodegradable waste is a loss of resources and results in GHG [27]. The Landfill Directive therefore aims to reduce the amount of biodegradable waste that is landfilled. Minimising waste to landfill is a key element of the EU strategy and it is important to see whether progress is made towards the targets going forward. Differences between countries in the amount of waste sent to landfill may be related to how their tax systems favour disposal by incineration rather than landfill. A number of Member States have a **landfill tax**. The limited capacity of landfills is another factor that can lead countries to impose a landfill tax [27]. 23 of 27 EU Member States have a landfill tax in EUR per tonne and 16 EU Member States of 27 have a landfill ban implemented [13]. Environmental taxes are principally effective in the case of domestic waste [53].

Extended Producer Responsibility (EPR) is an alternative regulatory tool highlighted by the EU (Directive 2008/98/EC) to support its goals of reducing waste generation and improving waste management [68]. EPR is a concept where manufacturers and importers of products should bear a significant degree of responsibility for the environmental impacts of their products throughout the product life-cycle, including upstream impacts inherent in the selection of materials for the products, impacts from manufacturers' production process itself, and downstream impacts from the use and disposal of the products [70]. In an analysis, Gupt and Sahay [47] show that the economic responsibility of the separate collection and recycling centres play an important role to the success of the environmental policies based on the extended producer responsibility. **Take-back responsibilities, regulatory provisions and financial flow** emerge as the three most significant parts of Extended Producer Responsibility. A deposit system accepts that the consumer pays an amount when purchasing a product that is refunded when it is returned or sent to a collection system. The literature regards the deposit system as a policy mix and shows its superiority [53]. Incentives are often more expensive to implement, but they avoid some of the potentially negative aspects of disincentives, such as an increase in illegal dumping to avoid taxes.

In addition, it has been shown that reimbursement (refund) systems or positive incentives can significantly increase the recycling of materials, be it printer cartridges, beverage packaging or end-of-life cars [82]. Studies point out that incentive pricing is effective (e.g. as waste weight progressive taxation). This system of taxation encourages individuals to minimize or reduce the amount of residual waste [53].

By implementing the **Single-use Plastic (SUP) Directive** [32], the EU aims to become a forerunner in the global fight against marine litter and plastic pollution by reducing the volume and impact of certain plastic products on the environment. The SUP directive is also one of the first directives which prohibit the placing of specific products on the market. Hence, if sustainable alternatives are available and affordable, SUP products cannot be placed on the markets of EU Member States (e.g. plates, plastic bags, cutlery, straws and stirrers, balloons, cups for beverages). The SUP Directive is the an important EU legislation to strengthen the transition to a CE by introducing measures ensuring separate collection for recycling of certain products such as fishing gear or beverage bottles [88]. From a governing perspective, coasts or coastal states can profit from spatial planning processes and multi-stakeholder assessments. Globally, each cubic meter of the sea will and is expected to meet numerous (often conflicting) requirements. Spatial planning processes considering the requirements of neighbouring countries and the global marine system helps to find more equitable and sustainable regulations for the use of and access to the oceans [96].

6 Outlook on Future, Innovative Waste Management Concepts

The way waste is managed affects the environment, the livelihood and health and well-being of the population, and the relationships between governments and citizens [51] on a local, national, and global level. The EU is committed to transitioning **from a linear to a circular economy** and EU waste law is a major part of EU's environmental policy and a key element in efforts to develop a sustainable and resource-efficient economy [63]. The increasing global consumption of energy and raw materials, accompanied by increasing waste generation and emissions to the environment, leads to growing worries about security and sustainability of the supply.

Due to the global supply and value chains, the complexity of the resource and emission problems to be solved is increasing. Raw materials and commodities are the backbone of the production and manufacturing value chain and play an important role as a basis of growth and wealth. Many countries are facing similar challenges and the complexity of the global waste management has increased over the past decades as a consequence of trends like population growth, urbanization, global-ization, economic and technological development. Strategies to tackle these issues are required. In this respect, the need to understand and see the physical value and the respective economy behind is an essential issue as well. The UNFC frame-work can assist to classify the physical economic values of resources systematically. In 2018, specifications for application of the UNFC framework to anthropogenic resources were published, however, a standardized classification methodology for different types of anthropogenic (secondary) resources (just like "waste materials") is essential but still missing [102].

The concept of a CE is a prevailing bridging concept to foster the fundamental links among resource use, emissions and waste to integrate economic and environmental policies [61]. However, preventing products and materials from becoming waste is the top priority in waste management by taking measures before a substance, material or product become waste [30]. **Waste prevention** strategies (products or services) can intensive the use with concepts like **reuse, sharing, smart product-design, multiple use and hence increase the performance and lifespan of products**. In particular, product design for a CE has come into focus as a new area of research in the broader field of sustainable design—extending product lifespans and fully recovering products and materials are key elements of this approach [80].

Further, **waste management is getting 'smarter'** focussing on solving waste management problems by using intelligent systems, sensors and mobile applica-tions. But the implementation of intelligent/smart waste management systems is still in its infancy [108]. Humans have created a variety of policies and technologies managing waste and minimize environmental and public health threats it poses [95]. In addition, digitization methods of **Industry 4.0** are developing rapidly in all sectors, especially for use in WM there are more and more applications used [79]. Our society is producing more, increasingly complex solid waste, and this waste is concentrated in cities [95]. Hence, homogeneity of the materials necessary for high-quality recy-cling must be achieved, which is increasingly done, e.g., in sorting plants through spectroscopic and visual analysis with coupled (automated) sorting. However, these analyses at best, record the main components of the materials (e.g. the polymer) and are not able to determine the more precise compositions (e.g. additives in plastics).

It is arguably an increasingly complex problem affecting the behaviour of citizens, manufacturers, product designers, and policy makers, and therefore encompassing infrastructure and managerial aspects of operationalizing the technologies [108].

In contrast to the state of the art, new or adapted technologies from other sectors are available to close material cycles. These technologies include those that directly examine materials and products, such as **artificial intelligence-assisted image recognition methods, sound-based methods, odour sensors and others**. Many methods require previously marked materials (QR, NFC, RFID). This tagging is associated with a database of any depth of information. These new technologies could be applied to specific waste fractions and streams by combining or replacing them with traditional state of the art methods and could make a significant contribution providing clean material loops. Further, technologies like RFID tags, NFC sensors,

GPS, etc. facilitate real-time data collection and inform effective decision-making in restoration activities [108].

Despite the ability of solutions improving the way resources are used and recycled, technology selections differ by context. Societies vary by technical capacity, regional aspects, income level, GDP, waste quantities and composition and often the best solution is neither the latest nor the most advanced technologically [51]. Summarized and considered more comprehensively, it is not (only) a technical solution that we need to tackle our resource and waste problem. A strong change in our consumer behaviour, far-reaching waste prevention measures and a noteworthy transformation of our production and economy system must be the answer to the challenges we are facing in our throw-away society.

References

1. Regulation No 450/2009 on active and intelligent materials and articles intended to come into contact with food, p 9
2. Al-Khatib IA, Kontogianni S, Nabaa HA, Al-Sari MI (2015) Public perception of hazardous-ness caused by current trends of municipal solid waste management. Waste Manag 36:323–330
3. Allesch A, Brunner PH (2016) Material flow analysis as a tool to improve waste management systems: the case of Austria. Environ Sci Technol 51:540–551
4. Allesch A, Laner D, Roithner C, Fazeni-Fraisl K, Lindorfer J, Moser S, Schwarz M (2019) Energie-und Ressourceneinsparung durch Urban Mining-Ansätze. Bundesministerium für Verkehr, Innovation und Technologie, Wien
5. Amasuomo E, Baird J (2016) The concept of waste and waste management. J Manag Sustain 6:88
6. Aznar-Sánchez JA, García-Gómez JJ, Velasco-Muñoz JF, Carretero-Gómez A (2018) Mining waste and its sustainable management: advances in worldwide research. Minerals 8:284
7. Baccini P, Pedraza A (2006) Die bestimmung von materialgehalten in gebäuden. Bauwerke als Ressourcennutzer und Ressourcenspender in der langfristigen Entwicklung urbaner Systeme–Ein Beitrag zur Exploration urbaner Lagerstätten. vdf Hochschulverlag AG an der ETH Zürich, Zürich, Switzerland, pp 103–132
8. Blengini GA (2009) Life cycle of buildings, demolition and recycling potential: a case study in Turin, Italy. Build Environ 44:319–330
9. Bringezu S, Schütz H, Pengue W, O'Brien M, Garcia F, Sims R, Howarth RW, Kauppi L, Swilling M, Herrick J (2014) Assessing global land use: balancing consumption with sustainable supply. United Nations Environment Programme Nairobi
10. Brunner PH, Rechberger H (2015) Waste to energy–key element for sustainable waste management. Waste Manag 37:3–12
11. Butera S, Christensen TH, Astrup TF (2015) Life cycle assessment of construction and demolition waste management. Waste Manag 44:196–205
12. Carattini S, Baranzini A, Lalive R (2018) Is taxing waste a waste of time? Evidence from a supreme court decision. Ecol Econ 148:131–151
13. Cewep (2021) Landfill Taxes and Bans
14. Chang N-B, Pires A, Martinho G (2011) Empowering systems analysis for solid waste management: challenges, trends, and perspectives. Crit Rev Environ Sci Technol 41:1449–1530
15. Chen DM-C, Bodirsky BL, Krueger T, Mishra A, Popp A (2020) The world's growing municipal solid waste: trends and impacts. Environ Res Lett 15:074021

16. Cheng K-L, Hsu S-C, Li W-M, Ma H-W (2018) Quantifying potential anthropogenic resources of buildings through hot spot analysis. Resour Conserv Recycl 133:10–20

17. Coelho A, de Brito J (2013) Environmental analysis of a construction and demolition waste recycling plant in Portugal-Part I: Energy consumption and CO_2 emissions. Waste Manag 33:1258–1267

18. Dahlbo H, Bachér J, Lähtinen K, Jouttijärvi T, Suoheimo P, Mattila T, Sironen S, Myllymaa T, Saramäki K (2015) Construction and demolition waste management–a holistic evaluation of environmental performance. J Clean Prod 107:333–341

19. Das S, Lee S-H, Kumar P, Kim K-H, Lee SS, Bhattacharya SS (2019) Solid waste management: scope and the challenge of sustainability. J Clean Prod 228:658–678

20. De Feo G, De Gisi S (2014) Using MCDA and GIS for hazardous waste landfill siting considering land scarcity for waste disposal. Waste Manag 34:2225–2238

21. De Medina-Salas L, Castillo-González E, Giraldi-Díaz M, Fernández-Rosales V, Welsh Rodríguez C (2020) A successful case in waste management in developing countries. J Pollut Effl Control 8:1–5

22. Deloitte Sustainability (2015) Study on data for a raw material system analysis: roadmap and test of the fully operational MSA for raw materials, Prepared for the European Commission, DG GROW. https://ec.europa.eu/jrc/en/scientific-tool/msa

23. Deloitte Sustainability (2017) Deloitte Sustainability, British Geological Survey, Bureau de Recherches Geologiques et Minieres, Netherlands Organisation for Applied Scientific Research (2017), Study on the review of the list of Critical Raw Materials—Criticality Assessment, Report prepared for the European Commission. https://publications.europa.eu/en/public ation-detail/-/publication/08fdab5f-9766-11e7-b92d-01aa75ed71a1/language-en. https://doi. org/10.2873/876644

24. Diaz LF (2017) Waste management in developing countries and the circular economy. SAGE Publications Sage UK, London, England

25. Edjabou ME, Jensen MB, Götze R, Pivnenko K, Petersen C, Scheutz C, Astrup TF (2015) Municipal solid waste composition: sampling methodology, statistical analyses, and case study evaluation. Waste Manag 36:12–23

26. EEA (2019) The European environment—state and outlook 2020. Knowledge for transition to a sustainable Europe. https://www.eea.europa.eu/publications/soer-2020. Accessed 02 June 2021

27. EEA (2020) 11. Waste. European Environment Agency

28. Ellen MacArthur Foundation (2021) What is a circular economy? https://ellenmacarthurfoun dation.org/topics/circular-economy-introduction/overview. Accessed 14 Dec 2021

29. EU Commission (2020) A new circular economy action plan—for a cleaner and more competitive Europe (COM(2020) 98 final)

30. EU Parliament (2008) Directive 2008/98/EC of the European Parliament and of the Council on waste and repealing certain Directives. Off J Eur Union

31. EU Parliament (2015) Understanding waste streams treatment of specific waste. European Parliamentary Research Service

32. EU Parliament (2019) Directive (EU) 2019/904 of the European Parliament and the council on the reduction of the impact of certain plastic products on the environment

33. European Commission (2022) Buildings and construction. https://ec.europa.eu/growth/ind ustry/sustainability/buildings-and-construction_de. Accessed 15 Jan 2022

34. EuroStat (2021a) Municipal waste statistics. https://ec.europa.eu/eurostat/statistics-explai ned/index.php?title=Municipal_waste_statistics#Municipal_waste_generation. Accessed 14 Dec 2021

35. EuroStat (2021b) Waste statistics. https://ec.europa.eu/eurostat/statistics-explained/index. php?title=Waste_statistics#Total_waste_generation. Accessed 14 Dec 2021

36. Ferronato N, Torretta V (2019) Waste mismanagement in developing countries: a review of global issues. Int J Environ Res Public Health 16:1060

37. Fischedick M, Roy J, Acquaye A, Allwood J, Ceron J-P, Geng Y, Kheshgi H, Lanza A, Perczyk D, Price L (2014) Industry. In: Climate change 2014: mitigation of climate change.

Contribution of working group III to the fifth assessment report of the intergovernmental panel on climate change. Technical report

38. Geisendorf S, Pietrulla F (2018) The circular economy and circular economic concepts—a literature analysis and redefinition. Thunderbird Int Bus Rev 60:771–782
39. Gharfalkar M, Court R, Campbell C, Ali Z, Hillier G (2015) Analysis of waste hierarchy in the European waste directive 2008/98/EC. Waste Manag 39:305–313
40. Ghisellini P, Cialani C, Ulgiati S (2016) A review on circular economy: the expected transition to a balanced interplay of environmental and economic systems. J Clean Prod 114:11–32
41. Gorg H (1997) Entwicklung eines Prognosemodells fur Bauabfalle als Baustein von Stoff-strombetrachtungen zur Kreislaufwirtschaft im Bauwesen. Verein zur Forderung des Instituts fur Wasserversorgung, Abwasserbeseitgung ….
42. Goyal S, Esposito M, Kapoor A (2018) Circular economy business models in developing economies: lessons from India on reduce, recycle, and reuse paradigms. Thunderbird Int Bus Rev 60:729–740
43. Graedel TE (2011) The prospects for urban mining, The Bridge
44. Gruhler K, Böhm R, Deilmann C, Schiller G (2002) Stofflich-energetische Gebäudesteckbriefe-Gebäudevergleiche und Hochrechnungen für Bebauungsstrukturen. DEU
45. Gu F, Guo J, Hall P, Gu X (2019) An integrated architecture for implementing extended producer responsibility in the context of Industry 4.0. Int J Prod Res 57:1458–1477
46. Guerrero LA, Maas G, Hogland W (2013) Solid waste management challenges for cities in developing countries. Waste Manag 33:220–232
47. Gupt Y, Sahay S (2015) Review of extended producer responsibility: a case study approach. Waste Manage Res 33:595–611
48. IPCC (2021) Climate change 2021. The physical science basis. Summary for policymakers. Working Group I contribution to the sixth assessment report, intergovernmental panel on cimate change
49. ISWA (2016) A roadmap for closing waste dumpsites. The world's most polluted places. International Solid Waste Association, Vienna, Austria
50. Jambeck JR, Geyer R, Wilcox C, Siegler TR, Perryman M, Andrady A, Narayan R, Law KL (2015) Plastic waste inputs from land into the ocean. Science 347:768–771
51. Kaza S, Yao L, Bhada-Tata P, Van Woerden F (2018) What a waste 2.0: a global snapshot of solid waste management to 2050. World Bank Publications
52. Khandelwal H, Dhar H, Thalla AK, Kumar S (2019) Application of life cycle assessment in municipal solid waste management: a worldwide critical review. J Clean Prod 209:630–654
53. Kirakozian A (2016) One without the other? Behavioural and incentive policies for household waste management. J Econ Surv 30:526–551
54. Kirchherr J, Reike D, Hekkert M (2017) Conceptualizing the circular economy: an analysis of 114 definitions. Resour Conserv Recycl 127:221–232
55. Kleemann F, Lederer J, Aschenbrenner P, Rechberger H, Fellner J (2016) A method for determining buildings' material composition prior to demolition. Build Res Inf 44:51–62
56. Kleemann F, Lederer J, Fellner J (2019) Urban mining—potentials and limitations for a cirucular economy, fact book for a circular economy CEC4europe. Circular economy coalition for Europe
57. Kudełko J (2018) Effectiveness of mineral waste management. Int J Min Reclam Environ 32:440–448
58. Kumar S, Smith SR, Fowler G, Velis C, Kumar SJ, Arya S, Rena, Kumar R, Cheeseman C (2017) Challenges and opportunities associated with waste management in India. R Soc Open Sci 4:160764
59. Lederer J, Gassner A, Kleemann F, Fellner J (2020) Potentials for a circular economy of mineral construction materials and demolition waste in urban areas: a case study from Vienna. Resour Conserv Recycl 161:104942
60. Letcher TM, Vallero DA (2019) Waste: a handbook for management. Academic Press

61. Luttenberger LR (2020) Waste management challenges in transition to circular economy–case of Croatia. J Clean Prod 256:120495
62. Maalouf A, Mavropoulos A, El-Fadel M (2020) Global municipal solid waste infrastructure: delivery and forecast of uncontrolled disposal. Waste Manag Res 38:1028–1036
63. Maitre-Ekern E (2021) Re-thinking producer responsibility for a sustainable circular economy from extended producer responsibility to pre-market producer responsibility. J Clean Prod 286:125454
64. Mathieux F, Ardente F, Bobba S, Nuss P, Blengini GA, Dias PA, Blagoeva D, de Matos CT, Wittmer D, Pavel C (2017) Critical raw materials and the circular economy. Publications Office of the European Union, Bruxelles, Belgium
65. McKelvey VE (1975) Concepts of reserves and resources
66. Menegaki M, Damigos D (2018) A review on current situation and challenges of construction and demolition waste management. Curr Opin Green Sustain Chem 13:8–15
67. Mian MM, Zeng X, Nasry ANB, Al-Hamadani SM (2017) Municipal solid waste management in China: a comparative analysis. J Mater Cycles Waste Manag 19:1127–1135
68. Niza S, Santos E, Costa I, Ribeiro P, Ferrão P (2014) Extended producer responsibility policy in Portugal: a strategy towards improving waste management performance. J Clean Prod 64:277–287
69. Nuss P, Blengini GA (2018) Towards better monitoring of technology critical elements in Europe: coupling of natural and anthropogenic cycles. Sci Total Environ 613:569–578
70. OECD (2006) Extended producer responsibility (EPR)— fact sheet: extended producer responsibility
71. OECD (2021) Looking to 2060: long-term global growth prospects—Bloomberg Brief. https://www.oecd.org/economy/lookingto2060long-termglobalgrowthprospects.htm. Accessed 04 Oct 2021
72. Pires A, Martinho G, Chang N-B (2011) Solid waste management in European countries: a review of systems analysis techniques. J Environ Manag 92:1033–1050. https://doi.org/10.1016/j.jenvman.2010.11.024
73. Ponnada MR, Kameswari P (2015) Construction and demolition waste management—a review. Safety 84:19–46
74. Prajapati P, Varjani S, Singhania RR, Patel AK, Awasthi MK, Sindhu R, Zhang Z, Binod P, Awasthi SK, Chaturvedi P (2021) Critical review on technological advancements for effective waste management of municipal solid waste-Updates and way forward. Environ Technol Innov 101749
75. Rahmasary AN, Robert S, Chang I-S, Jing W, Park J, Bluemling B, Koop S, van Leeuwen K (2019) Overcoming the challenges of water, waste and climate change in Asian cities. Environ Manag 63:520–535
76. Reck BK, Graedel TE (2012) Challenges in metal recycling. Science 337:690–695
77. Rentz O, Seemann A, Schultmann F (2001) Abbruch von Wohn-und Verwaltungsgebäuden–Handlungshilfe [Demolition of domestic and administrative buildings–recommendations for action]. Landesanstalt für Umweltschutz Baden-Würtemberg (LFU), Karlsruhe
78. Rodić L, Wilson DC (2017) Resolving governance issues to achieve priority sustainable development goals related to solid waste management in developing countries. Sustainability 9:404
79. Sarc R, Curtis A, Kandlbauer L, Khodier K, Lorber KE, Pomberger R (2019) Digitalisation and intelligent robotics in value chain of circular economy oriented waste management—a review. Waste Manag 95:476–492
80. Sauerwein M, Doubrovski E, Balkenende R, Bakker C (2019) Exploring the potential of additive manufacturing for product design in a circular economy. J Clean Prod 226:1138–1149
81. Schiller G, Müller F, Ortlepp R (2017) Mapping the anthropogenic stock in Germany: metabolic evidence for a circular economy. Resour Conserv Recycl 123:93–107
82. Schuyler Q, Hardesty BD, Lawson T, Opie K, Wilcox C (2018) Economic incentives reduce plastic inputs to the ocean. Mar Policy 96:250–255

83. Sharma HB, Vanapalli KR, Cheela VS, Ranjan VP, Jaglan AK, Dubey B, Goel S, Bhattacharya J (2020) Challenges, opportunities, and innovations for effective solid waste management during and post COVID-19 pandemic. Resour Conserv Recycl 162:105052

84. Sharma KD, Jain S (2020) Municipal solid waste generation, composition, and management: the global scenario. Soc Responsib J

85. Simoni M, Kuhn E, Morf LS, Kuendig R, Adam F (2015) Urban mining as a contribution to the resource strategy of the Canton of Zurich. Waste Manag 45:10–21

86. Singh J, Laurenti R, Sinha R, Frostell B (2014) Progress and challenges to the global waste management system. Waste Manag Res 32:800–812

87. Srivastava V, Ismail SA, Singh P, Singh RP (2015) Urban solid waste management in the developing world with emphasis on India: challenges and opportunities. Rev Environ Sci Bio/Technol 14:317–337

88. Syberg K, Nielsen MB, Clausen LPW, van Calster G, van Wezel A, Rochman C, Koelmans AA, Cronin R, Pahl S, Hansen SF (2021) Regulation of plastic from a circular economy perspective. Curr Opin Green Sustain Chem 100462

89. Thürer M, Tomašević I, Stevenson M (2017) On the meaning of 'waste': review and definition. Prod Plan Control 28:244–255

90. Tisserant A, Pauliuk S, Merciai S, Schmidt J, Fry J, Wood R, Tukker A (2017) Solid waste and the circular economy: a global analysis of waste treatment and waste footprints. J Ind Ecol 21:628–640

91. Tomić T, Schneider DR (2020) Circular economy in waste management–socio-economic effect of changes in waste management system structure. J Environ Manag 267:110564

92. Umweltbundesamt (2015) National and international waste classification. https://www.umw eltbundesamt.de/en/topics/waste-resources/waste-management/waste-types/waste-classific ation. Accessed 14 Dec 2021

93. UNECE (2021) United Nations framework classification for resources. Update 12019. https:// unece.org/DAM/energy/se/pdfs/UNFC/publ/UNFC_ES61_Update_2019.pdf. Accessed 27 Oct 2021

94. United Nations GA (2015) Resolution adopted by the General Assembly on 25 September 2015. United Nations, Washington

95. Vergara SE, Tchobanoglous G (2012) Municipal solid waste and the environment: a global perspective. Annu Rev Environ Resour 37:277–309

96. Visbeck M (2018) Ocean science research is key for a sustainable future. Nat Commun 9:1–4

97. WHO (2021) WHO Coronavirus (COVID-19) Dashboard, https://covid19.who.int/ (accessed 25.08.2021 2021).

98. Wiedinmyer C, Yokelson RJ, Gullett BK (2014) Global emissions of trace gases, particulate matter, and hazardous air pollutants from open burning of domestic waste. Environ Sci Technol 48:9523–9530

99. Wilson DC, Rodic L, Modak P, Soos R, Carpintero A, Velis K, Iyer M, Simonett O (2015) Global waste management outlook. UNEP United Nations Environment Programme

100. Wilson DC, Velis CA (2015) Waste management–still a global challenge in the 21st century: an evidence-based call for action. SAGE Publications Sage UK, London

101. Winkler K, Fuchs R, Rounsevell M, Herold M (2021) Global land use changes are four times greater than previously estimated. Nat Commun 12:1–10

102. Winterstetter A, Heuss-Assbichler S, Stegemann J, Ulrich K, Wäger P, Osmani M, Rechberger H (2021) The role of anthropogenic resource classification in supporting the transition to a circular economy. J Clean Prod 126753

103. Winterstetter A, Laner D, Rechberger H, Fellner J (2016) Integrating anthropogenic material stocks and flows into a modern resource classification framework: challenges and potentials. J Clean Prod 133:1352–1362

104. WMA (2002) Waste Management Act (Bundesgesetz über eine nachhaltige Abfallwirtschaft), BGBl. I Nr. 102/2002

105. Yıldız TD (2020) Waste management costs (WMC) of mining companies in Turkey: can waste recovery help meeting these costs? Resour Policy 68:101706

106. Zaman AU (2016) A comprehensive study of the environmental and economic benefits of resource recovery from global waste management systems. J Clean Prod 124:41–50
107. Zaman AU, Swapan MSH (2016) Performance evaluation and benchmarking of global waste management systems. Resour Conserv Recycl 114:32–41
108. Zhang A, Venkatesh VG, Liu Y, Wan M, Qu T, Huisingh D (2019) Barriers to smart waste management for a circular economy in China. J Clean Prod 240:118198
109. Zhang D, Liu X, Huang W, Li J, Wang C, Zhang D, Zhang C (2020) Microplastic pollution in deep-sea sediments and organisms of the Western Pacific Ocean. Environ Pollut 259:113948
110. Ziraba AK, Haregu TN, Mberu B (2016) A review and framework for understanding the potential impact of poor solid waste management on health in developing countries. Arch Public Health 74:1–11

Waste, Environment, and Sanitary Issues: Are They Really at Odds?

Maura Tomatis, Jasmine Rita Petriglieri, and Francesco Turci

1 Introduction

Assessment of waste health impact is an important scientific challenge. Waste may pose a risk to humans and the environment by two main processes: the primary release of a broad range of contaminants, including primary toxicants such as Pb, Cd, Hg, and PCBs, and the release of secondary toxicants that are produced during waste processing such as dioxins and difurans. Primary toxicants are related to the chemical composition of waste, while secondary toxicants depend on the waste management strategies that are adopted [1].

Health impact related to waste and waste management strategies is investigated through epidemiological studies. These studies are generally aimed at finding possible associations between the proximity to a disposal site or a treatment plant and some specific endpoints relevant for human health and environment. A number of relevant reviews have been published on this subject. However, despite some studies provide evidence of an association, the conclusion of most of these reviews are not definitive due to difficulties in interpreting data from primary studies, lack of accurate information on the exposure and control of potential confounders [2]. The main difficulty in the measurement of the impact of waste on public health arises from the variability of the exposure dose, which depends on several factors, and not only on the distance from the waste treatment site. In most of the available studies, however, the distance from the waste site is used as a proxy of exposure. Moving towards a

M. Tomatis · J. R. Petriglieri · F. Turci (✉)
G. Scansetti Interdepartmental Center for Studies On Asbestos and Other Toxic Particulates, University of Turin, V. P. Giuria 7, 10125 Torino, Italy
e-mail: francesco.turci@unito.it

M. Tomatis · F. Turci
Department of Chemistry, University of Turin, V. P. Giuria 7, 10125 Torino, Italy

J. R. Petriglieri
Department of Earth Sciences, University of Turin, V. Valperga Caluso 35, 10125 Torino, Italy

© The Author(s), under exclusive license to Springer Nature Switzerland AG 2023
M. Tribaudino et al. (eds.), *Minerals and Waste*, Earth and Environmental Sciences Library, https://doi.org/10.1007/978-3-031-16135-3_10

definition of exposome due to waste-generated chemicals and particulates is a key aspect for future assessment and management of risk.

Exposure to waste pollution results mainly from inhalation and ingestion, and different environmental matrices may be involved in the transport of contaminants. Air is the first environmental transport pathway for burning waste, but volatile products, such as dioxins, may contaminate food that becomes an indirect source of exposure [3]. Similarly, leachate from landfills can affect groundwater and soil and exposure occurs by ingestion [4]. Dermal contact represents a further possible way of exposure in occupational environments.

To further increase the complexity of a general approach to risk analysis of waste, it has to be taken into account that the impact of waste on public health is site-specific and the effect on humans and the environment are largely modulated by the different waste characteristics and waste management strategies. Waste characteristics depend on cultural, climatic, and socioeconomic conditions. It may vary between countries and seasonally, changing in quantity and composition over the course of the year [5]. Waste types may be conveniently categorized as (a) municipal, (b) agricultural, (c) electronic (e-waste), (d) hospital and (e) industrial waste [6]. Less common waste types include mining waste and waste from production and end of life of new technologies, such as engineered nanoparticles. Waste management strategies mostly depend on local governance and institutional capacity, while the waste characteristics are related to the consumption patterns and the industrialization level of the country that is investigated. For instance, waste is usually managed on a municipal or regional scale in industrialized countries, and technological methods such as mechanized collection, separation, and treatment are commonly employed. Open dumping is however still commonly used in less industrialized countries and emerging economies [7]. Improper disposal site construction and waste management causes diffuse pollution, such as the release of heavy metals in soils and waters. Open burning or uncontrolled incineration plants result in the emission of toxic volatile compounds, including dioxins and other persistent environmental pollutants (POPs) that might accumulate in the food chain [8]. Besides population resident near disposal or treatment sites, also waste workers are exposed to specific health risks. The occupational risks associated with waste management are again largely dependent on the availability of personal protective equipment (PPE) and the existence and enforcement of specific regulations of worker protection [2] (Fig. 1).

2   Municipal Solid Waste

Municipal solid waste (MSW) is usually defined as materials discarded from residential and commercial sources [9]. The mass of MSW produced in the world has been growing considerably for many decades and its amount is estimated to grow further. More than one billion metric tons of MSW are currently produced worldwide per year [10] and at least one third is not properly managed.

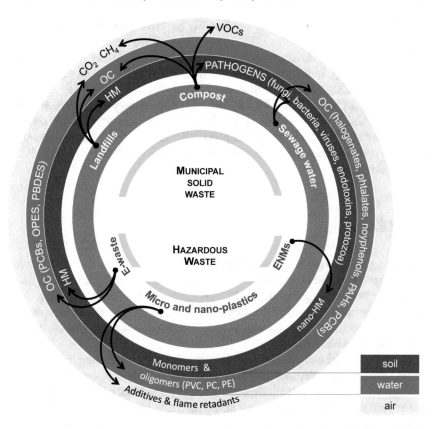

Fig. 1 Main hazardous substances in waste materials: HM = heavy metals; OC = organic compounds; VOC = volatile organic compounds; NPs = nanoparticles. Hazardous substance release into the environment depends on the waste characteristics and may vary seasonally and according to the waste management policies and practices enforced in the different geographic zones and countries

In many developed countries, MSW definition does not include industrial, medical, electronic, construction and demolition wastes which are collected separately. A comparison of MSW average compositions based on the income level of the countries shows that the major difference is in the relative amount of the organic fraction and in the percentage of recyclable paper waste. Organic fraction is higher in middle- and low-income countries than in the high-income ones, and ranges from 70 to 20%. Percentage of paper ranges from 23% in high-income countries to 7% in low-income countries. Plastic and household hazardous waste, including mineral oil, asbestos containing materials (ACM), and paints may also be present. The composition of MSW directly impacts on the feasibility of greener waste treatments. The composition indeed affects waste physical characteristics (moisture content, density, and calorific value) that in turn affect management options. High amount of organic fraction, for example, increases the waste humidity level and lower its calorific value, which in turn renders such MSW not suitable for production of thermal energy in waste-to-energy plants.

MSW is currently managed by two approaches: landfilling or conversion treatments. Landfilling is the most common practice of MSW management worldwide. Among MSW conversion strategies, incineration and composting are the two most diffuse. Different technological and safety levels may characterize the two approaches and a large technology gap between high- and low-income countries exists. Open dumping and open burning are still widely adopted in most developing countries and these uncontrolled MSW disposal practices often lead to considerable health risks.

2.1 Landfilling

Landfilling is a simple and economical method for the disposal of solid waste, as decomposition naturally occurs at the disposal site allowing transformation of the waste into stabilized and relatively inert materials. Landfills differ from dumpsites that are open and unregulated areas or holes in the ground with no environmental protection and disposal controls. Opposite to landfills, dumpsites often receive hazardous waste materials.

Landfills may be equipped with an impermeable layer separating the waste from the underlying soil, a drainage system for the leachate, and gas collection and ventilation system. Within these landfills, solid waste is placed into cells and covered with soil to mitigate odors, reduce water infiltration, and limit vector spreading, while waste decomposition proceeds. Landfill leachate is typically collected through a draining system running at the bottom of the landfill and is treated to remove organic material and suspended solids.

2.1.1 Hazardous Substances in Landfills

The main source of health and environmental hazard that might be generated by a landfill is represented by leaching waters and gas emission.

Leachate. Leachate is the aqueous effluent generated by rainwater percolating through the waste layers, biodegradation processes and the inherent water in waste. Leachate amount and composition vary considerably depending on landfilling technology, climate, waste composition and landfill age.

The amount of leachate produced is linked to precipitation, surface water run-off, infiltration of groundwater, and degree of waste compaction. Leachate production is larger in landfills built without engineered liners, waterproof covers, or leachate collection systems. Presence of poorly compacted materials and high moisture content promote leachate production. The climate affects leachates by influencing precipitation and water evaporation.

Waste composition directly influences the chemistry of leachate, albeit four groups of pollutants are always present in MSW landfill leachate [11, 12]. These are:

1. Macro-components containing inorganic cations: calcium (Ca^{2+}), magnesium (Mg^{2+}), sodium (Na^+), potassium (K^+), ammonium (NH_4^+), iron (Fe^{2+}), manganese (Mn^{2+}). Cations are commonly counterbalanced by inorganic anions, such as chloride (Cl^-), sulfate (SO_4^{2-}), and hydrogen carbonate (HCO_3^-).
2. Heavy metals: cadmium (Cd^{2+}), chromium (Cr^{3+}), copper (Cu^{2+}), lead (Pb^{2+}), nickel (Ni^{2+}), and zinc (Zn^{2+}).
3. Dissolved organic matter (DOM). DOM consists of a variety of biodegradable or non-biodegradable organic products ranging from small volatile acids to fulvic and humic-like compounds. The biodegradable fractions is composed of simple carbon-chained structures, that are originated from the initial decomposition processes of organic solid waste. The non-biodegradable fraction mainly consists of humic substances that derives from the condensation and polymerization of microbial degradation products. Dissolved organic matter in leachate is quantified as Chemical Oxygen Demand (COD) or Total Organic Carbon (TOC).
4. Xenobiotic organic compounds. This class includes highly toxic chemicals including benzene, toluene, ethylbenzene, xylenes, and halogenated hydrocarbons, such as tetrachloroethylene and trichloroethylene. The origin of xenobiotics in leachate is mainly from household or industrial chemicals, such as personal care products, pharmaceuticals or industrial products, pesticides. Xenobiotics are usually present in low concentrations, generally lower than 1 ppm.

Beside these four main classes, some other anions (borate, sulfide, arsenate, selenate) and metals (barium, lithium, mercury, and cobalt) may also be found in landfill leachate, although in very low concentrations.

The amount of each component varies according to the MSW composition, albeit with some exceptions. For example, heavy metal content in MSW largely varies among different landfills. Nonetheless, heavy metal average concentration in leachates is quite constant and usually low as less than 0.02% of heavy metals received at landfills is leached, due to the adsorptive capacity of the organic matter present in the waste. Similarly, sulfides and carbonates, which are commonly present in the landfills, form insoluble salts with heavy metals (Cd, Ni, Zn, Cu, and Pb) and limit their concentration in the leachate.

Landfill age is another key factor in defining the leachate composition [11, 13]. Leachate generation takes place in different phases, and several biological and chemical reactions occur. During a first, brief period, oxygen present in the void spaces of the freshly buried waste is rapidly consumed by aerobic decomposition with production of H_2O and CO_2. Most of the leachate produced during this aerobic phase results from the release of moisture. As oxygen is depleted, the environment becomes anaerobic. Organic biodegradable compounds, such as cellulose, undergo bacterial decomposition. Bacteria hydrolyze polymers to monosaccharides, which ferment to carboxylic acids and alcohols resulting in a decrease of pH. The acidic environment increases the solubility of many compounds. Acids and alcohols are subsequently converted to acetate, hydrogen, and carbon dioxide by acetogenic bacteria. Finally, methanogens convert the endproducts of the acetogenic reactions to methane and carbon. The pH increases as the carboxylic acids are consumed. This leads to a reduction of metal concentrations, as metallic cations precipitates as pH increases. Sulfate concentrations are also lowered by the microbial reduction of sulfate to sulfide. As a consequence of these processes, leachates with less than 5 years age have low pH and are mainly composed of low molecular weight organic matter, whereas leachates with more than 10 years age mostly contain non-biodegradable components with high molecular weight, such as humic acids.

Landfill gas. Landfill gas, also known as biogas, includes methane and carbon dioxide generated during organic matter decomposition. Landfills have been recognized as significant contributors to atmospheric methane, accounting for 6–18% of the global emissions. Biogas also contains other gases in traces, primarily sulfur compounds, responsible for odors, and volatile organic compounds (VOCs). VOCs may include a set of toxic compounds such as benzene, vinyl chloride, dichloromethane, chloroform, toluene, dichlorobenzene. The amount and composition of gas produced depends on the composition, amount, and age of waste accordingly with decomposition processes described above [14].

2.1.2 Environmental Impact

Even if the odors from landfill gaseous emissions are often the most disturbing factor for nearby population, the most severe environmental impact of landfills is related to leachate production that may percolate though the ground and pollute waters and soils [14], with detrimental effects on the food chain. Contamination of surface water by organic matter results in oxygen depletion in the waterbody and induces changes in the bottom fauna. Among the main constituents of leachate, ammonia produced during protein degradation, is highly toxic for aquatic organisms. Because the natural decomposition of waste continues for several years, without proper leachate treatments soil and water contamination may persist for decades, even after waste disposal has stopped.

Emission of gas generated from decomposition of organic matter starts one to two years after waste disposal into the landfill and continues for 15–25 years [14] upon disposal is completed. Emission methane and carbon dioxide and their accumulation in the atmosphere have a negative impact on climate change. In addition, because of the presence of flammable gas, landfill fires are relatively common incidents. These fires usually occur at low temperatures under anoxic conditions and may produce dioxins/furans, polynuclear aromatic hydrocarbons, and respirable particulate matter (PM) depending on waste composition.

Due to their high content in sulphurated compounds, the released gases may promote rain acidification. Acid rain, in turn, negatively affects soils and ecosystems, e.g., by inducing a reduction in photosynthesis, enzyme inhibition, and changes in synthetic pathways on plants.

2.1.3　Health Impact

In vitro toxicity studies

A number of bioassays are used to assess leachate toxicity [12, 15] and to evaluate the effectiveness of treatment used to reduce the amount of environmental toxicants. Environmental impact is assessed by means of bioassays on bacteria, algae, plants, invertebrates, and aquatic organisms. The impact on human health is investigated with tests on mammalian cells.

As an example, report of in vitro toxicity tests carried out of different landfill leachates are reported. Some researchers observed that leachate samples with high organic substances level were cytotoxic in different human cells lines [16, 17] inducing necrosis [18], apoptosis, shrunken morphologies, cytoplasmic vacuolations and detachment of the treated cells from the substratum [19]. Genotoxic effect of these leachates was also observed [19, 20] on lymphocytes, liver cells and, to a lesser extent, on osteosarcoma cells. COD reduction [21] and phenol removal [22] significantly decreased leachate genotoxicity.

Epidemiological studies

Several epidemiological studies on the health effects of landfills, reviewed by [2, 23–26], have been conducted in both high- and low-income countries, with contrasting results. Health risks from landfills are difficult to quantify, information on exposure is insufficient, populations are often exposed to low concentrations of toxic substances. In general a lack of characterization of the types of chemicals that are contained in leachates and the amount that can be released into the environment prevent causal correlations to be established [26]. Nonetheless, some health problems seem to be observed more frequently on populations residing near landfills and these include low birth weight, congenital anomalies, and respiratory diseases. In municipal waste workers and recyclers, musculoskeletal disorders, respiratory, gastro-intestinal and skin conditions have been observed [27].

Neonatal outcomes. There is some evidence of an increased risk of adverse birth and neonatal outcomes for residents near landfills and dumpsites [23]. A literature survey of studies carried out in the USA prior to 1998 indicated that women living near MSW disposal sites have an increased risk of delivering infants with birth defects (chromosome anomalies, heart/circulatory defects, neural tube defects). However, the studies were limited in number and conclusions regarding causality could not be draw [26]. Moreover, some of the examined landfills may have received hazardous wastes mixed with the nonhazardous wastes [25]. An increased risk of congenital anomalies was observed also in an UK study in 2005 [28]. The results refer to landfills operating between the early 1980s and the late 1990s, before the UK accepted the European Landfill Directive [1999/31/EC] to improve standards and reduce adverse effects on the environment. Limited evidence of an increase in infants with low birth weights has been found [26, 29].

Respiratory diseases: There is some evidence of an increased risk of respiratory diseases. About 80% of resident population that lived near a landfill, which served

New York City from 1948 to 1999, reported respiratory problems (asthma or respiratory symptoms). Several studies conducted in low-income countries in the 1980s to 1990s have hypothesized a relationship between working at open dumps and increased respiratory illness (symptoms of cough and breathlessness during working hours, reduction of pulmonary function) [25]. Studies carried out between 2000 and 2020 in Europe, North America [23], and South Africa [30] highlighted some evidence of an increased risk of respiratory diseases (mucosal irritation, shortness of breath, pneumonia, bronchitis, asthma and sleep apnea), mainly in young people [23]. Respiratory symptoms have been mostly associated to H_2S and PM2.5 emission.

Cancer. Although excesses in bladder, pancreas, liver, kidney, stomach, larynx and lung cancer, leukemia and non-Hodgkin lymphoma were reported in several studies [26, 27], the evidence for a causal relationship between residence near a landfill and cancer is still weak [2, 23, 24, 31].

2.2 Composting

Composting is a solid waste disposal technology where the organic fraction of the waste undergoes microbiological decomposition under carefully controlled conditions. The organic matter contained in MSW ranges from garden and food waste to mixed household wastes. The biodegradable fraction may vary from 50 to 90% depending upon countries [32]. Microbes metabolize the organic material and can reduce its volume by as much as 50%. The stabilized product is called compost and may be used in agriculture depending on its quality and maturity. Prime quality compost may be used for the production of agricultural products, secondary quality compost is suitable for landscape and forest improvement [33]. To make compost suitable for agricultural purposes, the quality of the incoming waste has to be carefully controlled, avoiding the contamination from hazardous and non-degradable wastes.

Composition and physical characteristics of compost vary according to the nature of the waste material and the conditions under which the decomposition takes place. Usually the aerobic composting process includes three stages. At the first stage, the temperature is moderate (40–45 °C) and the degradable soluble compounds are decomposed by mesophilic microorganisms. The metabolic activity of the bacteria results in a rapid increase of the temperature (50–60 °C) and the replacement of mesophiles by thermophilic microorganisms, which decompose polysaccharides, proteins and fats. The high temperature (55 °C) allows elimination of parasites and pathogens. At the final stage, the temperature decreases and mesophiles are restored to the dominant position. After decomposition, reorganization of the organic matter in stable molecules, such as humic acid (humification phase), stabilizes the compost [34].

2.2.1   Hazardous Substances in Compost

Bioaerosol emission and, although at lesser extent, release of heavy metals are the major sources of environmental and health risk posed by composting technology.

Bioaerosols: the term refers to all particles with a biological origin in suspension in the air. Bioaerosols may consist of both microorganisms (bacteria, fungi, virus, protozoa, algae, pollen) and biomolecules (toxins, debris from membranes). Among microorganisms, actinomycetes, *Aspergillus fumigatus*, and *Penicillium sp.* are some of the dominant species [35]. Bioaerosols are released during the necessary processes require to oxygenate the compost. Those involve vigorous movements of composting material such as shredding and compost pile turning. Microorganism and biomolecule concentration is significantly higher in the area surrounding the composting plants than in the environmental background, but usually returns to the background level about 250 m from the emission source [36]. However, some studies have found significant levels of actinomycetes and thermotolerant fungi at a distance of 600–1400 m from the site. Bioaerosol concentrations are highly variable ranging from 10^1 to 10^7 CFU/m^3 for bacteria and fungi, and 0.1 to 56 mg/m^3 for organic dust [37].

Heavy metals: Although health and environmental risk depends on availability rather than on concentration of pollutants [32], the presence of heavy metals (namely Cd, Cr, Cu, Pb, Ni and Zn) represents an obvious concern, if compost has to be further used for agricultural purposes. Heavy metals originate from contaminated waste and are not degraded throughout the composting process. Their amount usually increases during composting because of the degradation of the organic material with consequent waste weight loss. However, although the bioavailable fraction can differ from one metal to another, their availability is usually reduced as the compost matures. Indeed, increase of concentration of humic acids with respect to those of fulvic acids reduces metals availability as a result of complexation. Compost acidity also affects metal availability. Mature compost generally has a pH between 6 and 8 and neutral/alkaline pH values promote metal precipitation, opposite to acid pH that allows metal leaching.

Organic compounds: Compost may contain several organic contaminants, including polycyclic aromatic hydrocarbon (PAH), polychlorinated dibenzo-p-dioxins/polychlorinated dibenzofurans, polychlorinated biphenyls, and endocrine disruptors such as estrogens, testosterone, and phthalate esters. These compounds can be extremely persistent in the environment and may cause adverse health effects. However, most of these compounds are degraded during microbial decomposition and their level in compost are usually very low [32].

Volatile organic compounds (VOCs) are often detected in compost facilities. Most of VOCs occur where fresh material is stored or in the initial zone of the composting process, where crushing machines stir and facilitate desorption of volatile compounds. In a 2009 study, VOC concentrations were detected to be below the ACGIH exposure limits [38]. Emission of gas (CH_4, NH_3, CO_2) also takes place as a consequence of the degradation processes.

2.2.2   Environmental Impact

Contamination of soil with heavy metals is the major risk related to the use of poor-quality compost. However, in several countries regulations are enforced to limit the quantity of heavy metals and organic pollutants allowed in the compost [39].

Gas emissions from composting plants may contribute to eutrophication, acidification and photochemical oxidation processes.

The odor due to the VOC emissions is a negative impact of composting plants, which begins with the arrival of fresh material or is generated in aerobic fermentation in the initial phases of the decomposition. Sulfured compounds of intense odor are generated also in plants operating under anaerobic conditions [32].

However, composting also has many environmental advantages such as the reduction of the amount of material disposed to landfill, which in turn allows decreasing the total leachate production, and the increase of the calorific value of feedstock, that allows generating more energy, in case of waste-to-energy technology are used.

2.2.3   Health Impact

Bioaerosols have the potential to produce respiratory diseases [2, 27, 35, 38, 40]. The risk depends on the concentrations of bioaerosol components as well as personal exposure and health status.

Allergic alveolitis, allergic rhinitis, asthma, and upper airway irritation are the most frequent effects of prolonged exposure to bioaerosols. Some pathogens, including *Aspergillus fumigatus*, an opportunistic fungal microorganism, may cause invasive aspergillosis in immuno-compromised individuals. *A. fumigatus* and *thermophilic actinomycetes* can cause hypersensitivity pneumonitis, alveolitis and bronchial asthma. Exposure to endotoxins, components of the cell wall of Gram-negative bacteria, peptidoglycans in the wall of Gram-positive bacteria, $\beta(1-3)$-D-glucan in the cell wall of molds and mycotoxins may cause asthmatic or allergic episodes.

Workers of compost plants, who experience the highest levels of exposure to bioaerosol, are at highest risk. A significant increase of health complaints of the upper airways and the skin, as well as higher concentrations of specific antibodies against molds and actinomycetes were reported in cohorts of compost workers. Some evidence of gastrointestinal, and proinflammatory effects at high levels of exposure were also reported [37]. Compost workers also often suffer from eye irritation.

Few studies have investigated the impact on the health of residents living close to composting facilities, with contrasting results. Some of these studies did not find association between respiratory symptoms and place of residence [41] or with concentration of *Aspergillus fumigatus* [42]. Conversely, other studies [43] indicated an association between bioaerosol pollution and irritative respiratory symptoms and tiredness.

Currently, available studies do not provide quantitative dose–response estimates, and the identification of harmful levels of bioaerosol is not yet possible, nor the extrapolation of clear indications on the safety distance between composting plants and residential areas could be obtained [37]. Some review studies suggest that there is insufficient evidence to provide a quantitative association between health risk for residents and exposure to compost [40]. However, some authors recorded qualitative evidence of adverse health effects that justifies the adoption of precautionary approach, as recommended by the UK Environmental Agency [36].

2.3 Sewage Treatments

Sewage sludge is the solid residue obtained by the treatment of wastewater released from various sources, such as houses, industries, and medical facilities.

The composition of the sludge is variable and depends on wastewater origin, technology applied in treatment plants, and step of the treatment process [44]. Wastewaters from highly populated cities contains several toxic chemicals such as disinfectants, drugs along with their metabolites, endocrine disruptors, pesticides, herbicides, ammonia, heavy metals. To remove such a large spectrum of pollutants, various treatments are required to ensure that sewage pollution is effectively reduced to an acceptable environmental and health risk. Primary wastewater treatment usually involves sedimentation to remove settleable solids. The removed solid constitutes the primary sludge. Secondary wastewater treatment aims to remove organic material by microbiological degradation. Flocculated microorganisms form a biomass, which is concentrated to obtain the secondary sludge. Tertiary treatment may be carried out when high quality effluents are required, i.e., for drinking water. Tertiary treatments include ozone disinfection to remove pathogens, filtration or coagulation to reduce suspended solids, filtration on activated charcoal to eliminate organic contaminants, nitrification to decrease ammonia content and several other ad hoc treatments. The residues from tertiary treatment are incorporated into sludges from primary and secondary treatment. Resulting sludge is thus a heterogeneous mixture of microorganisms, organic (50%–70%) and inorganic (30%–50%) components with high moisture content [45]. Organic fraction contains proteins and peptides, lipids, polysaccharides, alongside micro-pollutants. Inorganic fraction includes minerals such as quartz, calcite and significant amounts of nitrogen (2.5–4.0%) and phosphorus (0.5%–2.5%) that are useful nutrients for plant growth. These characteristics make sewage sludge a good candidate to be used as fertilizer or soil improver. However, because of the presence of large amounts of pollutants and pathogens, sewage sludge needs to be treated before recycling or disposal. Several treatment technologies, aimed to destroy pathogens and eliminate offensive odors, may be employed depending on sludge characteristics [46]. The most common treatments are lime stabilization, which consists in addition of calcium oxide to strongly basify (pH > 12) the sludge. Extreme pH values drastically reduce microbiological risk and stop any aerobic and anaerobic processes. Sludges may also be digested to produce biogas (60–70% of CH_4, 30–40% of CO_2 and trace amounts of H_2S and NH_3). Digested sludge still contains a high quantity of nutrients as well as inorganic contaminants. Sludge that does not meet quality criteria for agricultural purposes [47] may be disposed of in landfill or by thermal treatment. Thermal treatment of sludges decomposes organic pollutants, neutralize pathogens, reduces the sludge volume, and recovers energy. Incineration (exothermic oxidation of solids generating CO_2, H_2O, and ash) is the most used technology. The residual ash can be used as a raw material for brick production. Other thermal treatments include pyrolysis (thermal conversion in an oxygen deficient environment at T 350–900 °C generating H_2, CO, CO_2 and CH_4, bio-oil, and char) and gasification (thermal conversion into H_2, CO, and at lesser extent CO_2 and CH_4 under partial oxidation at T 700–1000 °C) [44, 48].

2.3.1 Hazardous Substances in Sewage

The major concern for human health about wastewaters and sewage sludge arises from the presence of organic and inorganic pollutants, and pathogens [49].

Pathogens. Most of the microbes in sewage have fecal origin. Enteric bacteria (*E. coly, Salmonella enterica, Shigella dysenteriae*) are commonly found. *Legionella sp., Mycobacterium sp.,* and *Leptospira sp.* are some of the frequently found bacteria from the non-enteric class. Enteric viruses, resistant to disinfection treatments, protozoans and their oocytes are also often encountered [50].

Metallic elements. The main source of heavy metals and metalloids like As and Se is industrial wastewater. The total content varies within wide limits (from 0.5 to 2% of dry sludge). The amount of element, generally, decreases as follows: Zn > Cu > Cr > Ni > Pb > Cd. However, as previously mentioned, the toxicity is mainly related to availability instead of concentration. Mn, Ni and Zn are frequently present in the most mobile fractions, e.g., they are adsorbed on colloidal materials or on iron-manganese oxides, which are unstable in anoxic conditions, or associated with carbonate that easily dissolves at low pH. Cu typically forms complexes with organic compounds or sulfides and displays medium mobility. Cr is frequently found as insoluble sulfide with very low availability, as well as Fe that forms insoluble crystalline compounds, while Cd may be found in all different fractions [45]. Availability strongly depends on the sludge origin and treatment. Lime-stabilized sludge shows the largest range of leachable metals, while char from sludge pyrolysis has the lowest metal availability [51].

Heavy metals are also found in the solid residue that results from sludge thermal processing. Their distribution depends on thermal process, reaction time and moisture content of the sludge. In high temperature processes, As, Hg, Cd and Pb are mostly distributed in the fine fly ash particles, while Cr, Ni, Mn and Cu are mainly found in bottom ash or coarse ash [45].

Organic contaminants. Several organic contaminants are present in sludge and their concentrations are closely related with the amounts in wastewater. Contaminants observed in sludge include halogenated organic compounds, linear alkylbenzene sulphonates, phthalates; nonylphenols, polychlorinated biphenyls, polychlorinated dibenzo-dioxins and -furans, polybrominated diphenylethers [52]. The main sources of these substances are organic solvents, lubricants, resins, rubber, plasticizers, wood protectives, and some pesticides. These organic pollutants are often associated with heavy metals, *e.g.*, PAHs are associated with Hg, while PCBs are associated with Pb, Cd, and Cu [53].

In addition, emerging contaminants such as endocrine disruptors, pharmaceuticals and personal care products, microplastic and nanoparticles have attracted increasing attention in recent years [54].

2.3.2   Environmental Impact

Environmental risk of wastewaters and sludges arises from their potency to contaminate soil and water. Forests and marshlands have been used as sludge disposal in both Europe and the USA in the past. The detrimental effects on the environment include groundwater contamination, ecotoxic effects and risk of spreading pathogens. Accumulation of heavy metals and persistent organic contaminants in soil and vegetation may occur. Accumulation of nitrogen and phosphorus reduces biodiversity and harms the quality of water, promoting fast algae growth and decreasing oxygen content. Zinc accumulation negatively affects soil microbial activity [55].

Thermal processing of sludge may also impact on the environment. Presence of sulfur, nitrogen and chlorine in the sludge may lead to emission of hazardous compounds such as SO_x, NO_x, dioxins, furans, and other toxic chlorinated products [56]. Furthermore, incineration and gasification, which involves use of high temperatures, may volatilize heavy metal compounds, especially those associated with Cl because of their low boiling points.

2.3.3   Health Impact

Major risks arise from ingestion of contaminated water/food and inhalation of bioaerosol or gas. Health risk may be of biological or chemical origin.

Health risk posed by microorganisms present in wastewater and in sludges depends on their concentrations, persistence in the environment, infective dose, and host response. Most of the pathogens may survive for days or weeks in the soil and on crops that come in contact with wastewater or non-properly treated sludge. Intestinal worm diseases are easily transmitted to humans when untreated wastewater is used to water crops. A survey of the epidemiological literature up to 2003 indicates evidence of an increase of intestinal worm (mainly *Ascaris lumbricoides*) and bacterial infections (typhoid, cholera, *Helicobacter pylori*), and of diarrheal disease in consumers of vegetables irrigate with untreated wastewater. The risk significantly decreased with treated wastewater (fecal coliform guideline of ≤ 1000 FC/100 ml, ≤ 0.1 nematode eggs per liter). Risk of bacterial infections increases when sprinkler irrigation is used and the population is exposed to wastewater aerosols containing $>10^6$ total coliform (TC) per 100 ml.

Gastrointestinal symptoms (nausea, vomiting, diarrhea) and diseases of the airways (asthma and chronic bronchitis) were evidenced in sewage workers. Fatigue and headache have also been reported. Some studies suggest that respiratory symptoms and illness are related to exposure to bio-aerosols containing pro-inflammatory bacterial endotoxins and/or to gas like hydrogen sulfide and sulfur dioxide [57, 58]. A low increase of subclinical, but non clinical, hepatitis A was also observed [59]. Some studies report an increased risk of stomach cancer, and a few studies report an increased risk of cancer in the larynx, liver, prostate and of leukemia. However, no causation with chemical or biological agents commonly found in sewage treatment plants has been demonstrated so far [57].

Non-occupational exposures have been studied in a limited way. Odor is the main complaint of populations non-occupationally exposed to sewage sludge emitted gases. Some studies suggest that living in the proximity of a sewage treatment plant favors the occurrence of gastrointestinal symptoms (nausea, vomiting, and diarrhea). Ambient odors may produce health effects by direct irritation and/or toxicological routes. Air pollution by pathogens such as staphylococci has also been indicated as a possible cause [60]. Respiratory, gastrointestinal, and irritation symptoms have been reported in people living in lands where treated sewage sludge is applied as fertilizer [61].

3 Hazardous Waste

US-EPA has defined hazardous waste as "waste with properties that make it dangerous or capable of having a harmful effect on human health or the environment." Examples of hazardous waste include asbestos, brake fluid, printer toner, batteries, solvents, pesticides, lubricant oils, end-of-life products containing ozone-depleting substances, heavy metals, or organic pollutants such as polychlorinated dibenzo-p-dioxins and dibenzofurans (European council directive 2000/532/EC). Waste might also be classified as "hazardous" when it is: (i) reactive (able to trigger chemical reactions in particular conditions such as unused explosives or lithium/sulfur batteries which can give explosion when react with water); (ii) corrosive (able to corrode metal containers such as liquid with pH $\leq$ 2 or $\geq$ 12.5); (iii) inflammable (substances with a flash point under 60 °C, able to spark fires or blow up such as solvents and waste oils); (iv) toxic (able to cause death or injury when ingested, inhaled, or absorbed by the skin).

3.1 E-Waste

The term electronic waste (e-waste) refers to all equipment dependent on electrical currents or electromagnetic fields to work properly. E-waste includes a broad range of electronic devices from different sources (household appliances, medical devices, lighting equipment, monitoring and control instruments, mobile phones and computers, batteries, electric plastic casings, cathode-ray tubes etc.) and it is classified as hazardous waste [62]. However, opposite to other categories, e-waste also has significant potential for value recovery. Indeed, e-waste typically contains consistent amounts of copper, gold, palladium, and silver, alongside iron and steel. Besides metals, which may constitute up to 60% of total components, e-waste contains glass and plastics. E-waste production is continuously increasing, but only a little part (as low as 20%, in 2016) of global e-waste is properly recycled or disposed of, with a dramatic difference between high- and low-income countries [63]. E-waste management strategies include disposal to landfills or recycling with recovering of valuable materials. To this purpose, collected e-waste is shredded and processed to separate metallic from non-metallic fraction. Hydrometallurgical and pyrometallurgical methods are used for metal recovery [1, 62, 63]. Hydrometallurgical processes include solvent extraction, leaching, adsorption and ion exchange processes for selective dissolution of metal components by means of mineral acids and oxidants such as HCl, H_2SO_4, HNO_3/H_2O_2, $HClO_4$, NaClO and lixiviants as cyanide, thiosulfate, thiourea and halide. Pyrometallurgical methods consist of thermal treatments through conflagration, smelting in furnaces or with alkali chemicals, sintering, plasma melting and other solid–liquid–gas reactions at high temperature. The discards are disposed of in landfills or incinerated.

Unfortunately, e-waste is often processed and recycled by informal businesses in emerging economies of Asia, Africa, and South America. Informal recycling is carried out either on waste dumping sites or in recycling shops. Many operators are children and adolescents who often live at the dumping sites or nearby. Informal recycling traditionally involves manual and primitive techniques, including cutting, shredding and acid leaching of electronic components, melting of electronic boards or burning cable wires to extract metals, cleaning with hazardous solvents and finally, burning of residual materials [64].

3.1.1 Hazardous Substances in E-Waste

E-waste pollution arises from both primary and secondary toxicants, mainly released during waste combustion, which is often performed in open fires without environmental controls to protect workers health and the environment.

Heavy metals. Several metals can be found in e-waste depending on origin, age, and manufacturer of the electronic devices. The most frequent metals are Pb, Cu, Cr, Zn, Cd, Se, Fe, Ni, Mn and Sb. They come from different sources. Pb is a common component of screens of old television, Hg and Ba are present in fluorescent lamps, Cd may be present in printer drums and in batteries, Be in power supply boxes, Cr in data tapes and floppy-disks, Zn in cathode ray tubes (CRT) [63, 65]. Exposure to metals takes place during e-waste desoldering, shredding, melting, and burning. Pyrometallurgical processes, for example, may cause emissions of fumes containing low melting point metals such as Cu, Cd, and Pb. Metal concentrations are generally higher at e-waste sites with respect to the surrounding environment. High amounts of Pb are often found in recycling plants [66]. Heavy metals can be also released in the leachate of e-waste landfills, when percolating waters are not properly collected and treated. Some studies provided convincing evidences that a significant amount of Pb can be released by electronic devices that containing color CRT or printer wiring boards with Pb-bearing solder [67].

Organic pollutants. a wide range of organic compounds is present in or is generated during recycling of e-waste. The most commonly found organic pollutants are: (i) polychlorinated biphenyls (PCBs), that were used as coolants, dielectrics, and lubricants in old electrical equipment before their ban; (ii) flame retardants such as polybrominated diphenyl ethers (PBDEs) and organophosphate ethers (OPEs) added to electronic goods to meet flammability standards; (iii) secondary combustion products such as dioxins, furans, PAHs and phenols.

The release of flame retardants into the environment typically occurs during shredding of flame-retardant plastics. Emission of dioxins and brominated/chlorinated dibenzofurans can occur during pyrometallurgical treatments of halogen-containing plastics. Proper design of the technological process may significantly reduce the emissions. Usually, high temperature treatments must be followed by rapid cooling of the gases and the use of a gas cleaning system is required. Similar concerns arise from e-waste incineration procedures [67].

Besides heavy metals and organic pollutants, particulate matter generated during sorting, dismantling, and burning of e-waste further contributes to environmental pollution. Other substances used during dismantling and recycling activities may increase the occupational risk. Hydrometallurgical processes, for example, may exposure workers to acid fumes, acid solutions, and cleaning solvents.

A brief overview of the main hazardous chemicals in e-waste and their effects on human health is reported in Table 1

3.1.2 Environmental Impact

Accumulation of organic and inorganic contaminants in soils and waters is the most common environmental scenario related to an incorrect e-waste management. Severe pollution from e-waste has been well documented in Ghana, India, and China [85]. High levels of BFRs, PAHs and PCBs were observed in the surroundings of e-waste disposal/treatment areas. Large contamination by heavy metals (Cu, Ni, Pb, Zn) of dust from roads adjacent recycling sites, and of soil and freshwater has been also observed [67]. Pollutant concentration in residential areas near the treatment facilities is often elevated compared to background levels, suggesting potential atmospheric contamination of the surrounding environment.

Long-term intake of contaminated water and food by dairy ruminants causes bioaccumulation of several hazardous substances. PCBs for example-accumulate in adipose tissue, liver, and fatty portions of meat. They are also excreted in edible products such as eggs and milk. Accumulation of PBDEs and PCBs in aquatic species from e-waste sites has been demonstrated [65].

Overall, the negative environmental impact of e-waste treatment can be reduced significantly by enforcing a correct waste management, which includes recycling of

Table 1 Potential hazardous chemicals in e-waste and their effects on human health (adapted from [62, 63, 65, 68, 69])

	Main toxicant sources	Route of exposure	Health effects	Mechanism of action	References
Organic contaminants					
Brominated flame retardants:	Flame retardant plastics	Ingestion, inhalation, and transplacental		Bioaccumulation in fatty tissues	
Polybrominated diphenyl ethers (PBDEs)			Thyroid problem, alteration of neurobehavioral development	Disruption of thyroid hormone balance, disruption of second messenger communications, alteration of neurotransmitter systems	[70, 71]
Polybrominated biphenyls (PBBs)			Skin disorders, alteration of thyroid function, reproductive development disorder in animals; probably carcinogenic to humans (Group 2A, IARC)	Alteration of hepatic metabolism, interaction with several cellular receptors	[72]
Polychlorinated biphenyls (PCBs)	Dielectric fluids, lubricants and cooling in, generators capacitors and transformers	Ingestion, inhalation, dermal contact, and transplacental	Chloracne and other dermal alterations, anti-estrogenic activity, thyroid hormone alteration immunosuppression; liver, lung and thyroid toxicity. Carcinogenic to humans (Group 1, IARC)	Multiple modes of action: receptor binding, gene expression, protein-activity and cellular replication changes, induction of oxidative stress	[72]

(continued)

Table 1 (continued)

	Main toxicant sources	Route of exposure	Health effects	Mechanism of action	References
Polychlorinated dibenzodioxins (PCDDs) and dibenzofurans (PCDFs)	Waste incineration a T < 800 °C (dioxin production depends on temperature, O_2 and CO_2 concentrations, and on the chlorine content of waste)	Ingestion, inhalation, dermal contact, and transplacental	Harmful potential varies depending on the number and structural position of chlorine atoms. Chloracne and other dermal alterations. Disturbances in the nervous, immune, reproductive, and endocrine system	Accumulation in the adipose tissue. Binding to Ah receptor with deregulation of expression of a large number of genes	[73, 74]
Polyaromatic hydrocarbons (PAHs)	Waste incineration	Ingestion, inhalation, or dermal contact	Carcinogenicity, mutagenicity, teratogenicity	Induction of oxidative stress and genotoxic damage, Interaction with Ah receptor	[75]
Heavy metals					
Lead (Pb)	CRTs (4–22% of Pb), television sets, PC monitors, batteries, light bulbs, lamps	Ingestion, inhalation or dermal contact	Lead poisoning: weakness, irritability, asthenia, nausea, abdominal pain, and anaemia. Chronic exposure: renal toxicity, neurotoxicity mainly in children. Cardiovascular effects and changes in endocrine and immune functions. Inorganic lead compounds are probably carcinogenic to humans (Group 2A; IARC)	ROS generation and inhibition of antioxidant defenses, interference with DNA repair; PKC activation with upregulation of 'immediate early response' genes and consequent induction of cell proliferation	[76]

(continued)

Table 1 (continued)

	Main toxicant sources	Route of exposure	Health effects	Mechanism of action	References
Hexavalent chromium (Cr)	Anti-corrosion coatings, data tapes, floppy disks	Ingestion, inhalation, dermal contact	Lung cancer. Dermatitis	DNA damage	[73, 77]
Cadmium (Cd)	Batteries, older CRTs, infrared detectors, semi-conductor chips, ink or toner photocopying machines, mobile phones	Ingestion, inhalation	Kidney damage. Renal toxicity. Bone disease (osteomalacia and osteoporosis). Possibly reproductive damage, and lung emphysema. Cadmium compounds cause cancer of lung, kidney and prostate (Group 1; IARC)	Induction of oxidative stress and mitochondrial damage, inhibition of DNA repair, deregulation of cell proliferation and disturbance of tumor-suppressor functions	[73, 78]
Mercury (Hg)	CRTs, thermostats, sensors, monitors, cold cathode fluorescent lamps	Ingestion, inhalation or dermal contact	Toxicity varies with the form of Hg. Massive acute exposure to Hg vapor: bronchitis and SNC symptoms (tremors); chronic exposure: neurological dysfunction. Prenatal exposure to organic mercury: neurodevelopment delays and cognitive deficits of children	Alteration of the tertiary and quaternary structure of proteins and binding with sulfhydryl groups, interference with DNA transcription and protein synthesis, disruption of subcellular elements in the central nervous system	[79]

(continued)

Table 1 (continued)

	Main toxicant sources	Route of exposure	Health effects	Mechanism of action	References
Zinc (Zn)	CRTs, metal coatings, batteries	Ingestion, inhalation	Increased risk of Cu deficiency (anemia, neurological abnormalities), acute respiratory distress syndrome		[80]
Nickel (Ni)	Batteries, electron gun in CRTs	Ingestion, inhalation or dermal contact, and transplacental	Dermatitis, asthma and chronic bronchitis. Ni (II) compounds are carcinogenic to humans (Group 1; IARC)	Induction of oxidative stress, oxidative DNA damage, chromosomal alterations, inhibition of DNA repair	[73, 81]
Barium (Ba)	CRTs (2–9% Ba), fluorescent lamps	Ingestion, inhalation or dermal contact	Increase blood pressures, changes in heart rhythm, stomach irritation, muscle weakness, benign pneumoconiosis	Blocking of the K+ channels of the Na-K pump in cell membranes	[82]
Beryllium (Be)	Power supply boxes, computers, x-ray machines, ceramic components	Ingestion, inhalation, and transplacental	Acute exposure: inflammatory obstructive lung disease. Be and Be compounds cause lung cancer	Several molecular mechanisms, possibly interrelated: ROS generation, chronic inflammatory response, proto-oncogenes and apoptosis activation	[73, 83]

(continued)

Table 1 (continued)

	Main toxicant sources	Route of exposure	Health effects	Mechanism of action	References
Copper (Cu)	Television and PC board, mobile phone, DVD player scrap	Ingestion, inhalation or dermal contact	Chronic inhalation: irritation of t nose, mouth and eyes, headaches. Chronic intake: stomachaches, vomiting and diarrhea, liver cirrhosis	Oxidative damage	[84]

glass and valuable metals, incineration of hazardous substances and deposition of the residue in sanitary landfills.

3.1.3 Health Impact

Occupational, direct local exposure of people living near disposal/recycling sites and indirect exposure of the general population to hazardous chemicals of e-waste and combustion products are well documented, mainly when informal disposal or recycling are carried out [64, 86]. E-waste is informally managed in many countries, including China, Ghana, Nigeria, India, Thailand, the Philippines, and Vietnam). Most of the Chinese studies [68, 87–89] evidence that high levels of heavy metals and persistent organic pollutants are found in the environment near e-waste sites and that workers engaged in e-waste recycling and nearby residents are often contaminated [90]. PCBs and PBDEs were detected in serum from residents and in hair from e-waste workers. Serum level of thyroid stimulating hormone (TSH), frequency of micro-nucleated and binucleated lymphocyte cells, and oxidative stress markers in urine were found to be significantly higher in e-waste workers with respect to the control group. Overexposure to Ba, Cd, Pb, and Y has been also documented in workers of facilities processing CRTs in US and Europe. Overexposure to Hg was observed during recycling of alkaline batteries and fluorescent lamps, and central nervous system and respiratory symptoms have been related to these exposures [91]. Furthermore, high levels of Cd and Pb in blood of children, and of Cr in umbilical cord blood of neonates have been reported [67]. Although a clear association between exposure to e-waste toxicants and health effects was not always found, several studies report a significant increase of risk for both malignant and non-malignant pathologies.

Neonatal outcomes and growth. Spontaneous abortions, stillbirths, and premature births, reduced birthweights and birth lengths, likely associated with exposures to persistent organic pollutants, have been reported in people exposed to e-waste in China. Some studies carried out in the same area evidenced a reduction of physical growth indicators (weight, height, and body-mass index) in children living in the e-waste recycling town. Researchers related these adverse outcomes with the exposure to heavy metals [86, 92], mainly Pb [93].

Neurodevelopment. Some studies report an association between blood levels of Cd, Mn, and Pb and behavioral abnormalities, sensory processing difficulties, attention deficit and hyperactivity in children [92].

Respiratory diseases. decrease in lung function (Forced Vital Capacity, FVC, and Forced Expiratory Volume, FEV) and asthma were observed in children living in e-waste recycling areas and associated with a high blood Pb concentration [93]. Lung damage is also associated with other compounds commonly found in e-waste, such as polycyclic aromatic hydrocarbons, Cr(IV), Cd, Ni, As, and Li [86, 92].

Cardiovascular diseases. Although a cause-effect relationship could not be established, some studies report high prevalence of cardiovascular diseases (hypertension in adults, and vascular inflammation and lipid disorders in children) in e-waste areas, probably linked with heavy metal exposure [92].

Cancer. several compounds present in e-waste are carcinogenic (Cr(IV), Cd and Be) [73], probably carcinogenic (some polychlorinated and polybrominated compounds, and some polycyclic aromatic hydrocarbon) or possibly carcinogenic (polybrominated diphenyl ethers, metallic nickel, and some polycyclic aromatic hydrocarbons) [72]. In agreement, some occupational studies revealed a slight increase of cancer risk, mainly related to exposure to Cr and Cd, in e-waste Chinese workers [91].

3.2 Micro and Nanoplastics

Plastics are synthetic materials typically made of organic polymers and other chemical additives, such as phthalates, bisphenols, and flame retardants, that give plastic products peculiar properties. Polyethylene (PE), polypropylene (PP), polycarbonate (PC), polystyrene (PS), and polyvinyl chloride (PVC) are the most widely produced plastics with many applications in everyday life, including packaging, construction, electronics, and medical equipment. The growing demand of plastic products has led to global production of more than 300 million tons a year (367 million tons in 2020 [94]), with China being the leading plastics producer worldwide. When not properly managed, plastic products and consequently plastic waste can pollute the environment for decades. It is estimated that about 6.3 billion tons of plastic waste had been generated worldwide from the 1950s to 2015 [95] and that the current worldwide flow of plastic waste is of 5 million tons [94]. A significant amount of plastic waste is currently disposed of directly into the environment or by applying improper reuse or recycling methods. Several plastics (e.g., PVC, PS) are particularly difficult to recycle and their end-of life increases the risk of release of toxic contaminants in the environment. Many plastics may be chemically harmful in some contexts, either because they are toxic per se or because they absorb other pollutants. Plastic recycling often involves waste-to-energy technology with a where plastic is burned, and the thermal energy reconverted in electricity. However, plastic incineration can generate priority pollutants and greenhouse gases. Only 20–25% of the plastic waste is appropriately recycled and incinerated [95, 96]. The United Nations' Globally Harmonized System of Classification and Labelling of Chemicals states that more than 50% of the chemical ingredients of plastics are hazardous [97], and they may be released into the environment after they are discarded.

After entering the environment, interactions between plastic waste and environmental components (UV light, water) can disintegrate and alter large pieces of plastics into smaller plastic debris. Fragmentation and (photo)degradation of plastic waste generates micro- or nano sized plastic particles (MNPs). It has been recently estimated that at least 78% of the priority pollutants listed by the EPA and 61% listed by the European Union are associated with plastic debris [98]. In addition, primary microplastic particles are intentionally produced and added to consumer products, mainly cosmetics and in agriculture, when plastic-containing sludges are used ad fertilizers. According to the diameter of plastic fragments or particles, microplastics

(MPs) are plastic particles with a diameter less than 5 mm, while nanoparticles (NPs) range in diameter below 100 nm [96]. Any types of MNPs, in terms of types, sizes, and shapes, have the potential to contaminate all kind of waste [99].

MNPs are widespread pollutants in oceans, soils, biota, and Earth's atmosphere. They can be observed in both marine and terrestrial ecosystems and be ingested and accumulated by animals along the food chain. Even if the investigation on the effects plastics that is ingested/inhaled is quite recent, studies in humans and mussels have found that MNP debris can reach cells and tissues where they can cause harm [98].

There are currently no guidelines or regulations that cover MNPs production and MNPs waste management in a comprehensive manner. Some reviews demand for the development of regulatory measures on unintentional release of MNPs, including measures to increase the capture of MNPs at all relevant stages of products' life cycle [100].

3.2.1   Hazardous Substances in Microplastics and Nanoplastics

Main concern about MNPs toxicity is related to both release of hazardous substances (monomers and additives used in plastic manufacturing) and adsorption of toxic chemicals from the environment.

Monomers and oligomers that form the polymeric chains can leach during MNPs use and disposal [101]. Some monomers have adverse effects on human health. PC, PS and PVC are common examples of plastics that have been shown to release toxic monomers. Vinyl chloride used in the production of PVC is a well-established human carcinogen [102]. Bisphenol A (BPA), used in the PC production and as an additive in other plastics, and styrene, used in the PS production, are suspected endocrine disrupting chemicals. BPA has been detected in urine, blood, breast milk and tissue samples [103].

Similarly, some of the additives incorporated into the basic polymers to improve their performance are associated with adverse health effects in animals and humans [104]. The most common additives include plasticizers, flame retardants, antioxidants, acid scavengers, light and heat stabilizers, lubricants, pigments, and antistatic agents [105]. Nonylphenol and brominated flame-retardants are endocrine disruptors and carcinogens. Phthalates, used to make plastics more durable, have been associated with developmental anomalies [106]. Cd-, Ba-, Ca-, and Zn-dicarboxylates and organotin, used as heat stabilizers, are toxic to aquatic organisms [107].

In addition, MNPs may adsorb environmental pollutants, such as heavy metals, persistent organic and hydrophobic organic pollutants. For instance, Cr shows a higher potential for adsorption onto microplastics with respect to other heavy metals [108] or phenanthrene and pyrene that easily adsorb on PE and PS [109].

3.2.2 Environmental Impact

The detection of solid nano- and microplastic particles, fragments and fibers, virtually in every matrix and location worldwide is part of the global plastic pollution problem [100]. MNPs impact marine and terrestrial ecosystems, including oceans, rivers, air, drinking waters, sediments, and often enter food chain. MNPs-based environmental pollution is expected to increase due to the degradation of plastic materials that is dispersed in the environment, especially in the oceans. Moreover, during their life cycle, MNPs undergo abiotic and biotic transformations which creates new potentially toxic chemicals.

Primary MNPs derive from consumer products in which plastic fibers or particles are added to confer special properties. MNPs are easily released from these products into the environment during the product life cycle. Another significant source of MNPs is plastic debris that degrades from the larger pieces of plastics due to UV radiation, physical wear, and biodegradation in the environment. Microplastic fibers of PS and PP are some of the most common types of MNPs detected in the environment [110].

MNPs have been found to be present in most of the marine environment, where they can also be absorbed and bioaccumulated by marine animals, and in soil as well as in terrestrial invertebrate (e.g., earthworms) that live on the surface and deep layers of the soil [96, 111]. Recent studies evidenced that remote areas are impacted by MNPs that can migrate via atmospheric transport [112]. Plastic debris generated from different processes have a different impact on the environment. Storms and heavy rains in the nearby of cities generate microplastics with larger particle size distribution, that are deposited during wet events. Smaller microplastics are deposited during dry weather and can be transported for long distances by wind. This latter form of microplastic represent the major part of microplastic mass on the planet [113]. No conclusive data on microplastics mobilization through terrestrial ecosystems are still available. Evidences suggest percolation from the surface to deeper layers, resuspension into the atmosphere, and transport into ground and surface waters as the main processes involved in the MNPs transport cycle [112]. In streams, microplastics can be deposited and resuspended, buried in sediments, or exported to downstream ecosystems [114]. Little information is available on microplastics movement through food chains, but some evidence advises a trophic type of transfer [115].

The impact of microplastic on ecosystem is far more extended than direct toxicological effects. Larger plastic fragments, *e.g.*, bottles, bags, and floating pontoons, can transport species to new habitats, where they might prompt ecological damages and endanger endemic species [98].

3.2.3 Health Impact

Humans can be exposed to MNPs by ingestion and inhalation [116, 117]. Drinking of polluted freshwater and ingestion of marine food, where plastic bioaccumulation occurs, are the main ways of oral exposure. Microplastic have also been found in table salt, beer and other food items. There is a lack of information on the fate of MNPs in the human gastrointestinal tract, but more than 90% of ingested plastic is expected to be excreted via feces. Factors affecting fate of ingested particles and potential health effects are size, polymer type, and presence of additive chemicals that could be leached in biological fluids [118].

Cellular studies on mammalian intestinal cells showed negligible toxicity pattern, with a low dose-dependent cytotoxicity, negligible ROS production and DNA damage or chromosomal aberration, nor alteration of cell membrane integrity and fluidity, of polystyrene beads of 50–100 nm size. Adverse effects of MNPs ingestion have only been observed in experimental studies on aquatic organisms and in few studies on rodents exposed to polystyrene microparticles [119]. These studies highlighted damage of gut barrier function, alteration of intestinal microbiota, reduction of triglyceride and total cholesterol levels and decrease in some key gene expressions related to lipogenesis and triglyceride synthesis in liver [116]. No data on effects on humans of ingested plastics are available to date.

Exposure through inhalation implies that airborne MNPs are dispersed from industrial emissions, clothes and house furniture, materials in buildings, waste incineration, landfills, and sewage sludge. Potential health hazard of airborne MNPs to the general population is still a subject of debate, while studies on occupational hazard have evidenced an increased risk of irritation of the respiratory tract, dyspnea, coughing, and reduced lung capacity among workers in nylon flock, polyester, polyolefin, and polyamide fiber plants. A loss of lung functionality was also observed following exposure to PVC dust in manufacturing plants [120]. Plastic fibers have been found in human lung tissues [121]. In vitro tests have demonstrated that plastic fibers are extremely durable in physiological fluids. Polypropylene, polyethylene and polycarbonate fibers did not dissolve or modify their characteristics in a synthetic extracellular lung fluid after 180 days of incubation [122].

Available studies on mammalian cells mainly focused on the effect of micro- and nanosized PS beads. Reduction of cell viability, increase of ROS generation, and expression of pro-inflammatory cytokines (IL-6 and IL-8) have been observed in bronchial epithelial cells exposed to high dose (1000 μg/ml) of negatively charged PS. A significant dose dependent cytotoxicity, increasing with decreasing particle size, was also evidenced in lung epithelial cells. To date more research is needed for the risk assessment of the impact of microplastics on human health.

3.3 Engineered Nanomaterials

Nanomaterials have been defined as "*materials containing particles, in an unbound state or as an aggregate or as an agglomerate and where, for 50% or more of the particles in the number size distribution, one or more external dimensions is in the size range 1–100 nm*" (2011/696/EU).

Engineered nanomaterials (ENMs) are used in a wide range of products including plastics, textiles, medicines, cosmetics, clothing, sunscreens, electronics, and their production has been increasing progressively over the past years [123]. ENMs produced at industrial scale are currently nanostructures of metals or metal oxides, including Ag, Au, TiO_2, ZnO, and SiO_2, or carbon, including C60 fullerenes, carbon nanotubes (CNTs), graphene and graphene oxide. Despite the increasing attention posed to the environmental fate of ENMs, and to the potential exposure and health effects during their production and use, little is known about ENMs in waste. Under current European REACH (Registration, Evaluation, Authorization and Restriction of Chemicals) regulation, an exposure assessment for the waste life stage has to be carried only if more than 10 ton per year of the registered substance are produced or imported and the substance is classified as dangerous according to Regulation (EC) No 1272/2008. Based on the exposure assessment (i) a proper waste management approach, (ii) a list of physical and chemical properties that may affect waste treatment options, and (iii) precautions required for waste treatment have to be specified. Nonetheless, waste containing ENMs is currently managed along with conventional waste stream, without sufficient knowledge of the associated risks and impacts on the environment [124].

A classification of nano-waste, based on the potential health risk, recognized five risk classes [125]. The most important parameter that was considered for the classification is the potential to release of ENMs from the matrix when the ENM-containing product is put in contact with air or water. ENMs firmly embedded in a solid matrix, e.g., sintered into solar panels or memory chips, where ENMs release is unlikely to occur, were classified as low risk, considering their negligible contribution to ENMs exposure. Loosely bonded ENMs (e.g., nanoparticles in liquid suspension) were classified as high-risk nano-waste. The other contributing factor to the classification of ENMs is their (potential) toxicity. Even if a scientifically-sound nano-waste classification is required for determining the appropriate ways of treating and disposing of various types of nano-waste, the current lack of conclusive information on the toxicity of nanomaterials to humans and the environment is mainly preventing the practical implementation of nano-waste classification.

Begin included in the conventional waste stream, nano-waste management is processed, recycled, and incinerated with technologies used to manage conventional waste. The key factors governing the fate of ENMs during incineration are the temperature of the furnace and the chemical properties of the embedded nanomaterials. If furnace T is higher than the melting point of the ENM, a complete degradation of the nanomaterial is expected. This makes incineration the most suitable treatment for carbon-based ENMs. Nanomaterials composed of metal and metal oxides show much higher melting points and are usually more persistent. For instance, transition metal oxides, such as SiO_2 and TiO_2, are not significantly removed during incineration of waste and undergo complex physical and chemical transformations, depending on specific material properties of ENM and surrounding matrix [126].

3.3.1   Sources of ENMs in Waste

Main sources of ENMs in waste are construction materials that might contain titanium dioxide, zinc oxide, silicon dioxide and aluminum oxide nanoparticles, insulation materials, and paints [127, 128], and wastewaters from several manufacturing
activities [129, 130]. Among these activities, chemical industry, such as the chlo-
rate and electroplating industries, produce wastewater containing large amounts of
nanoparticles (mainly TiO_2, CeO_2, SiO_2, Al_2O_3, ZnO, Fe(0) and iron oxides) which
readily adsorb heavy metals. Likewise, household wastewater may contain a significant amount of nanoparticles (mainly TiO_2, ZnO, and Ag,) from cosmetics, clothing,
and catering utensils [131]. Wastewater treatment usually removes most of the nano-
fraction, which is trapped in sludge. Sludge is usually incinerated or disposed of
in landfill. Landfills also receive MSW that may contain a wide range of ENM-
containing products and the potential presence of ENMs in the draining waters of
the landfill should be considered in the environmental risk evaluation.

3.3.2   Environmental Impact

ENMs release into the environment can arise from compost and sewage sludge application to cultivated field. A study on compost containing Ag NPs and Ag doped TiO_2
NPs demonstrated that a negligible amount of nanoparticles is released in the ground.
This finding suggests that ENM-sludge leaching may not likely pollute ground and
ground waters, but indicates that ENMs may accumulate in the agricultural top-soil
[132]. In agreement, another study detected TiO_2 oxide NPs in the top-soil of an
agricultural site that had been amended with compost over a long-period of time
[133].

   Other sources of potential emissions of nanomaterials from waste are landfill
leachates [123]. Zn, Ag, Cu, Ni associated to the colloidal fraction of leachate
were detected. Nanoparticles of carbonates, quartz, and iron oxide, likely from
natural sources, have also been detected. During disposal in landfills, ENMs may
undergo physico-chemical transformation such as aggregation, dissolution, complexation, reduction/oxidation and sulfidation. These complex and largely unpredictable
processes may affect ENMs stability and mobility. Leachate characteristics, such as
pH and ionic strength, also influence these processes, also taking into account that
landfill leachate characteristics change over time with landfill age. Low pH values at
the early stages and higher pH values for older and more degraded waste differentiate
aggregation and chemical process that might modify ENMs. A study on the fate of
CNT in landfill leachate showed a significant CNT aggregation and little CNT transport in the early acidic leachate, but a significant transport capacity was achieved
in older leachate where higher levels of electrolytes were found. Studies on ZnO,
TiO_2, and Ag ENMs found that the dispersion and dissolution of NPs increased after
interaction with leachate components. However, in leachate water suspension, ENMs
mainly aggregate, forming larger particles that are easily separated in the solid waste
following leachate treatment [127]. However, several metal and metalloid NPs were

observed in leachate collected from uncontrolled dump site of construction materials. The migration of nanomaterials following rains to aquatic systems due to the absence of landfill liners and leachate collection systems, and the dispersion by wind erosion should be regarded as high environmental pollution routes [134].

A minor risk of nanoparticles emission is associated with incineration. Indeed, incineration gases (and particles) are convoyed to emission control technologies that remove contaminants before flue gas is released into the atmosphere. Emission control technologies for municipal waste incinerators have continually improved over the years, and today the efficiency of the removal of solid emissions, including nanoparticles, is very high [135, 136].

ENMs release from waste may pose a threat for the ecosystem. ENMs may influence plant growth and both enhancive and inhibitive effects have been documented. NPs can be taken up by plant roots and transported through plant vascular systems. ENMs may also affect microbial systems. A number of studies on ENM toxicity have been devoted to investigating the effects on NPs on plants and microorganisms, although most of the studies were designed to test conditions different from those found in landfills and composting facilities. ENMs may exert toxic effects on soil and aquatic microorganisms [123, 137, 138]. The bactericidal properties of Ag NPs suggest a potential toxic effects on soil and freshwater biota [139, 140]. Ag NPs may be internalized in the cell, may cause ROS generation, damage to cellular components, activation of antioxidant enzymes and depletion of antioxidant molecules, binding of proteins, and damage to the cell membrane [141].

Acute toxicity of ZnO NPs to bacteria, algae, aquatic and terrestrial invertebrates has been observed, although the extent of ZnO toxicity is dependent on the species investigated and the physico-chemical properties of the specific nanomaterial [142, 143]. ROS generation, cell membrane damage with enhanced membrane permeability, internalization of NPs and uptake of dissolved zinc ions are likely responsible for the observed toxic effects [144].

Cu-based NPs show a well-established antimicrobial effects and this could pose a risk to soil biota, especially when they are released in large amount and for long period of time [145].

3.3.3 Health Impact

A limited number of studies evaluate the health effect on humans of the exposure to ENMs released from waste. Exposure involves both workers in waste management facilities and residents in proximity of disposal sites. ENM exposure in occupational environment may take place during all steps of the waste management or recycling process, particularly during activities that involve handling, abrasion, shredding, milling, and compaction of waste [146, 147]. Occupational exposure limit (OEL), established to protect workers from adverse health effects caused by nanomaterial inhalation, have been established for titanium dioxide, carbon- and silver-based nanomaterials, zinc oxide and other metal oxides [148]. A number of human and animal studies suggest that inhalation of metal and carbon ENMs may cause airway inflammation and exacerbation of respiratory symptoms, mainly in patients with chronic airway diseases. The mechanisms underlying this response is still unclear, but likely involves oxidative cellular damage [149]. Some experimental studies have been devoted to investigating cellular effects of waste that contains ENMs. A comparative study on toxicity of particulate matter that was obtained from the combustion of plastic and paper waste, with or without ENMs, showed that the presence of Ag, TiO_2, NiO, fullerene, Fe_2O_3 and quantum dots do not significantly modify the geno- and cytotoxicity of the particulate matter on lung epithelial cells [150]. Residual ash from Al-based nanocomposites was not cytotoxic [151], but incineration changes surface chemistry and size distribution of nanoclays, which elicited differential cellular responses with respect to pristine material [152]. Generally speaking, the amount of scientific information about the toxic effects and the physico-chemical modification that ENMs can undergo during waste treatment is currently limited and further studies on the potential toxicity after waste treatment processes are required to properly assess potential health effect of nano-waste and derived nanoparticles.

References

1. Ahirwar R, Tripathi AK (2021) E-waste management: a review of recycling process, environmental and occupational health hazards, and potential solutions. Environ Nanotechnol Monitor Manage 15:100409
2. Giusti L (2009) A review of waste management practices and their impact on human health. Waste Manage 29:2227–2239. https://doi.org/10.1016/j.wasman.2009.03.028
3. Xu P, Chen Z, Wu L, Chen Y, Xu D, Shen H, Han J, Wang X, Lou X (2019) Health risk of childhood exposure to PCDD/Fs emitted from a municipal waste incinerator in Zhejiang, China. Sci Total Environ 689:937–944
4. Vaccari M, Vinti G, Tudor T (2018) An analysis of the risk posed by leachate from dumpsites in developing countries. Environments 5
5. Vergara SE, Tchobanoglous G (2012) Municipal solid waste and the environment: a global perspective. Annu Rev Environ Resour 37:277–309
6. Alam P, Ahmade K (2013) Impact of solid waste on health and the environment. Int J Sustain Dev Green Econ (IJSDGE) 2:165–168
7. Ferronato N, Torretta V (2019) Waste mismanagement in developing countries: a review of global issues. Int J Environ Res Pub Health 16. https://doi.org/10.3390/ijerph16061060
8. Wiedinmyer C, Yokelson RJ, Gullett BK (2014) Global emissions of trace gases, particulate matter, and hazardous air pollutants from open burning of domestic waste. Environ Sci Technol 48:9523–9530. https://doi.org/10.1021/es502250z
9. Williams PT (2005) Waste treatment and disposal. John Wiley & Sons
10. Hoornweg D, Bhada-Tata P (2012) What a waste: waste management around the world. World Bank, Washington, DC, pp 9–15
11. Kjeldsen P, Barlaz MA, Rooker AP, Baun A, Ledin A, Christensen TH (2002) Present and long-term composition of MSW landfill leachate: a review. Crit Rev Environ Sci Technol 32:297–336. https://doi.org/10.1080/10643380290813462
12. Luo HW, Zeng YF, Cheng Y, He DQ, Pan XL (2020) Recent advances in municipal landfill leachate: a review focusing on its characteristics, treatment, and toxicity assessment. Sci Total Environ 703. https://doi.org/10.1016/j.scitotenv.2019.135468
13. Kulikowska D, Klimiuk E (2008) The effect of landfill age on municipal leachate composition. Biores Technol 99:5981–5985. https://doi.org/10.1016/j.biortech.2007.10.015

14. Vaverkova MD (2019) Landfill impacts on the environment-review. Geosciences 9. https://doi.org/10.3390/geosciences9100431

15. Ghosh P, Thakur IS, Kaushik A (2017) Bioassays for toxicological risk assessment of landfill leachate: a review. Ecotoxicol Environ Saf 141:259–270. https://doi.org/10.1016/j.ecoenv.2017.03.023

16. Baderna D, Maggioni S, Boriani E, Gemma S, Molteni M, Lombardo A, Colombo A, Bordonali S, Rotella G, Lodi M, Benfenati E (2011) A combined approach to investigate the toxicity of an industrial landfill's leachate: chemical analyses, risk assessment and in vitro assays. Environ Res 111:603–613. https://doi.org/10.1016/j.envres.2011.01.015

17. Ghosh P, Gupta A, Thakur IS (2015) Combined chemical and toxicological evaluation of leachate from municipal solid waste landfill sites of Delhi, India. Environ Sci Pollut Res Int 22:9148–9158. https://doi.org/10.1007/s11356-015-4077-7

18. Talorete T, Limam A, Kawano M, Ben Rejeb Jenhani A, Ghrabi A, Isoda H (2008) Stress response of mammalian cells incubated with landfill leachate. Environ Toxicol Chem 27:1084–1092. https://doi.org/10.1897/06-648.1

19. Alimba CG, Gandhi D, Sivanesan S, Bhanarkar MD, Naoghare PK, Bakare AA, Krishnamurthi K (2016) Chemical characterization of simulated landfill soil leachates from Nigeria and India and their cytotoxicity and DNA damage inductions on three human cell lines. Chemosphere 164:469–479. https://doi.org/10.1016/j.chemosphere.2016.08.093

20. Ghosh P, Das MT, Thakur IS (2014) Mammalian cell line-based bioassays for toxicological evaluation of landfill leachate treated by Pseudomonas sp. ISTDF1. Environ Sci Pollut Res Int 21:8084–8094. https://doi.org/10.1007/s11356-014-2802-2

21. Ghosh P, Swati and Thakur IS, (2014) Enhanced removal of COD and color from landfill leachate in a sequential bioreactor. Bioresour Technol 170:10–19. https://doi.org/10.1016/j.biortech.2014.07.079

22. Wang G, Lu G, Yin P, Zhao L, Yu QJ (2016) Genotoxicity assessment of membrane concentrates of landfill leachate treated with Fenton reagent and UV-Fenton reagent using human hepatoma cell line. J Hazard Mater 307:154–162. https://doi.org/10.1016/j.jhazmat.2015.12.069

23. Vinti G, Bauza V, Clasen T, Medlicott K, Tudor T, Zurbrugg C, Vaccari M (2021) Municipal solid waste management and adverse health outcomes: a systematic review. Int J Environ Res Public Health 18. https://doi.org/10.3390/ijerph18084331

24. Porta D, Milani S, Lazzarino AI, Perucci CA, Forastiere F (2009) Systematic review of epidemiological studies on health effects associated with management of solid waste. Environ Health 8. https://doi.org/10.1186/1476-069x-8-60

25. Cointreau S (2006) Occupational and environmental health issues of solid waste management: special emphasis on middle-and lower-income countries. Urban Papers 2

26. Vrijheid M (2000) Health effects of residence near hazardous waste landfill sites: a review of epidemiologic literature. Environ Health Perspect 108:101–112. https://doi.org/10.2307/3454635

27. WHO (2015) Waste and humans health: evidences and needs. WHO meeting Report, World Health Organization, Bonn, D

28. Palmer SR, Dunstan FD, Fielder H, Fone DL, Higgs G, Senior ML (2005) Risk of congenital anomalies after the opening of landfill sites. Environ Health Perspect 113:1362–1365. https://doi.org/10.1289/ehp.7487

29. Rushton L (2003) Health hazards and waste management. Br Med Bull 68:183–197. https://doi.org/10.1093/bmb/ldg034

30. Tomita A, Cuadros DF, Burns JK, Tanser F, Slotow R (2020) Exposure to waste sites and their impact on health: a panel and geospatial analysis of nationally representative data from South Africa, 2008–2015. The Lancet Planet Health 4:e223–e234

31. Mattiello A, Chiodini P, Bianco E, Forgione N, Flammia I, Gallo C, Pizzuti R, Panico S (2013) Health effects associated with the disposal of solid waste in landfills and incinerators in populations living in surrounding areas: a systematic review. Int J Public Health 58:725–735. https://doi.org/10.1007/s00038-013-0496-8

32. Farrell M, Jones DL (2009) Critical evaluation of municipal solid waste composting and potential compost markets. Bioresour Technol 100:4301–4310. https://doi.org/10.1016/j.bio rtech.2009.04.029
33. Hargreaves JC, Adl MS, Warman PR (2008) A review of the use of composted municipal solid waste in agriculture. Agr Ecosyst Environ 123:1–14. https://doi.org/10.1016/j.agee. 2007.07.004
34. Azim K, Soudi B, Boukhari S, Perissol C, Roussos S, Alami IT (2018) Composting parameters and compost quality: a literature review. Org Agric 8:141–158
35. Wery N (2014) Bioaerosols from composting facilities-a review. Front Cell Infect Microbiol 4. https://doi.org/10.3389/fcimb.2014.00042
36. Agency UE (2010) Composting and Potential Health Effects from Bioaerosols: Our Interim Guidance for Permit Applicants. UK Environment Agency, Bristol, UK
37. Pearson C, Littlewood E, Douglas P, Robertson S, Gant TW, Hansell AL (2015) Exposures and health outcomes in relation to bioaerosol emissions from composting facilities: a systematic review of occupational and community studies. J Toxicol Environ Health B Crit Rev 18:43–69. https://doi.org/10.1080/10937404.2015.1009961
38. Domingo JL, Nadal M (2009) Domestic waste composting facilities: a review of human health risks. Environ Int 35:382–389. https://doi.org/10.1016/j.envint.2008.07.004
39. Amlinger F, Pollak, M., Favoino (2004) Heavy metals and organic compounds from wastes used as organic fertilisers_Annex 2: compost quality definition. Legislation and standards. Working Group: compost legislation and standard
40. Robertson S, Douglas P, Jarvis D, Marczylo E (2019) Bioaerosol exposure from composting facilities and health outcomes in workers and in the community: a systematic review update. Int J Hyg Environ Health 222:364–386. https://doi.org/10.1016/j.ijheh.2019.02.006
41. Cobb N, Sullivan P, Etzel RA (1995) Pilot study of health complaints associated with commercial processing of mushroom compost in southeastern Pennsylvania. J Agromed 2:13–25
42. Browne ML, Ju CL, Recer GM, Kallenbach LR, Melius JM, Horn EG (2001) A prospective study of health symptoms and Aspergillus fumigatus spore counts near a grass and leaf composting facility. Compost Sci & Utilization 9:241–249
43. Herr CEW, zur Nieden A, Jankofsky M, Stilianakis NI, Boedeker RH, Eikmann TF (2003) Effects of bioaerosol polluted outdoor air on airways of residents: a cross sectional study. Occup Environ Med 60:336–342. https://doi.org/10.1136/oem.60.5.336
44. Gao NB, Kamran K, Quan C, Williams PT (2020) Thermochemical conversion of sewage sludge: a critical review. Prog Energy Combust Sci 79. https://doi.org/10.1016/j.pecs.2020. 100843
45. Udayanga WC, Veksha A, Giannis A, Lisak G, Chang VW-C, Lim T-T (2018) Fate and distribution of heavy metals during thermal processing of sewage sludge. Fuel 226:721–744
46. Fytili D, Zabaniotou A (2008) Utilization of sewage sludge in EU application of old and new methods—a review. Renew Sustain Energy Rev 12:116–140. https://doi.org/10.1016/j.rser. 2006.05.014
47. CEC (2000) Council Directive of 27 April 2000 on Working Document on Sludge—third draft. In: Environment ECD (ed) Brussels (B)
48. Raheem A, Sikarwar VS, He J, Dastyar W, Dionysiou DD, Wang W, Zhao M (2018) Opportunities and challenges in sustainable treatment and resource reuse of sewage sludge: a review. Chem Eng J 337:616–641. https://doi.org/10.1016/j.cej.2017.12.149
49. Fijalkowski K, Rorat A, Grobelak A, Kacprzak MJ (2017) The presence of contaminations in sewage sludge—the current situation. J Environ Manage 203:1126–1136. https://doi.org/ 10.1016/j.jenvman.2017.05.068
50. Jia S, Zhang X (2020) Biological HRPs in wastewater. High-risk pollutants in wastewater, Elsevier, pp 41–78
51. Richards BK, Steenhuis TS, Peverly JH, McBride MB (2000) Effect of sludge-processing mode, soil texture and soil pH on metal mobility in undisturbed soil columns under accelerated loading. Environ Pollut 109:327–346. https://doi.org/10.1016/S0269-7491(99)00249-3

52. Choudri B, Al-Awadhi T, Charabi Y, Al-Nasiri N (2020) Wastewater treatment, reuse, and disposal-associated effects on environment and health. Water Environ Res 92:1595–1602
53. Lazzari L, Sperni L, Bertin P, Pavoni B (2000) Correlation between inorganic (heavy metals) and organic (PCBs and PAHs) micropollutant concentrations during sewage sludge composting processes. Chemosphere 41:427–435. https://doi.org/10.1016/S0045-6535(99)00289-1
54. Clarke BO, Smith SR (2011) Review of "emerging" organic contaminants in biosolids and assessment of international research priorities for the agricultural use of biosolids. Environ Int 37:226–247. https://doi.org/10.1016/j.envint.2010.06.004
55. Manzetti S, van der Spoel D (2015) Impact of sludge deposition on biodiversity. Ecotoxicology 24:1799–1814. https://doi.org/10.1007/s10646-015-1530-9
56. Yoshida H, Christensen TH, Scheutz C (2013) Life cycle assessment of sewage sludge management: a review. Waste Manage Res 31:1083–1101. https://doi.org/10.1177/0734242x13504446
57. Thorn J, Kerekes E (2001) Health effects among employees in sewage treatment plants: a literature survey. Am J Ind Med 40:170–179
58. Muzaini K, Yasin SM, Ismail Z and Ishak AR (2021) Systematic review of potential occupational respiratory hazards exposure among sewage workers. Front Public Health 9
59. Glas C, Hotz P, Steffen R (2001) Hepatitis A in workers exposed to sewage: a systematic review. Occup Environ Med 58:762–768
60. Jaremków A, Szałata Ł, Kołwzan B, Sówka I, Zwoździak J and Pawlas K (2017) Impact of a sewage treatment plant on health of local residents: gastrointestinal system symptoms. Pol J Environ Stud 26
61. Keil A, Wing S, Lowman A (2011) Suitability of public records for evaluating health effects of treated sewage sludge in North Carolina. N C Med J 72:98–104
62. Kumar A, Holuszko M, Espinosa DCR (2017) E-waste: an overview on generation, collection, legislation and recycling practices. Resour Conserv Recycl 122:32–42
63. Ilankoon I, Ghorbani Y, Chong MN, Herath G, Moyo T, Petersen J (2018) E-waste in the international context—a review of trade flows, regulations, hazards, waste management strategies and technologies for value recovery. Waste Manage 82:258–275
64. Ádám B, Göen T, Scheepers PT, Adliene D, Batinic B, Budnik LT, Duca R-C, Ghosh M, Giurgiu DI, Godderis L (2021) From inequitable to sustainable e-waste processing for reduction of impact on human health and the environment. Environ Res 194:110728
65. Frazzoli C, Orisakwe OE, Dragone R, Mantovani A (2010) Diagnostic health risk assessment of electronic waste on the general population in developing countries' scenarios. Environ Impact Assess Rev 30:388–399
66. Okeme J, Arrandale V (2019) Electronic waste recycling: occupational exposures and work-related health effects. Curr Environ Health Rep 6:256–268
67. Tsydenova O, Bengtsson M (2011) Chemical hazards associated with treatment of waste electrical and electronic equipment. Waste Manage 31:45–58
68. Li W, Achal V (2020) Environmental and health impacts due to e-waste disposal in China—a review. Sci Total Environ 737:139745
69. Perkins DN, Brune Drisse MN, Nxele T, Sly PD (2014) E-waste: a global hazard. Ann Glob Health 80:286–295. https://doi.org/10.1016/j.aogh.2014.10.001
70. McDonald TA (2002) A perspective on the potential health risks of PBDEs. Chemosphere 46:745–755. https://doi.org/10.1016/s0045-6535(01)00239-9
71. Wu Z, He C, Han W, Song J, Li H, Zhang Y, Jing X, Wu W (2020) Exposure pathways, levels and toxicity of polybrominated diphenyl ethers in humans: a review. Environ Res 187:109531. https://doi.org/10.1016/j.envres.2020.109531
72. IARC (2016) Polychlorinated biphenyls and polybrominated biphenyls international agency for research on cancer, Lyon, F.
73. IARC (2012) Arsenic, metals, fibres, and dusts: review of human carcinogens. International Agency for Research on Cancer, Lyon, F.

74. Schecter A, Birnbaum L, Ryan JJ, Constable JD (2006) Dioxins: an overview. Environ Res 101:419–428. https://doi.org/10.1016/j.envres.2005.12.003
75. IARC (1983) Polynuclear aromatic compounds, Part 1: chemical, environmental and experimental data. International Agency for Research on Cancer, Lyon, F.
76. IARC (2006) Inorganic and organic lead compounds. International Agency for Research on Cancer, Lyon, F.
77. Costa M, Klein CB (2006) Toxicity and carcinogenicity of chromium compounds in humans. Crit Rev Toxicol 36:155–163. https://doi.org/10.1080/10408440500534032
78. Genchi G, Sinicropi MS, Lauria G, Carocci A, Catalano A (2020) The effects of cadmium toxicity. Int J Environ Res Public Health 17. https://doi.org/10.3390/ijerph17113782
79. Bernhoft RA (2012) Mercury toxicity and treatment: a review of the literature. J Environ Public Health 2012:460508. https://doi.org/10.1155/2012/460508
80. Agnew UM, Slesinger TL (2022) Zinc toxicity. StatPearls, Treasure Island (FL)
81. Das KK, Reddy RC, Bagoji IB, Das S, Bagali S, Mullur L, Khodnapur JP, Biradar MS (2018) Primary concept of nickel toxicity—an overview. J Basic Clin Physiol Pharmacol 30:141–152. https://doi.org/10.1515/jbcpp-2017-0171
82. Oskarsson A (2022) Barium. Elsevier, Handbook on the toxicology of metals, pp 91–100
83. Strupp C (2011) Beryllium metal II. a review of the available toxicity data. Ann Occup Hyg 55:43–56. https://doi.org/10.1093/annhyg/meq073
84. Gaetke LM, Chow CK (2003) Copper toxicity, oxidative stress, and antioxidant nutrients. Toxicology 189:147–163
85. Sepúlveda A, Schluep M, Renaud FG, Streicher M, Kuehr R, Hagelüken C, Gerecke AC (2010) A review of the environmental fate and effects of hazardous substances released from electrical and electronic equipments during recycling: examples from China and India. Environ Impact Assess Rev 30:28–41
86. Grant K, Goldizen FC, Sly PD, Brune M-N, Neira M, van den Berg M, Norman RE (2013) Health consequences of exposure to e-waste: a systematic review. Lancet Glob Health 1:e350–e361
87. Song Q, Li J (2015) A review on human health consequences of metals exposure to e-waste in China. Environ Pollut 196:450–461
88. Xu X, Zeng X, Boezen HM, Huo X (2015) E-waste environmental contamination and harm to public health in China. Front Med 9:220–228
89. Cai K, Song Q, Yuan W, Ruan J, Duan H, Li Y, Li J (2020) Human exposure to PBDEs in e-waste areas: a review. Environ Pollut 115634
90. Awasthi AK, Wang M, Awasthi MK, Wang Z, Li J (2018) Environmental pollution and human body burden from improper recycling of e-waste in China: a short-review. Environ Pollut 243:1310–1316
91. Ceballos DM, Dong Z (2016) The formal electronic recycling industry: challenges and opportunities in occupational and environmental health research. Environ Int 95:157–166
92. Parvez SM, Jahan F, Brune M-N, Gorman JF, Rahman MJ, Carpenter D, Islam Z, Rahman M, Aich N, Knibbs LD (2021) Health consequences of exposure to e-waste: an updated systematic review. The Lancet Planet Health 5:e905–e920
93. Zeng X, Huo X, Xu X, Liu D, Wu W (2020) E-waste lead exposure and children's health in China. Sci Total Environ 734:139286
94. Plastic Europe (2022) Brussels, Belgium
95. Geyer R, Jambeck JR, Law KL (2017) Production, use, and fate of all plastics ever made. Sci Adv 3. https://doi.org/10.1126/sciadv.1700782
96. Jiang B, Kauffman AE, Li L, McFee W, Cai B, Weinstein J, Lead JR, Chatterjee S, Scott GI, Xiao S (2020) Health impacts of environmental contamination of micro- and nanoplastics: a review. Environ Health Prev Med 25:29. https://doi.org/10.1186/s12199-020-00870-9
97. Lithner D, Larsson A, Dave G (2011) Environmental and health hazard ranking and assessment of plastic polymers based on chemical composition. Sci Total Environ 409:3309–3324. https://doi.org/10.1016/j.scitotenv.2011.04.038

98. Rochman CM, Browne MA, Halpern BS, Hentschel BT, Hoh E, Karapanagioti HK, Rios-Mendoza LM, Takada H, Teh S, Thompson RC (2013) Classify plastic waste as hazardous. Nature 494:169–171. https://doi.org/10.1038/494169a

99. Hussain CM, Keçili R (2019) Modern environmental analysis techniques for pollutants. Elsevier

100. Mitrano DM, Wohlleben W (2020) Microplastic regulation should be more precise to incentivize both innovation and environmental safety. Nat Commun 11:5324. https://doi.org/10.1038/s41467-020-19069-1

101. Peng J, Wang J, Cai L (2017) Current understanding of microplastics in the environment: occurrence, fate, risks, and what we should do. Integr Environ Assess Manag 13:476–482

102. IARC (2012) Chemical agents and related occupations. International Agency for Research on Cancer, Lyon, F.

103. Halden RU (2010) Plastics and health risks. Annu Rev Public Health 31:179–194

104. Gunaalan K, Fabbri E, Capolupo M (2020) The hidden threat of plastic leachates: a critical review on their impacts on aquatic organisms. Water Res 184:116170

105. Hahladakis JN, Velis CA, Weber R, Iacovidou E, Purnell P (2018) An overview of chemical additives present in plastics: migration, release, fate and environmental impact during their use, disposal and recycling. J Hazard Mater 344:179–199

106. Meeker JD, Sathyanarayana S, Swan SH (2009) Phthalates and other additives in plastics: human exposure and associated health outcomes. Philos Trans Roy Soc B: Biol Sci 364:2097–2113

107. Turner A, Filella M (2021) Hazardous metal additives in plastics and their environmental impacts. Environ Int 156:106622

108. Liao Y-l and Yang J-y (2020) Microplastic serves as a potential vector for Cr in an in-vitro human digestive model. Sci Total Environ 703:134805

109. Wang W, Wang J (2018) Comparative evaluation of sorption kinetics and isotherms of pyrene onto microplastics. Chemosphere 193:567–573

110. Hu L, Chernick M, Lewis AM, Ferguson PL, Hinton DE (2020) Chronic microfiber exposure in adult Japanese medaka (Oryzias latipes). PLoS ONE 15:e0229962. https://doi.org/10.1371/journal.pone.0229962

111. Bouwmeester H, Hollman PC, Peters RJ (2015) Potential health impact of environmentally released micro- and nanoplastics in the human food production chain: experiences from nanotoxicology. Environ Sci Technol 49:8932–8947. https://doi.org/10.1021/acs.est.5b01090

112. Rochman CM, Hoellein T (2020) The global odyssey of plastic pollution. Science 368:1184–1185. https://doi.org/10.1126/science.abc4428

113. Brahney J, Hallerud M, Heim E, Hahnenberger M, Sukumaran S (2020) Plastic rain in protected areas of the United States. Science 368:1257–1260. https://doi.org/10.1126/science.aaz5819

114. Hoellein TJ, Shogren AJ, Tank JL, Risteca P, Kelly JJ (2019) Microplastic deposition velocity in streams follows patterns for naturally occurring allochthonous particles. Sci Rep 9:1–11

115. Provencher JF, Ammendolia J, Rochman CM, Mallory ML (2019) Assessing plastic debris in aquatic food webs: what we know and don't know about uptake and trophic transfer. Environ Rev 27:304–317. https://doi.org/10.1139/er-2018-0079

116. Chang X, Xue Y, Li J, Zou L, Tang M (2020) Potential health impact of environmental micro- and nanoplastics pollution. J Appl Toxicol 40:4–15. https://doi.org/10.1002/jat.3915

117. Wright SL, Kelly FJ (2017) Plastic and human health: a micro issue? Environ Sci Technol 51:6634–6647. https://doi.org/10.1021/acs.est.7b00423

118. Smith M, Love DC, Rochman CM, Neff RA (2018) Microplastics in seafood and the implications for human health. Curr Environ Health Rep 5:375–386. https://doi.org/10.1007/s40572-018-0206-z

119. Lusher AH P, Mendoza-Hill J (2017) Microplastics in fisheries and aquaculture: status of knowledge on their occurrence and implications for aquatic organisms and food safety

120. Prata JC (2018) Airborne microplastics: consequences to human health? Environ Pollut 234:115–126. https://doi.org/10.1016/j.envpol.2017.11.043

121. Pauly JL, Stegmeier SJ, Allaart HA, Cheney RT, Zhang PJ, Mayer AG, Streck RJ (1998) Inhaled cellulosic and plastic fibers found in human lung tissue. Cancer Epidemiol Preven Biomarkers 7:419–428
122. Law B, Bunn W, Hesterberg T (1990) Solubility of polymeric organic fibers and manmade vitreous fibers in Gambles solution. Inhalation Toxicol 2:321–339
123. Part F, Berge N, Baran P, Stringfellow A, Sun W, Bartelt-Hunt S, Mitrano D, Li L, Hennebert P, Quicker P (2018) A review of the fate of engineered nanomaterials in municipal solid waste streams. Waste Manage 75:427–449
124. OECD (2016) Nanomaterials in waste streams
125. Musee N (2011) Nanowastes and the environment: potential new waste management paradigm. Environ Int 37:112–128
126. Ounoughene G, LeBihan O, Debray B, Chivas-Joly C, Longuet C, Joubert A, Lopez-Cuesta J-M and Le Coq L (2017) Thermal disposal of waste containing nanomaterials: first investigations on a methodology for risk management. J Phys Conf Ser IOP Publishing, 012024
127. Manžuch Z, Akelytė R, Camboni M, Carlander D, García RP, Kriščiūnaitė G, Baun A, Kaegi R (2021) Study on the product lifecycles, waste recycling and the circular economy for nanomaterials. European Chemical Agency, Helsinky, FL
128. Jones W, Gibb A, Goodier C, Bust P, Song M, Jin J (2019) Nanomaterials in construction–what is being used, and where? Proceed Inst Civ Eng Constr Mater 172:49–62
129. Brar SK, Verma M, Tyagi R, Surampalli R (2010) Engineered nanoparticles in wastewater and wastewater sludge–evidence and impacts. Waste Manage 30:504–520
130. Kunhikrishnan A, Shon HK, Bolan NS, El Saliby I, Vigneswaran S (2015) Sources, distribution, environmental fate, and ecological effects of nanomaterials in wastewater streams. Crit Rev Environ Sci Technol 45:277–318
131. Liu W, Weng C, Zheng J, Peng X, Zhang J, Lin Z (2019) Emerging investigator series: treatment and recycling of heavy metals from nanosludge. Environ Sci Nano 6:1657–1673
132. Stamou I, Antizar-Ladislao B (2016) The impact of silver and titanium dioxide nanoparticles on the in-vessel composting of municipal solid waste. Waste Manage 56:71–78
133. Yang Y, Wang Y, Westerhoff P, Hristovski K, Jin VL, Johnson M-VV, Arnold JG (2014) Metal and nanoparticle occurrence in biosolid-amended soils. Sci Total Environ 485:441–449
134. Oliveira ML, Izquierdo M, Querol X, Lieberman RN, Saikia BK, Silva LF (2019) Nanoparticles from construction wastes: a problem to health and the environment. J Clean Prod 219:236–243
135. Buonanno G, Morawska L (2015) Ultrafine particle emission of waste incinerators and comparison to the exposure of urban citizens. Waste Manage 37:75–81
136. Johnson DR (2016) Nanometer-sized emissions from municipal waste incinerators: a qualitative risk assessment. J Hazard Mater 320:67–79
137. Roma J, Matos AR, Vinagre C, Duarte B (2020) Engineered metal nanoparticles in the marine environment: a review of the effects on marine fauna. Marine Environ Res 105110
138. Bondarenko O, Juganson K, Ivask A, Kasemets K, Mortimer M, Kahru A (2013) Toxicity of Ag, CuO and ZnO nanoparticles to selected environmentally relevant test organisms and mammalian cells in vitro: a critical review. Arch Toxicol 87:1181–1200
139. Courtois P, Rorat A, Lemiere S, Guyoneaud R, Attard E, Levard C, Vandenbulcke F (2019) Ecotoxicology of silver nanoparticles and their derivatives introduced in soil with or without sewage sludge: a review of effects on microorganisms, plants and animals. Environ Pollut 253:578–598
140. McGillicuddy E, Murray I, Kavanagh S, Morrison L, Fogarty A, Cormican M, Dockery P, Prendergast M, Rowan N, Morris D (2017) Silver nanoparticles in the environment: sources, detection and ecotoxicology. Sci Total Environ 575:231–246
141. McShan D, Ray PC, Yu H (2014) Molecular toxicity mechanism of nanosilver. J Food Drug Anal 22:116–127
142. Ma H, Williams PL, Diamond SA (2013) Ecotoxicity of manufactured ZnO nanoparticles–a review. Environ Pollut 172:76–85

143. Du J, Tang J, Xu S, Ge J, Dong Y, Li H, Jin M (2020) ZnO nanoparticles: recent advances in ecotoxicity and risk assessment. Drug Chem Toxicol 43:322–333

144. Sirelkhatim A, Mahmud S, Seeni A, Kaus NHM, Ann LC, Bakhori SKM, Hasan H, Mohamad D (2015) Review on zinc oxide nanoparticles: antibacterial activity and toxicity mechanism. Nano-micro Lett 7:219–242

145. Bakshi M and Kumar A (2021) Copper based nanoparticles in the soil-plant environment: assessing their applications, interactions, fate, and toxicity. Chemosphere 130940

146. Roes L, Patel MK, Worrell E, Ludwig C (2012) Preliminary evaluation of risks related to waste incineration of polymer nanocomposites. Sci Total Environ 417:76–86

147. Köhler AR, Som C, Helland A, Gottschalk F (2008) Studying the potential release of carbon nanotubes throughout the application life cycle. J Clean Prod 16:927–937

148. Rodríguez-Ibarra C, Déciga-Alcaraz A, Ispanixtlahuatl-Meráz O, Medina-Reyes EI, Delgado-Buenrostro NL, Chirino YI (2020) International landscape of limits and recommendations for occupational exposure to engineered nanomaterials. Toxicol Lett 322:111–119

149. Leikauf GD, Kim S-H, Jang A-S (2020) Mechanisms of ultrafine particle-induced respiratory health effects. Exp Mol Med 52:329–337

150. Vejerano EP, Ma Y, Holder AL, Pruden A, Elankumaran S, Marr LC (2015) Toxicity of particulate matter from incineration of nanowaste. Environ Sci Nano 2:143–154

151. Chivas-Joly C, Longuet C, Pourchez J, Leclerc L, Sarry G, Lopez-Cuesta J-M (2019) Physical, morphological and chemical modification of Al-based nanofillers in by-products of incinerated nanocomposites and related biological outcome. J Hazard Mater 365:405–412

152. Stueckle TA, White A, Wagner A, Gupta RK, Rojanasakul Y, Dinu CZ (2019) Impacts of organomodified nanoclays and their incinerated byproducts on bronchial cell monolayer integrity. Chem Res Toxicol 32:2445–2458

Printed in the United States
by Baker & Taylor Publisher Services